DATE DUE

BRODART, INC. Cat. No. 23-221

Elementary Functional Analysis

Mathematical Analysis by Georgi E. Shilov

Revised English editions translated and edited by
Richard A. Silverman

Volume 1. **Elementary Real and Complex Analysis**

Volume 2. **Elementary Functional Analysis**

Georgi E. Shilov

Elementary Functional Analysis

Revised English edition translated and edited by
Richard A. Silverman

The MIT Press Cambridge, Massachusetts, and London, England

This book was set in Monotype Baskerville,
printed on Decision Offset,
and bound in Columbia Millbank Vellum
by Halliday Lithograph Corp.
in the United States of America.

Library of Congress Cataloging in Publication Data

Shilov, Georgii Evgen'evich.
Elementary functional analysis.

Translation of v. 2 of Matematicheskii analiz.
Bibliography: p.
1. Functional analysis. I. Silverman, Richard A., ed. II. Title.
QA320.S4713 1974 515'.7 73-21882
ISBN 0-262-19122-9

Contents

Preface

Volume 2 of my series of books on mathematical analysis, entitled *Elementary Functional Analysis*, follows the same general principles as discussed in the preface to Volume 1 (*Elementary Real and Complex Analysis*, The MIT Press, 1973). Volume 2 begins with a long key chapter on the basic structures of mathematical analysis. Here, in turn, I treat linear spaces, metric spaces, normed linear spaces, Hilbert spaces, and finally normed algebras. The "standard models" for all these spaces are sets of functions (hence the term "functional analysis"), rather than sets of points in a finite-dimensional space (as in Volume 1). The study of the normed linear space of bounded sequences and of linear functionals on this space leads one naturally to the basic ideas of generalized limits and Toeplitz's theorem. Normed algebras figure in the theory of operators acting in a normed linear space. Among topics treated here are the "operational calculus" of analytic functions defined on the spectrum of a linear operator and theorems of the type of the Fredholm alternative.

Chapter 2 is devoted to differential equations, and contains the basic theorems on existence and uniqueness of solutions of ordinary differential equations for functions taking values in a Banach space. The solution of the linear equation with constant (operator) coefficients is written in general form in terms of the exponential of the operator. In the finite-dimensional case this leads to explicit formulas for the solutions not only of first-order equations, but also of higher-order equations and systems of equations. The method of "variation of constants" is applied to the linear equation with variable (operator) coefficients. Chapter 3 presents a theory of curvature for curves in a multidimensional space, and borrows freely from the material in Chapters 1 and 2.

The rest of the book is essentially a brief introduction to Fourier analysis and some of its ramifications. In Chapter 4, on orthogonal expansions, a key role is played by Fourier series, with emphasis on various kinds of convergence and summability for such series. Chapter 5, on Fourier transforms, besides presenting the usual real theory also deals with problems in the complex domain, in particular with problems involving the Laplace transform.

As in Volume 1, each chapter is accompanied by a set of problems. Hints and answers to these problems are given at the end of the book. Starred sections contain ancillary material that can be omitted on first reading.

G. E. S.

1 Basic Structures of Mathematical Analysis

Mathematical structures have already been discussed in the first volume of this course (Vol. 1, Sec. 2.5).† We now say a few more words about structures, with special reference to the structures encountered in analysis. The objects of mathematical analysis are numbers and functions, together with operations defined on them, and from the most general point of view, the relations between these objects are described by the theory of sets. In fact, numbers and functions make up various kinds of sets, and the relations of set inclusion, as well as the operations of taking unions, intersections, and complements, allow us to describe certain general properties of these sets. We arrive at the basic structures of analysis by imposing further conditions on these sets, stated in the form of various systems of axioms, corresponding to the properties or operations used in classical mathematical analysis. This gives rise to the following mathematical structures:

(1) *Linear spaces*, where linear operations, namely addition of elements and multiplication of elements by numbers, are defined;
(2) *Metric spaces*, where the operation of passage to the limit is axiomatized, with the help of the concept of distance;
(3) *Normed linear spaces* (or *Banach spaces*), where both linear operations and passage to the limit are considered;
(4) *Hilbert spaces*, where the concept of a scalar product is axiomatized, which allows us to deal not only with lengths of vectors but also with angles between vectors;
(5) *Normed algebras*, where the operations already considered are supplemented by the operation of multiplying one element by another.

Imposing the extra condition that the number of dimensions be finite, we are led to three kinds of linear spaces:

(1′) Finite-dimensional *affine* (i.e., nonmetrized) linear spaces;
(3′) Finite-dimensional normed linear spaces;
(4′) *Euclidean spaces*, i.e., finite-dimensional Hilbert spaces.

There are a large number of structures lying between the basic structures just enumerated. Despite their great importance, we will say nothing about these intermediate structures (topological spaces, partially ordered spaces, etc.) at this time.

The following diagram illustrates schematically the basic structures which will now be studied in detail:

† We refer to G. E. Shilov, *Mathematical Analysis, Volume 1. Elementary Real and Complex Analysis* (translated by R. A. Silverman), The MIT Press, Cambridge, Mass. (1973), henceforth cited by the abbreviation "Vol. 1."

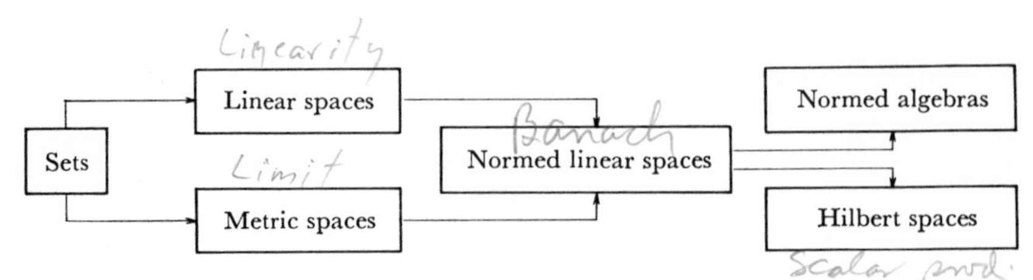

The arrows indicate particularization, i.e., transition from a general structure to a more special one.

1.1. Linear Spaces†

1.11. To construct the axiomatics of a linear space, we start from the properties of n-dimensional real space R_n (Vol. 1, Sec. 2.61), avoiding the expression of elements in coordinate form and replacing the field R of real numbers by an arbitrary field K (Vol. 1, Sec. 1.22). Thus by a *linear space* $\mathbf{K}$ over a field K is meant a set of objects (elements) $x, y, z, \ldots$ called *vectors* for which there are defined operations of addition and multiplication by numbers (from the field K)‡ satisfying the following axioms:

(a) $x+y=y+x$ for every $x, y \in \mathbf{K}$;
(b) $(x+y)+z=x+(y+z)$ for every $x, y, z \in \mathbf{K}$;
(c) $\mathbf{K}$ contains an element 0, called the *zero vector*, such that $x+0=x$ for every $x \in \mathbf{K}$;
(d) For every $x \in \mathbf{K}$ there exists an element $y \in \mathbf{K}$, called the *negative* of x, such that $x+y=0$;
(e) $\alpha(x+y)=\alpha x+\alpha y$ for every $x, y \in \mathbf{K}$ and $\alpha \in K$;
(f) $(\alpha+\beta)x=\alpha x+\beta x$ for every $x \in \mathbf{K}$ and $\alpha, \beta \in K$;
(g) $1 \cdot x=x$ for every $x \in \mathbf{K}$;
(h) $\alpha(\beta x)=(\alpha\beta)x$ for every $x \in \mathbf{K}$ and $\alpha, \beta \in K$.

If the field K is the field R of real numbers, the space $\mathbf{K}$ is called a *real* linear space and is denoted by $\mathbf{R}$. If K is the field C of complex numbers, the space $\mathbf{K}$ is called a *complex* linear space and is denoted by $\mathbf{C}$.

1.12. Axioms a–d are precisely the same as the addition axioms for real numbers (Vol. 1, Sec. 1.21). Therefore all the consequences of the addition axioms for real numbers deduced in Vol. 1, Sec. 1.3 continue to hold for an

† A detailed exposition of the theory of linear spaces is given in G. E. Shilov, *Linear Algebra* (translated by R. A. Silverman), Prentice-Hall, Inc., Englewood Cliffs, N. J. (1971), henceforth cited by the abbreviation "Lin. Alg.".

‡ More exactly, given any two vectors $x, y \in \mathbf{K}$, there is a rule leading to a (unique) element $x+y \in \mathbf{K}$, called the *sum* of x and y, and given any number $\alpha \in K$ and any vector $x \in \mathbf{K}$, there is a rule leading to a (unique) element $\alpha x \in \mathbf{K}$, called the *product* of α and x.

arbitrary linear space, namely uniqueness of the zero vector, uniqueness of the negative element for every $x \in \mathbf{K}$, and unique solvability of the equation $a+x=b$ (the latter guarantees that the operation of subtraction can be well defined).

The operation of multiplication of elements of a linear space by one another is not defined, and in fact Axioms e–h are only externally similar to the multiplication axioms for real numbers given in Vol. 1, Sec. 1.22. Therefore only certain of the theorems of Vol. 1, Sec. 1.4 can be carried over to linear spaces, namely the following propositions (with almost no change in the proofs):

a. THEOREM. *The formula*

$$0 \cdot x = 0$$

holds for every $x \in \mathbf{K}$.†

b. THEOREM. *If* $\alpha x = 0$, *then either* $\alpha = 0$ *or* $x = 0$.

Proof. If $\alpha \neq 0$, then, by Axioms g and h,

$$x = \frac{1}{\alpha}\alpha x = \frac{1}{\alpha} \cdot 0 = 0.$$

But $0 = 0 \cdot y$, where $y \in \mathbf{K}$ is arbitrary, and hence

$$x = \left(\frac{1}{\alpha} \cdot 0\right) y = 0 \cdot y = 0,$$

by Axiom h and Theorem 1.12a. ▮‡

c. THEOREM. *The formula*

$$-x = (-1)x$$

holds for every $x \in \mathbf{K}$.

These theorems are the analogues of Vol. 1, Theorems 1.47a, 1.47b, and 1.49, respectively.

1.13. Examples of linear spaces. We now describe four kinds of linear spaces over the field R of real numbers:

a. The real numbers themselves with the ordinary operations of arithmetic.

† In the right-hand side of the equation, 0 denotes the zero vector, and in the left-hand side the number 0 from the field K.

‡ The symbol ▮ means Q.E.D. and indicates the end of a proof. Theorem 1.12a refers to the (unique) theorem in Sec. 1.12a, Example 1.13i to the example in Sec. 1.13i, etc.

b. The n-dimensional real space R_n (often called n-dimensional *coordinate space*).

c. The space $R(E)$ of all functions defined on some set E and taking real values, with the usual operations of addition and multiplication by real numbers (as defined for numerical functions in Vol. 1, Sec. 4.31b).

d. The space $\mathbf{R}(E)$ of all functions defined on E and taking vector values (in the real space $\mathbf{R}$), with the operations

$$(x+y)(t)=x(t)+y(t), \qquad (\alpha x)(t)=\alpha x(t) \qquad (t \in E)$$

of addition and multiplication by real numbers which are natural for vector functions.

Note that each of the above examples is a generalization of the example which precedes it.

Replacing the field of real numbers by an arbitrary field K, we get four examples of linear spaces over the field K:

e. The field K itself.

f. The n-dimensional coordinate space K_n over the field K, consisting of all n-tuples $(\alpha_1,\ldots,\alpha_n)$ of elements of the field K, with the operations of addition and multiplication by numbers (in K) defined by the rules

$$(\alpha_1,\ldots,\alpha_n)+(\beta_1,\ldots,\beta_n)=(\alpha_1+\beta_1,\ldots,\alpha_n+\beta_n),$$
$$\beta(\alpha_1,\ldots,\alpha_n)=(\beta\alpha_1,\ldots,\beta\alpha_n).$$

g. The space $K(E)$ of all functions defined on some set E and taking values in the field K, with the usual operations of addition and multiplication by numbers (as defined for numerical functions).

h. The space $\mathbf{K}(E)$ of all functions defined on E and taking vector values (in the space $\mathbf{K}$), with the operations of addition and multiplication by numbers (in K) which are natural for vector functions.

The following example, although not comprised in the above list, is one of the most important spaces in analysis:

i. The space $R^s(M)$ of all continuous real functions defined on a metric space M.†

The last example does not generalize to the case of functions taking values in an arbitrary field, since the concept of continuity is in general not defined for such functions (continuity of a function requires a metric in the space where it takes its values, and we have not introduced a metric in an arbitrary

† The superscript s stands for the German word "stetig" = "continuous."

field K). Thus, for the time being, we can generalize Example 1.13i only to functions taking values in the n-dimensional real space R_n, where both linear operations and the concept of continuity have been defined (Vol. 1, Sec. 5.81).

j. The space $R_n^s(M)$ of all continuous functions defined on a metric space M and taking values in the n-dimensional real space R_n.

k. A special case ($n=2$) of the last example, meriting special mention, is the space $C^s(M)$ of all continuous complex-valued functions on a metric space M.

A useful restriction of Example 1.13i will be considered in Sec. 1.23a.

1.14.a. The vectors $x_1, \ldots, x_n$ of a linear space **K** are said to be *linearly dependent* if there exist constants $\alpha_1, \ldots, \alpha_n$ in the field K, not all equal to zero, such that

$$\alpha_1 x_1 + \cdots + \alpha_n x_n = 0. \tag{1}$$

On the other hand, if it follows from (1) that $\alpha_1 = \cdots = \alpha_n = 0$, then the vectors $x_1, \ldots, x_n$ are said to be *linearly independent.*

b. A linear space **K** is said to be *n-dimensional* if **K** contains n linearly independent vectors, while every $n+1$ vectors in **K** are linearly dependent. We say that **K** is *infinite-dimensional* if **K** contains n linearly independent vectors for every $n=1, 2, \ldots$

c. By a *basis* of an n-dimensional space **K** we mean any set of n linearly independent vectors in **K**. If $f_1, \ldots, f_n$ is a basis and x is any vector in **K**, then the $n+1$ vectors $x, f_1, \ldots, f_n$ are now linearly dependent, and hence there exist numbers $\alpha_0, \alpha_1, \ldots, \alpha_n$ in the field K, not all equal to zero, such that

$$\alpha_0 x + \alpha_1 f_1 + \cdots + \alpha_n f_n = 0.$$

Here $\alpha_0 \neq 0$, since otherwise the vectors $f_1, \ldots, f_n$ would be linearly dependent. Dividing by α_0 and writing $\beta_j = -\alpha_j/\alpha_0$ ($j=1, \ldots, n$), we get the following *expansion of the vector f with respect to the basis* $f_1, \ldots, f_n$:

$$x = \beta_1 f_1 + \cdots + \beta_n f_n.$$

Such an expansion *is necessarily unique*, since otherwise we would again get linear dependence between the vectors $f_1, \ldots, f_n$ (why?).

d. The n-dimensional real space R_n (Vol. 1, Sec. 2.61) is an n-dimensional linear space in the sense of the above definition. In fact, the vectors

$$e_1 = (1, 0, \ldots, 0), \ldots, e_n = (0, 0, \ldots, 1)$$

are obviously linearly independent, while every $n+1$ vectors

$$y_1=(x_1^{(1)},\ \ldots,\ x_n^{(1)}),\ \ldots, y_{n+1}=(x_1^{(n+1)},\ \ldots,\ x_n^{(n+1)})$$

are linearly dependent, as already noted in Vol. 1, Sec. 2.64.

In the same way, the space K_n (Example 1.13f) is n-dimensional in the sense of the above definition.

e. Let Ω be an infinite set on the real line $-\infty<x<\infty$, and let $P(\Omega)$ be the linear space of all polynomials $p(x)=a_0+a_1x+\cdots+a_nx^n$ (of all degrees) defined on Ω, with coefficients in an arbitrary field K and with the usual operations. Then $P(\Omega)$ is a linear space over the field K.

THEOREM. *The functions* $1, x, \ldots, x^n$ *are linearly independent for every* $n=1, 2, \ldots$

Proof. Suppose a relation

$$\alpha_0+\alpha_1x+\cdots+\alpha_nx^n\equiv 0$$

holds on Ω. Successively setting x equal to (distinct) values $x_0, x_1, \ldots, x_n$ (in Ω), we get a system

$$\alpha_0+\alpha_1x_0+\cdots+\alpha_nx_0^n=0,$$
$$\alpha_0+\alpha_1x_1+\cdots+\alpha_nx_1^n=0,$$
$$\cdots$$
$$\alpha_0+\alpha_1x_n+\cdots+\alpha_nx_n^n=0$$

of equations in the unknowns $\alpha_0, \alpha_1, \ldots, \alpha_n$, with a nonzero determinant, in fact, the Vandermonde determinant (Lin. Alg., Sec. 1.55c). It follows that

$$\alpha_0=\alpha_1=\cdots=\alpha_n=0. \quad \blacksquare$$

COROLLARY. *The space* $P(\Omega)$ *is infinite-dimensional.*

Proof. An immediate consequence of the definition in Sec. 1.14b. ∎

f. THEOREM. *The space* $R^s(M)$ *of all continuous real functions defined on an infinite metric space* M *is infinite-dimensional, and similarly for the space* $C^s(M)$ *of all continuous complex functions on* M.

Proof. We need only exhibit n linearly independent functions in the space $R^s(M)$ for every $n=1, 2, \ldots$ Let $t_1, \ldots, t_n$ be distinct points of the space M, and let

$$d=\min_{j,k}\rho(t_j,t_k)$$

where ρ is the metric in M. Let $y=\varphi(x)$ be a continuous function of a real argument x, equal to 1 if $x=0$ and 0 if $|x|\geqslant d$. The function $\rho(t_j,t)$ is a continuous function of t (Vol. 1, Sec. 5.12b), and hence the same is true of the function

$$x_j(t)=\varphi(\rho(t_j,t))$$

(Vol. 1, Sec. 5.15a). By construction, the function $x_j(t)$ equals 1 if $t=t_j$ and 0 if $t=t_k$, $k\neq j$. Suppose a relation

$$\alpha_1 x_1(t)+\cdots+\alpha_n x_n(t)\equiv 0$$

holds on M. Then, setting $t=t_j$, we find that $\alpha_j=0$ ($j=1,\ldots,n$), so that the functions $x_1(t),\ldots,x_n(t)$ are linearly independent. ∎

g. A subset $\mathbf{E}\subset\mathbf{K}$ is called a *subspace* of the linear space $\mathbf{K}$ if $x\in\mathbf{E}$, $y\in\mathbf{E}$ implies $x+y\in\mathbf{E}$ and $\alpha x\in\mathbf{E}$ for every number $\alpha\in K$. Every linear space $\mathbf{K}$ has two trivial subspaces, one consisting of the single (unique) zero element 0, the other coinciding with the whole space $\mathbf{K}$. All other subspaces of $\mathbf{K}$ are called *proper subspaces.*

h. Direct sums. A linear space $\mathbf{K}$ is called the *direct sum* of its subspaces $\mathbf{L}_1,\ldots,\mathbf{L}_m$ if for every $x\in\mathbf{K}$ there exists an expansion

$$x=x_1+\cdots+x_m,\qquad x_1\in\mathbf{L}_1,\ldots,x_m\in\mathbf{L}_m,$$

and if this expansion is unique, i.e., if

$$x=x_1+\cdots+x_m=y_1+\cdots+y_m,\qquad x_j\in\mathbf{L}_j, y_j\in\mathbf{L}_j\qquad(j=1,\ldots,m)\tag{2}$$

implies $x_1=y_1,\ldots,x_m=y_m$.

The condition (2) for the uniqueness of the expansion of an arbitrary element x can be replaced by the simpler condition for the uniqueness of the expansion of zero, i.e., if there exists an expansion

$$0=x_1+\cdots+x_m,\qquad x_1\in\mathbf{L}_1,\ldots,x_m\in\mathbf{L}_m,\tag{3}$$

then $x_1=\cdots=x_m=0$ (Lin. Alg., Sec. 2.45).

Thus the space R_n is the direct sum of the n one-dimensional subspaces "generated" by any n linearly independent vectors. Moreover, R_n can also be represented in various different ways as a direct sum of subspaces not all of dimension 1. Quite generally, given any subspace $\mathbf{L}\subset R_n$, there is another subspace $\mathbf{M}\subset R_n$ such that the direct sum of $\mathbf{L}$ and $\mathbf{M}$ is the whole space R_n (Lin. Alg., Sec. 2.46).

i. Factor spaces. Given a subspace $\mathbf{L}$ of a linear space $\mathbf{K}$, an element $x\in\mathbf{K}$ is said to be *comparable* with an element $y\in\mathbf{K}$ (more exactly, *comparable relative to* $\mathbf{L}$) if $x-y\in\mathbf{L}$. The set of all elements $y\in\mathbf{K}$ comparable with a given element $x\in\mathbf{K}$ is called a *class*, and is denoted by $\mathbf{X}$. The class $\mathbf{X}$ contains the element x itself, and any two elements $y \in \mathbf{X}$, $z \in \mathbf{X}$ are comparable with each other. Moreover, if $u \notin \mathbf{X}$, then u is not comparable with any element of $\mathbf{X}$. Therefore two classes either have no elements in common or else coincide completely.

The whole space $\mathbf{K}$ can be represented as a union of nonintersecting classes $\mathbf{X}$, $\mathbf{Y}$, ... This set of classes will be denoted by $\mathbf{K}/\mathbf{L}$. Linear operations can be introduced in the set $\mathbf{K}/\mathbf{L}$ as follows: Given two classes $\mathbf{X}$, $\mathbf{Y}$ and two numbers α, β of the field K, we wish to define the class $\mathbf{Z} = \alpha\mathbf{X} + \beta\mathbf{Y}$. To this end, we choose arbitrary elements $x \in \mathbf{X}$, $y \in \mathbf{Y}$ and find the class $\mathbf{Z}$ containing the element $z = \alpha x + \beta y$. It can be shown that $\mathbf{Z}$ is uniquely defined and that the operations in question satisfy the axioms of Sec. 1.11. The zero element of the space $\mathbf{K}/\mathbf{L}$ is the class containing the zero element of the space $\mathbf{K}$ and hence coinciding with the subspace $\mathbf{L}$ itself. The negative of the class $\mathbf{X}$ is the class consisting of the negatives of the elements of the class $\mathbf{X}$.† The resulting linear space $\mathbf{K}/\mathbf{L}$ is called the *factor space of the space* $\mathbf{K}$ *relative to the subspace* $\mathbf{L}$.

j. Morphisms of linear spaces. Let $\mathbf{X}$ and $\mathbf{Y}$ be two linear spaces over the same field K. Then a mapping $y = \omega(x)$ of $\mathbf{X}$ into $\mathbf{Y}$ is called a *morphism* (synonymously, a *homomorphism* or *linear operator*) of $\mathbf{X}$ into $\mathbf{Y}$ if

$$\omega(\alpha_1 x_1 + \alpha_2 x_2) = \alpha_1 \omega(x_1) + \alpha_2 \omega(x_2)$$

for every pair of elements x_1, $x_2 \in \mathbf{X}$ and every pair of numbers α_1, $\alpha_2 \in K$.

A morphism ω is called an *epimorphism* if it maps the space $\mathbf{X}$ onto the whole space $\mathbf{Y}$. A morphism ω mapping $\mathbf{X}$ onto part (or all) of $\mathbf{Y}$ in a one-to-one fashion (so that $x_1 \neq x_2$ implies $\omega(x_1) \neq \omega(x_2)$) is called a *monomorphism*. A morphism ω which is both an epimorphism and a monomorphism, i.e., which maps $\mathbf{X}$ onto all of $\mathbf{Y}$ in a one-to-one fashion, is called an *isomorphism* (in keeping with the general definition of an isomorphism of structures given in Vol. 1, Sec. 2.52). The notation for a morphism ω of $\mathbf{X}$ into $\mathbf{Y}$ is $\omega: \mathbf{X} \to \mathbf{Y}$.

Let $\mathbf{X}$ be a subspace of the space $\mathbf{Y}$. Then the mapping ω which assigns to every element $x \in \mathbf{X}$ the same element $x \in \mathbf{Y}$ is a monomorphism $\omega: \mathbf{X} \to \mathbf{Y}$, while the mapping ω' which assigns to every element $u \in \mathbf{Y}$ the class $\mathbf{U} \in \mathbf{Y}/\mathbf{X}$ containing u is an epimorphism $\omega': \mathbf{Y} \to \mathbf{Y}/\mathbf{X}$.

THEOREM. *Every n-dimensional space* $\mathbf{K}_n$ *over a field* K *is isomorphic to the n-dimensional coordinate space* K_n.

Proof. Let $f_1, \ldots, f_n$ be a system of n linearly independent vectors of the space $\mathbf{K}_n$. Then for every $x \in \mathbf{K}_n$ there exists a (unique) expansion $x =$

† The proofs of these assertions are given in Lin. Alg., Sec. 2.48c.

$\xi_1 f_1 + \cdots + \xi_n f_n$ (Sec. 1.14c). Assigning the vector $x' = (\xi_1, \ldots, \xi_n) \in K_n$ to the vector $x \in \mathbf{K}_n$, we get a one-to-one mapping $\omega: \mathbf{K}_n \to K_n$. It is easily verified that ω preserves linear operations and hence is an isomorphism. ▌

Thus, for example, R_n/R_m $(n<m)$ is isomorphic to R_{n-m}.

k. Direct products. Given two linear spaces **X** and **Y**, we can form the *direct product* $\mathbf{P}(\mathbf{X},\mathbf{Y})$ of X and Y, i.e., the set of all ordered pairs (x,y) with $x \in \mathbf{X}$, $y \in \mathbf{Y}$ (cf. Vol. 1, Sec. 2.82). Linear operations in the direct product are introduced by applying them to the separate components of the ordered pairs:

$$\alpha_1(x_1,y_1) + \alpha_2(x_2,y_2) = (\alpha_1 x_1 + \alpha_2 x_2, \alpha_1 y_1 + \alpha_2 y_2).$$

The validity of the axioms of Sec. 1.11 is easily verified. It is obvious that the space $\mathbf{P}(\mathbf{X},\mathbf{Y})$ contains the two subspaces

$$\mathbf{X}^* = \{(x,y): y=0\}, \qquad \mathbf{Y}^* = \{(x,y): x=0\}$$

which are isomorphic (as just defined) to the spaces **X** and **Y**, respectively. Note that

$$(x,y) = (x,0) + (0,y)$$

for every element $(x,y) \in \mathbf{P}(\mathbf{X},\mathbf{Y})$. This representation of (x,y) as a sum of terms belonging to $\mathbf{X}^*$ and $\mathbf{Y}^*$ is unique, by the definition of addition in $\mathbf{P}(\mathbf{X},\mathbf{Y})$ and equality of elements in $\mathbf{P}(\mathbf{X},\mathbf{Y})$. Thus the direct product of two spaces **X** and **Y** is the direct sum of its subspaces $\mathbf{X}^*$ and $\mathbf{Y}^*$, which are isomorphic to **X** and **Y**, respectively.

1.15. Linear operators

a. In the context of analysis, morphisms of linear spaces are most often called "linear operators." Thus by a *linear operator mapping a linear space* **X** *into a linear space* **Y** we mean a mapping $\mathbf{A}: \mathbf{X} \to \mathbf{Y}$ such that

$$\mathbf{A}(\alpha_1 x_1 + \alpha_2 x_2) = \alpha_1 \mathbf{A} x_1 + \alpha_2 \mathbf{A} x_2$$

for every $x_1, x_2 \in \mathbf{X}$ and $\alpha_1, \alpha_2 \in K$ (it is assumed that **X** and **Y** are over the same field K). If $\mathbf{X}=\mathbf{Y}$, **A** is said to be an *operator (acting) in the space* **X**.

b. The operator assigning the zero vector of the space **Y** to every vector $x \in \mathbf{X}$ is obviously a linear operator mapping **X** into **Y**. This operator is called the *zero operator*, denoted by **0**.

c. The operator which assigns to every vector $x \in \mathbf{X}$ the same vector $x \in \mathbf{X}$ is a linear operator acting in **X**. This operator is called the *unit operator* (or *identity operator*), denoted by **E**.

d. When the space $\mathbf{Y}$ is one-dimensional, the operator $\mathbf{A}$ is called a *linear functional*. This term is preferred in cases where $\mathbf{X}$ is an infinite-dimensional space; in the finite-dimensional case $\mathbf{A}$ is more often called a *linear form*.

e. Given two linear operators $\mathbf{A}_1$ and $\mathbf{A}_2$ mapping a space $\mathbf{X}$ into a space $\mathbf{Y}$, the operator $\mathbf{A}_1+\mathbf{A}_2$, called the *sum* of the operators $\mathbf{A}_1$ and $\mathbf{A}_2$, is defined by the formula

$$(\mathbf{A}_1+\mathbf{A}_2)x=\mathbf{A}_1x+\mathbf{A}_2x,$$

while the operator $\alpha\mathbf{A}_1$ where $\alpha \in K$, called the *product of the operator* $\mathbf{A}_1$ *and the number* α, is defined by the formula

$$(\alpha\mathbf{A}_1)x=\alpha(\mathbf{A}_1x).$$

Note that $\mathbf{A}_1+\mathbf{A}_2$ and $\alpha\mathbf{A}_1$ are both linear operators mapping $\mathbf{X}$ into $\mathbf{Y}$.

f. It is easily verified that the operations of addition of operators and multiplication of operators by numbers satisfy the axioms of Sec. 1.11, which characterize operations in a linear space. Hence *the set of all linear operators mapping a linear space* $\mathbf{X}$ *into a linear space* $\mathbf{Y}$ *is itself a linear space*. The zero element of this space is just the zero operator $\mathbf{0}$ of Sec. 1.15b.

g. Multiplication of operators. Let $\mathbf{B}$ be a linear operator mapping a space $\mathbf{X}$ into a space $\mathbf{Y}$, and let $\mathbf{A}$ be a linear operator mapping a space $\mathbf{Y}$ into a space $\mathbf{Z}$ (where all the spaces are over the same field K). Then the operator $\mathbf{P}=\mathbf{AB}$, called the *product of the operators* $\mathbf{A}$ *and* $\mathbf{B}$ (in that order), is defined as the operator mapping $\mathbf{X}$ into $\mathbf{Z}$ such that

$$\mathbf{P}x=(\mathbf{AB})x=\mathbf{A}(\mathbf{B}x)$$

for every $x \in \mathbf{X}$ (note that first the operator $\mathbf{B}$ acts on the vector $x \in \mathbf{X}$ and then the operator $\mathbf{A}$ acts on the resulting vector $\mathbf{B}x \in \mathbf{Y}$). Clearly $\mathbf{P}=\mathbf{AB}$ is a linear operator mapping the space $\mathbf{X}$ into the space $\mathbf{Y}$. Moreover, we have the formulas

$$\alpha(\mathbf{AB})=(\alpha\mathbf{A})\mathbf{B}=\mathbf{A}(\alpha\mathbf{B}),$$
$$\mathbf{A}(\mathbf{B}_1+\mathbf{B}_2)=\mathbf{AB}_1+\mathbf{AB}_2,$$
$$(\mathbf{A}_1+\mathbf{A}_2)\mathbf{B}=\mathbf{A}_1\mathbf{B}+\mathbf{A}_2\mathbf{B},$$
$$\mathbf{A}(\mathbf{BC})=(\mathbf{AB})\mathbf{C},$$

expressing the associative and distributive laws for multiplication of operators. Here $\alpha \in K$ is any number, $\mathbf{A}$, $\mathbf{A}_1$, and $\mathbf{A}_2$ are operators mapping a space $\mathbf{Y}$ into a space $\mathbf{Z}$, while $\mathbf{B}$, $\mathbf{B}_1$, and $\mathbf{B}_2$ are operators mapping a space $\mathbf{X}$ into a space $\mathbf{Y}$, and $\mathbf{C}$ is an operator mapping a space $\mathbf{W}$ into a space $\mathbf{X}$; the expressions on the left and right are operators mapping $\mathbf{X}$ into $\mathbf{Z}$ in the first three formulas, and an operator mapping $\mathbf{W}$ into $\mathbf{Z}$ in the last formula. Furthermore, if $\mathbf{E_X}$ is the unit operator in the space $\mathbf{X}$ and $\mathbf{E_Y}$ the unit operator in the space $\mathbf{Y}$, then

$$\mathbf{E_Y B} = \mathbf{B E_X} = \mathbf{B}$$

for any operator $\mathbf{B}$ mapping $\mathbf{X}$ into $\mathbf{Y}$.

Operators acting in a space $\mathbf{X}$, i.e., mapping a space $\mathbf{X}$ into itself, can be multiplied by one another in any order, and the result is again an operator acting in $\mathbf{X}$. But this multiplication is in general noncommutative, so that $\mathbf{AB} \neq \mathbf{BA}$ for many pairs of operators $\mathbf{A}$ and $\mathbf{B}$. As for the powers of an operator $\mathbf{A}$ acting in a space $\mathbf{X}$, they are defined as follows:

$$\mathbf{A}^0 = \mathbf{E_X},\ \mathbf{A}^1 = \mathbf{A},\ \mathbf{A}^2 = \mathbf{A}\cdot\mathbf{A},\ \ldots,\ \mathbf{A}^{k+1} = \mathbf{A}^k\cdot\mathbf{A},\ \ldots$$

h. Let

$$p(\lambda) = \sum_{k=0}^{n} a_k \lambda^k$$

be any polynomial with coefficients in a field K, and let $\mathbf{A}$ be any operator acting in a space $\mathbf{X}$. Then with $p(\lambda)$ we can associate the "operator polynomial"

$$p(\mathbf{A}) = \sum_{k=0}^{n} a_k \mathbf{A}^k,$$

which is also a linear operator acting in the same space $\mathbf{X}$ as $\mathbf{A}$ itself. Note that if

$$p(\lambda) = p_1(\lambda) + p_2(\lambda),$$

then

$$p(\mathbf{A}) = p_1(\mathbf{A}) + p_2(\mathbf{A}),$$

while if

$$p(\lambda) = p_1(\lambda) p_2(\lambda),$$

then

$$p(\mathbf{A}) = p_1(\mathbf{A}) p_2(\mathbf{A})$$

(Lin. Alg., Sec. 6.25e).

i. Given an operator $\mathbf{A}$ mapping a space $\mathbf{Y}$ into a space $\mathbf{X}$ and an operator $\mathbf{B}$ mapping $\mathbf{X}$ into $\mathbf{Y}$, suppose $\mathbf{AB} = \mathbf{E_X}$. Then the operator $\mathbf{A}$ is called a *left inverse* of $\mathbf{B}$, while $\mathbf{B}$ is called a *right inverse* of $\mathbf{A}$. If $\mathbf{B}$ maps $\mathbf{X}$ into itself, then, among all the operators acting in **X**, **B** may have both a left inverse, say **A**, and a right inverse, say **C**. If **B** is such an operator, these inverses necessarily coincide, since

$$\mathbf{A}=\mathbf{A}\mathbf{E}_{\mathbf{X}}=\mathbf{A}(\mathbf{B}\mathbf{C})=(\mathbf{A}\mathbf{B})\mathbf{C}=\mathbf{E}_{\mathbf{X}}\mathbf{C}=\mathbf{C}. \tag{4}$$

Moreover (4) implies that *every left (or right) inverse of the operator* **B** coincides with **A** = **C**. This uniquely defined operator **A** = **C** is called the *inverse* of the operator **B**, and is denoted by $\mathbf{B}^{-1}$.

In infinite-dimensional spaces there are operators with left inverses (even infinitely many left inverses) and no right inverses at all, or, conversely, right inverses and no left inverses at all (Lin. Alg., Sec. 4.76).

j. Given an operator **A** acting in a space **X**, a subspace $\mathbf{X}'\subset\mathbf{X}$ is said to be *invariant under the operator* **A** if $x\in\mathbf{X}'$ implies $\mathbf{A}x\in\mathbf{X}'$. A nonzero vector $f\in\mathbf{X}$ is said to be an *eigenvector* of an operator **A** (acting in the space **X**) if

$$\mathbf{A}f=\lambda f \quad (\lambda\in K). \tag{5}$$

The number λ figuring in (5) is called the *eigenvalue* of **A** corresponding to the eigenvector f. An eigenvector f obviously "generates" a one-dimensional invariant subspace, consisting of all vectors of the form αf, $\alpha\in K$.

k. A linear combination of eigenvectors of an operator **A**, all belonging to the same eigenvalue λ, is obviously itself an eigenvector of **A** with the same eigenvalue λ. It follows that *the set of all eigenvectors of the operator* **A** *corresponding to a given eigenvalue* λ *is a subspace of the space* **X**. This subspace is called the *eigenspace* (or *characteristic subspace*) of **A** corresponding to the eigenvalue λ.

l. THEOREM. *The eigenvectors* $f_1, f_2, \ldots, f_n$ *of the operator* **A** *corresponding to distinct eigenvalues* $\lambda_1, \lambda_2, \ldots, \lambda_n$ *are linearly independent.*

Proof. We use induction on the integer n. The theorem is obviously true for $n=1$. Assuming that the theorem is true for any $n-1$ eigenvectors of **A**, we now show that it remains true for any n eigenvectors of **A**. Suppose, to the contrary, that $f_1, f_2, \ldots, f_n$ are linearly dependent, so that there is a linear relation

$$\alpha_1 f_1+\alpha_2 f_2+\cdots+\alpha_n f_n=0 \tag{6}$$

between the eigenvectors $f_1, f_2, \ldots, f_n$, with $\alpha_1\neq 0$, say. Applying the operator **A** to this relation, we get

$$\alpha_1\lambda_1 f_1+\alpha_2\lambda_2 f_2+\cdots+\alpha_n\lambda_n f_n=0. \tag{7}$$

Multiplying (6) by λ_n and subtracting the result from (7), we find that

$$\alpha_1(\lambda_1-\lambda_n)f_1+\alpha_2(\lambda_2-\lambda_n)f_2+\cdots+\alpha_{n-1}(\lambda_{n-1}-\lambda_n)f_{n-1}=0,$$

which, by the induction hypothesis, implies that all the coefficients

$$\alpha_1(\lambda_1-\lambda_n),\ \alpha_2(\lambda_2-\lambda_n),\ \ldots,\ \alpha_{n-1}(\lambda_{n-1}-\lambda_n)$$

vanish, in particular that

$$\alpha_1(\lambda_1-\lambda_n)=0,$$

contrary to the assumption that $\alpha_1 \neq 0$, $\lambda_1 \neq \lambda_n$. This contradiction shows that the eigenvectors $f_1, f_2, \ldots, f_n$ must be linearly independent. ∎

1.16. Examples of linear operators

a. Let $A=\|a_{jk}\|$ be an $m\times n$ matrix (i.e., a matrix with m rows and n columns), made up of elements of the field K. Let $\mathbf{K}_n$ be an n-dimensional linear space with basis $e_1, \ldots, e_n$, and let $\mathbf{K}_m$ be an m-dimensional linear space with basis $f_1, \ldots, f_m$ (where both $\mathbf{K}_n$ and $\mathbf{K}_m$ are over the field K). Suppose that with every vector

$$x=\sum_{k=1}^{n}\xi_k e_k \in \mathbf{K}_n$$

we associate a vector

$$y=\sum_{j=1}^{n}\eta_j f_j \in \mathbf{K}_m,$$

in accordance with the rule

$$\eta_j=\sum_{k=1}^{n}a_{jk}\xi_k \qquad (j=1,\ldots,m).$$

This gives a linear operator mapping the space $\mathbf{K}_n$ into the space $\mathbf{K}_m$.

b. The continuous analogue of the operator in the preceding example is the operator

$$y(t)=\mathbf{A}x(t)=\int_a^b K(t,s)x(s)\,ds. \tag{8}$$

Here $x(t)$ is an element of the space $R^s(a,b)$,† $K(t,s)$ is a real function of two variables defined on the rectangle $a\leqslant t\leqslant b$, $c\leqslant s\leqslant d$, and $y(t)$ is a function defined on the interval $a\leqslant t\leqslant b$. It is easily verified that $y(t)$ is continuous on $[a,b]$ if $K(t,s)$ is continuous on the rectangle $a\leqslant t\leqslant b$, $c\leqslant s\leqslant d$. In this case, $\mathbf{A}$ is a linear operator mapping the space $R^s(a,b)$ into the space $R^s(c,d)$.

The operator (8) is called the *Fredholm (integral) operator*. Such operators will be considered in more detail in Sec. 1.98.

† I.e., the space of all continuous real functions on the interval $[a,b]$.

c. The *integration operator*

$$\mathbf{I}x(t) = \int_a^t x(\tau)\, d\tau \qquad (a \leqslant t \leqslant b)$$

is a special case of (8), mapping the space $R^s(a,b)$ into itself.

d. The expression

$$F(x) = \int_a^b f(\tau) x(\tau)\, d\tau,$$

involving a fixed (continuous) function $f(t)$, is an example of a linear functional defined on the space $R^s(a,b)$.

1.17. Linear operators in finite-dimensional spaces

a. We now find the general form of a linear operator $\mathbf{A}$ mapping an n-dimensional space $\mathbf{K}_n$ into an m-dimensional space $\mathbf{K}_m$. Let $e_1, \ldots, e_n$ be a basis in the space $\mathbf{K}_n$ and $f_1, \ldots, f_m$ a basis in the space $\mathbf{K}_m$. Applying the operator $\mathbf{A}$ to the basis vectors $e_1, \ldots, e_n$, we get

$$\begin{aligned} \mathbf{A}e_1 &= a_{11} f_1 + \cdots + a_{m1} f_m, \\ \mathbf{A}e_2 &= a_{12} f_1 + \cdots + a_{m2} f_m, \\ &\ldots \\ \mathbf{A}e_n &= a_{1n} f_1 + \cdots + a_{mn} f_m, \end{aligned} \tag{9}$$

where the a_{ij} are certain numbers in the field K. Thus, for fixed bases $e_1, \ldots, e_n$ and $f_1, \ldots, f_m$ in the spaces $\mathbf{K}_n$ and $\mathbf{K}_m$, the operator $\mathbf{A}$ gives rise to an $m \times n$ matrix†

$$A = \left\| \begin{matrix} a_{11} & a_{12} & \cdots & a_{1n} \\ a_{21} & a_{22} & \cdots & a_{2n} \\ \cdot & \cdot & \cdots & \cdot \\ a_{m1} & a_{m2} & \cdots & a_{mn} \end{matrix} \right\|,$$

where the kth column consists of the components of the vector $\mathbf{A}e_k$ in the basis $f_1, \ldots, f_m$.

Now let

$$x = \sum_{k=1}^{n} \xi_k e_k$$

† Note the distinction between the symbol $\mathbf{A}$ (boldface Roman) for an *operator* and the corresponding symbol A (lightface Italic) for the *matrix* of $\mathbf{A}$.

be any vector in $\mathbf{K}_n$, and let

$$\mathbf{A}x = \sum_{j=1}^{m} \eta_j f_j \in \mathbf{K}_m.$$

Then

$$\sum_{j=1}^{m} \eta_j f_j = \mathbf{A}x = \sum_{k=1}^{n} \xi_k \mathbf{A} e_k = \sum_{k=1}^{n} \xi_k \sum_{j=1}^{m} a_{jk} f_j = \sum_{j=1}^{m} \left(\sum_{k=1}^{n} a_{jk} \xi_k \right) f_j,$$

and hence

$$\eta_j = \sum_{k=1}^{n} a_{jk} \xi_k \qquad (j = 1, \ldots, m). \tag{10}$$

But (10) is just the rule figuring in Example 1.16a. Therefore Example 1.16a actually gives the general form of a linear operator mapping the space $\mathbf{K}_n$ into the space $\mathbf{K}_m$.

If $\mathbf{A}$ maps $\mathbf{K}_n$ into itself, then $m = n$ and the matrix A is square. If $\mathbf{A}$ maps $\mathbf{K}_n$ into a one-dimensional space $\mathbf{K}_1$, then $m = 1$ and the matrix of $\mathbf{A}$ reduces to

$$A = \| a_1 \;\; a_2 \;\; \cdots \;\; a_n \|.$$

In this case, the action of $\mathbf{A}$ is expressed by the formula

$$\mathbf{A}x = \sum_{k=1}^{n} a_k \xi_k$$

(where reference to the single basis vector in $\mathbf{K}_1$ is omitted), and represents a linear form (Sec. 1.15d).

b. We now find the matrix analogues of the algebraic operations on linear operators described in Sec. 1.15e. Let $\mathbf{A}_1$ and $\mathbf{A}_2$ be two linear operators mapping a space $\mathbf{X}$ with basis $e_1, \ldots, e_n$ into a space $\mathbf{Y}$ with basis $f_1, \ldots, f_m$. Moreover, let $A_1 = \|a_{ij}^{(1)}\|$ be the matrix of the operator $\mathbf{A}_1$ and $A_2 = \|a_{ij}^{(2)}\|$ the matrix of the operator $\mathbf{A}_2$ relative to these bases, so that

$$\mathbf{A}_1 e_j = \sum_{i=1}^{m} a_{ij}^{(1)} f_i, \qquad \mathbf{A}_2 e_j = \sum_{i=1}^{m} a_{ij}^{(2)} f_i \qquad (j = 1, \ldots, n).$$

Then, for arbitrary $\alpha_1, \alpha_2 \in K$,

$$(\alpha_1 \mathbf{A}_1 + \alpha_2 \mathbf{A}_2) e_j = \sum_{i=1}^{m} (\alpha_1 a_{ij}^{(1)} + \alpha_2 a_{ij}^{(2)}) f_i,$$

i.e., the matrix corresponding to the linear operator $\alpha_1 \mathbf{A}_1 + \alpha_2 \mathbf{A}_2$ is just $\|\alpha_1 a_{ij}^{(1)} + \alpha_2 a_{ij}^{(2)}\|$. Thus the matrix corresponding to the sum of two operators is obtained by element-by-element addition of the matrices of the operators themselves, while the matrix corresponding to the product of an operator and a number is obtained by multiplying every element of the matrix of the operator by the given number.

c. In particular, it follows that *the linear space* $\mathbf{L}(\mathbf{K}_n,\mathbf{K}_m)$ *of all linear operators mapping an n-dimensional space* $\mathbf{K}_n$ *into an m-dimensional space* $\mathbf{K}_m$ *is isomorphic to the nm-dimensional coordinate space* K_{nm} (recall Theorem 1.14j).

d. Next we find the matrix corresponding to the product of two operators. Choose a basis $e_1, \ldots, e_n$ in an n-dimensional space $\mathbf{X}$, a basis $f_1, \ldots, f_m$ in an m-dimensional space $\mathbf{Y}$, and a basis $g_1, \ldots, g_q$ in a q-dimensional space $\mathbf{Z}$. Let $\mathbf{B}$ be an operator mapping $\mathbf{X}$ into $\mathbf{Y}$, with $m \times n$ matrix $B = \|b_{jk}\|$, so that

$$\mathbf{B}e_k = \sum_{j=1}^{m} b_{jk} f_j \qquad (k=1,\ldots,n),$$

and let $\mathbf{A}$ be an operator mapping $\mathbf{Y}$ into $\mathbf{Z}$, with $q \times m$ matrix $A = \|a_{ij}\|$, so that

$$\mathbf{A}f_j = \sum_{i=1}^{q} a_{ij} g_i \qquad (j=1,\ldots,m).$$

Then for the product $\mathbf{P} = \mathbf{AB}$ we get

$$\begin{aligned}\mathbf{AB}e_k = \mathbf{A}(\mathbf{B}e_k) = \mathbf{A}\left(\sum_{j=1}^{m} b_{jk} f_j\right) &= \sum_{j=1}^{m} b_{jk}\mathbf{A}f_j \\ = \sum_{j=1}^{m}\sum_{i=1}^{q} b_{jk} a_{ij} g_i &= \sum_{i=1}^{q}\left(\sum_{j=1}^{m} a_{ij} b_{jk}\right) g_i .\end{aligned}$$

Hence the elements p_{ik} of the matrix P of the operator $\mathbf{P} = \mathbf{AB}$ are just

$$p_{ik} = \sum_{j=1}^{m} a_{ij} b_{jk} \qquad (i=1,\ldots,q;\ k=1,\ldots,n). \tag{11}$$

The matrix $P = \|p_{ik}\|$ obtained from the matrices $A = \|a_{ij}\|$ and $B = \|b_{jk}\|$ by formula (11) is called the *product of the matrices A and B* (in that order). Thus we can multiply a $q \times m$ matrix by an $m \times n$ matrix, and the result is a $q \times n$ matrix.

If $\mathbf{X} = \mathbf{Y} = \mathbf{Z}$, then A and B are square $n \times n$ matrices, and their product AB is also an $n \times n$ matrix.

e. Let $\mathbf{A}$ be an operator acting in an n-dimensional space $\mathbf{K}_n$, and let $\|a_{jk}\|$ be the matrix of $\mathbf{A}$ in some basis $e_1, \ldots, e_n$. Then the eigenvalues of the operator $\mathbf{A}$ (Sec. 1.15k) are just the roots of the *characteristic equation*†

† As usual, the determinant of a matrix A is denoted by $|A|$ or by $\det A$.

$$\begin{vmatrix} a_{11}-\lambda & a_{12} & \cdots & a_{1n} \\ a_{21} & a_{22}-\lambda & \cdots & a_{2n} \\ \cdot & \cdot & \cdots & \cdot \\ a_{n1} & a_{n2} & \cdots & a_{nn}-\lambda \end{vmatrix} = 0 \qquad (12)$$

(Lin. Alg., Sec. 4.94). If λ_0 is a root of equation (12), then the components of the corresponding eigenvector

$$f = \sum_{k=1}^{n} \xi_k e_k$$

are just the solutions of the (nontrivially compatible) homogeneous linear system

$$\begin{aligned} &(a_{11}-\lambda_0)\xi_1 + a_{12}\xi_2 + \cdots + a_{1n}\xi_n = 0, \\ &a_{21}\xi_1 + (a_{22}-\lambda_0)\xi_2 + \cdots + a_{2n}\xi_n = 0, \\ &\cdots \\ &a_{n1}\xi_1 + a_{n2}\xi_2 + \cdots + (a_{nn}-\lambda_0)\xi_n = 0. \end{aligned}$$

f. Following Lin. Alg., Chapter 6, we now describe the structure of an arbitrary linear operator **A** acting in an n-dimensional real or complex space.

Given any operator **A** acting in a complex space $\mathbf{C}_n$, we can represent $\mathbf{C}_n$ as a direct sum of subspaces invariant under **A**, in each of which the matrix of the operator **A** takes the form of a "Jordan block"

$$\begin{Vmatrix} \lambda & 1 & & & \\ & \lambda & 1 & & \\ & & \ddots & & \\ & & & \lambda & 1 \\ & & & & \lambda \end{Vmatrix} \qquad (13)$$

in a suitably chosen basis (the elements off the indicated diagonals are all zero). The basis of the space $\mathbf{C}_n$ obtained by combining the bases of all these invariant subspaces is called the *Jordan basis* of the operator **A**, and the matrix of **A** in this basis (a quasi-diagonal matrix with diagonal blocks of the form (13)) is called the *Jordan matrix* of the operator **A**. The numbers λ and the sizes of the Jordan blocks (13) are invariants of the operator **A**, i.e., they do not depend on the choice of the Jordan basis. In fact, the numbers λ are the roots of equation (12), and the sizes of the Jordan blocks can be found from the "elementary divisors" of the operator **A** (Lin. Alg., Sec. 6.44).

In particular, if all the roots of equation (12) are simple, the Jordan matrix of an operator **A** acting in a complex space $\mathbf{C}_n$ takes the diagonal form

$$\begin{Vmatrix} \lambda_1 & & & \\ & \lambda_2 & & \\ & & \ddots & \\ & & & \lambda_n \end{Vmatrix}. \tag{14}$$

***g.** Similarly, given any operator **A** acting in a real space $\mathbf{R}_n$, we can represent $\mathbf{R}_n$ as a direct sum of subspaces invariant under **A**, in each of which the matrix of the operator **A** is either of the form (13) or of the form

$$\begin{Vmatrix} \sigma & \tau & 1 & 0 & & & & & & \\ -\tau & \sigma & 0 & 1 & & & & & & \\ & & \sigma & \tau & 1 & 0 & & & & \\ & & -\tau & \sigma & 0 & 1 & & & & \\ & & & & & \ddots & & & & \\ & & & & & & \sigma & \tau & 1 & 0 \\ & & & & & & -\tau & \sigma & 0 & 1 \\ & & & & & & & & \sigma & \tau \\ & & & & & & & & -\tau & \sigma \end{Vmatrix} \tag{15}$$

(a "real Jordan block") in a suitably chosen basis. The basis of the space $\mathbf{R}_n$ obtained by combining the bases of all these invariant subspaces is called the *real Jordan basis* of the operator **A**, and the matrix of **A** in this basis (a quasi-diagonal matrix with diagonal blocks of the form (13) or (15)) is called the *real Jordan matrix* of the operator **A**. The numbers λ, σ, τ and the sizes of the Jordan blocks (13) and (15) do not depend on the choice of the real Jordan basis; the numbers λ and $\sigma+i\tau$ are the roots of equation (12), and the sizes of the Jordan blocks (13) and (15) can be found from the real elementary divisors of the operator **A** (Lin. Alg., Sec. 6.64).

***h.** In the case of a real space $\mathbf{R}_n$, if $\lambda=\sigma+i\tau$ is an imaginary (i.e., nonreal) root of equation (12), then so is the complex conjugate $\bar{\lambda}=\sigma-i\tau$. Suppose all the roots of (12) are simple, and let $\sigma_1 \pm i\tau_1, \ldots, \sigma_k \pm i\tau_k$ be the imaginary roots, while $\lambda_{2k+1}, \ldots, \lambda_n$ are the real roots. Then the real Jordan matrix of the operator **A** takes the form

$$\begin{Vmatrix} \sigma_1 & \tau_1 & & & & & & \\ -\tau_1 & \sigma_1 & & & & & & \\ & & \ddots & & & & & \\ & & & \sigma_k & \tau_k & & & \\ & & & -\tau_k & \sigma_k & & & \\ & & & & & \lambda_{2k+1} & & \\ & & & & & & \ddots & \\ & & & & & & & \lambda_n \end{Vmatrix}. \tag{16}$$

***i.** In a complex space, the Jordan form of every operator **A** with a *Hermitian* matrix $\|a_{jk}\|$ in some basis, so that $\bar{a}_{jk}=a_{kj}$ $(j, k=1, \ldots, n)$, takes the diagonal form (14), and the corresponding numbers λ_j (the eigenvalues of **A**) turn out to be real (Lin. Alg., Sec. 9.34). In a real space, the real Jordan matrix of every operator **A** with a *symmetric* matrix $\|a_{jk}\|$ in some basis, so that $a_{jk}=a_{kj}$ $(j, k=1, \ldots, n)$, also takes diagonal form (Lin. Alg., Sec. 9.45). However, if **A** has an *antisymmetric* matrix $\|a_{jk}\|$ in some basis, so that $a_{jk}=-a_{kj}$ $(j, k=1, \ldots, n)$, then the real Jordan matrix of **A** is of the form (16), where the numbers $\sigma_1, \ldots, \sigma_k, \lambda_{2k+1}, \ldots, \lambda_n$ are all zero (Lin. Alg., Sec. 9.46).

1.18. Algebras

a. A linear space **U** over a field K is called an *algebra* (more exactly, an algebra over K) if the elements of **U** are equipped with an operation of *multiplication*, denoted by $x \cdot y$ (or simply xy), which satisfies the following conditions:

(a) $\alpha(xy)=(\alpha x)y=x(\alpha y)$ for every $x, y \in \mathbf{U}$ and every $\alpha \in K$;
(b) $(xy)z=x(yz)$ for every $x, y, z \in \mathbf{U}$;
(c) $(x+y)z=xz+yz$ for every $x, y, z \in \mathbf{U}$;
(d) $x(y+z)=xy+xz$ for every $x, y, z \in \mathbf{U}$.

The first two conditions are called *associative* laws, the second two *distributive* laws.

b. In general, multiplication may not be commutative, i.e., the relation $xy=yx$ may fail to hold for certain pairs of elements $x, y \in \mathbf{U}$. If multiplication is commutative, so that $xy=yx$ for every pair of elements $x, y \in \mathbf{U}$, then the algebra **U** is said to be *commutative*.

c. An element $e \in \mathbf{U}$ is said to be the *unit* of the algebra **U** if $ex=xe=x$ for every $x \in \mathbf{U}$. Given any element $x \in \mathbf{U}$, the element $y \in \mathbf{U}$ is said to be the *inverse* of x if $xy=yx=e$. An element $x \in \mathbf{U}$ with an inverse is said to be *invertible*.†

d. A subspace $\mathbf{V} \subset \mathbf{U}$ is called a *subalgebra* of the algebra **U** if $x \in \mathbf{V}$, $y \in \mathbf{V}$ implies $xy \in \mathbf{V}$.

e. A subspace $\mathbf{J} \subset \mathbf{U}$ is called a *left ideal* of the algebra **U** if $x \in \mathbf{U}$, $y \in \mathbf{J}$ implies $xy \in \mathbf{J}$, and a *right ideal* of **U** if $y \in \mathbf{J}$, $z \in \mathbf{U}$ implies $yz \in \mathbf{J}$. An ideal which is both a left and a right ideal is called a *two-sided ideal*. In a commutative algebra there is obviously no distinction between left, right, and two-sided ideals.

† The inverse of an invertible element x is usual denoted by x^{-1}.

Every algebra **U** contains two trivial two-sided ideals, one called the *null ideal* consisting of the single element 0, the other coinciding with the whole algebra. All other ideals in **U** are called *proper ideals.*

f. Let **J** be a two-sided ideal of an algebra **U**. Then in the factor space **U**/**J** we can introduce an operation of multiplication for the classes **X**, **Y**, . . . in addition to the linear operations of Sec. 1.14i. In fact, given two classes **X** and **Y**, we choose arbitrary elements $x \in \mathbf{X}$, $y \in \mathbf{Y}$ and interpret **XY** as the class containing the product xy. It can be shown that this definition is well posed (in the sense that the class **XY** is independent of the choice of elements $x \in \mathbf{X}, y \in \mathbf{Y}$) and that the resulting operation of multiplication converts the space **U**/**J** itself into an algebra (Lin. Alg., Sec. 6.23b). This algebra is called the *factor algebra of the algebra* **U** *relative to the ideal* **J**.

g. Morphisms of algebras. Given two algebras **U** and **V** over the same field K, a mapping $\omega: \mathbf{U} \to \mathbf{V}$ of the algebra **U** into the algebra **V** is called a *morphism* (or *homomorphism*) of the algebra **U** into the algebra **V** if it is a morphism of the linear space **U** into the linear space **V** as defined in Sec. 1.14j and if moreover

$$\omega(x_1 x_2) = \omega(x_1)\omega(x_2) \tag{17}$$

for every pair of elements x_1, $x_2 \in \mathbf{U}$. A morphism ω which is an epimorphism, monomorphism, or isomorphism of the space **U** into the space **V** (Sec. 1.14j) is called an *epimorphism, monomorphism*, or *isomorphism* of the algebra **U** into the algebra **V**, provided the extra condition (17) is satisfied.

Thus the mapping $\omega: \mathbf{U} \to \mathbf{U}/\mathbf{J}$ which assigns to every element $x \in \mathbf{U}$ the class $\mathbf{X} \in \mathbf{U}/\mathbf{J}$ containing x is an epimorphism of the algebra **U** onto the algebra **U**/**J**.

h. Given any morphism $\omega: \mathbf{U} \to \mathbf{V}$, the set of elements $x \in \mathbf{U}$ for which $\omega(x) = 0$ is a two-sided ideal **J** in the algebra **U**. From a knowledge of ω, we can now define a morphism $\tilde{\omega}$ of the algebra **U**/**J** into the algebra **V**, by assigning the element $\omega(x) \in \mathbf{V}$ to the class $\mathbf{X} \in \mathbf{U}/\mathbf{J}$, where x is an arbitrary element of the class **X**. Then $\tilde{\omega}$ is a monomorphism (Lin. Alg., Sec. 6.25d). If the original morphism ω is an epimorphism of the algebra **U** into the algebra **V**, then the morphism $\tilde{\omega}$ is an isomorphism.

1.19. Examples of algebras and their morphisms

a. The set $\mathscr{P}$ of all polynomials

$$p(\lambda) = \sum_{k=0}^{n} a_k \lambda^k$$

(of all degrees) in λ with coefficients in a given field K, subject to the usual operations of addition and multiplication for polynomials, is obviously an algebra, in fact a commutative algebra with a unit.

b. The set $\mathscr{F}(G)$ of all analytic functions $f(\lambda)$ defined on an open set G of the complex plane forms a complex algebra with the usual operations of addition and multiplication for functions (Vol. 1, Sec. 4.72). This algebra is also commutative and has a unit. Consider the differentiation operator on $\mathscr{F}(G)$, i.e., the operator assigning to each function $f(\lambda) \in \mathscr{F}(G)$ its derivative $f'(\lambda)$. This operator is obviously linear and satisfies Leibniz's rule

$$[f(\lambda)g(\lambda)]^{(j)} = \sum_{i=0}^{j} \frac{j!}{i!\,(j-i)!} f^{(i)}(\lambda)g^{(j-i)}(\lambda). \tag{18}$$

c. The linear space of all linear operators acting in a linear space $\mathbf{K}$ forms an algebra (with the usual operations of addition and multiplication of operators), which is in general noncommutative.

d. By a *spectrum*, denoted by S, we mean any finite set of numbers $\lambda_1, \ldots, \lambda_m$ (from a field K), where each number λ_k $(k=1, \ldots, m)$ is assigned a positive integer r_k, called its *multiplicity*. By a *jet*, denoted by f, we mean any set of $r=r_1+\cdots+r_m$ numbers (from the field K), denoted by

$$f_{(j)}(\lambda_k) \qquad (j=0, \ldots, r_{k-1};\, k=1, \ldots, m). \tag{19}$$

Finally, we let $\mathscr{J}(S)$ denote the set of all jets on a given spectrum S.

We now use the following rules to introduce operations of addition and multiplication in $\mathscr{J}(S)$:

$$(f+g)_{(j)}(\lambda_k) = f_{(j)}(\lambda_k) + g_{(j)}(\lambda_k),$$
$$(\alpha f)_{(j)}(\lambda_k) = \alpha f_{(j)}(\lambda_k),$$
$$(fg)_{(j)}(\lambda_k) = \sum_{i=0}^{j} \frac{j!}{i!\,(j-i)!} f_{(i)}(\lambda_k)g_{(j-i)}(\lambda_k)$$

$(j=0, \ldots, r_{k-1};\, k=1, \ldots, m)$. For $j=0$ the last formula is replaced by

$$(fg)_{(0)}(\lambda_k) = f_{(0)}(\lambda_k)g_{(0)}(\lambda_k).$$

With these operations, the set $\mathscr{J}(S)$ becomes an algebra of dimension r over the field K (Lin. Alg., Sec. 6.71).

e. Given a linear operator $\mathbf{A}$ acting in an n-dimensional complex space $\mathbf{C}_n$, the set of all polynomials $p(\mathbf{A})$ in the operator $\mathbf{A}$ (Sec. 1.15h) with the natural operations of addition and multiplication of operators forms a complex algebra, which we denote by $\mathscr{P}(\mathbf{A})$. It turns out that this algebra is isomorphic to the algebra of jets $\mathscr{J}(S_{\mathbf{A}})$ of the preceding example, where $S_{\mathbf{A}}$ is the spectrum of the operator $\mathbf{A}$, i.e., the set of (distinct) eigenvalues λ_k of $\mathbf{A}$, each assigned a multiplicity r_k (a positive integer) equal to the maximum size of all the Jordan blocks of **A** (Sec. 1.17f) with the number λ_k along their diagonals. This isomorphism is established as follows: With every jet (19) defined on $S_{\mathbf{A}}$ we associate the operator $f(\mathbf{A})$ whose matrix in the Jordan basis of the operator **A** has the same quasi-diagonal structure as the matrix of **A** itself, except that every $p \times p$ Jordan block

$$\begin{Vmatrix} \lambda_k & 1 & 0 & \cdots & 0 \\ 0 & \lambda_k & 1 & \cdots & 0 \\ \cdot & \cdot & \cdot & \cdots & \cdot \\ 0 & 0 & 0 & \cdots & \lambda_k \end{Vmatrix} \qquad (p \leqslant r_k) \tag{20}$$

of the matrix of **A** is replaced by the following block of the same size:

$$\begin{Vmatrix} f_{(0)}(\lambda_k) & f_{(1)}(\lambda_k) & \frac{1}{2!}f_{(2)}(\lambda_k) & \cdots & \frac{1}{(p-1)!}f_{(p-1)}(\lambda_k) \\ 0 & f_{(0)}(\lambda_k) & f_{(1)}(\lambda_k) & \cdots & \frac{1}{(p-2)!}f_{(p-2)}(\lambda_k) \\ \cdot & \cdot & \cdot & \cdots & \cdot \\ 0 & 0 & 0 & \cdots & f_{(0)}(\lambda_k) \end{Vmatrix} \tag{21}$$

(Lin. Alg., Secs. 6.81–6.84). The operator $f(\mathbf{A})$ is actually of the form $p(\mathbf{A})$, where the polynomial $p(\lambda)$ satisfies the conditions

$$p^{(j)}(\lambda_k) = f_{(j)}(\lambda_k) \qquad (j=0, \ldots, r_{k-1};\ k=1, \ldots, m), \tag{22}$$

$p^{(j)}(\lambda)$ being the jth derivative of the polynomial $p(\lambda)$.

f. Again let **A** be a linear operator acting in an n-dimensional complex space $\mathbf{C}_n$, with spectrum $S_{\mathbf{A}}$ consisting of the distinct eigenvalues $\lambda_1, \ldots, \lambda_m$ of **A**. Consider the mapping ω of the algebra $\mathscr{F}(S_{\mathbf{A}})$ of all functions $f(\lambda)$ analytic on $S_{\mathbf{A}}$, i.e., analytic on some open set containing $S_{\mathbf{A}}$,† into the algebra of jets $\mathscr{J}(S_{\mathbf{A}})$ such that every function $f(\lambda) \in \mathscr{F}(S_{\mathbf{A}})$ is assigned the jet consisting of the numbers

$$f_{(j)}(\lambda_k) = f^{(j)}(\lambda_k) \qquad (j=0, \ldots, r_{k-1};\ k=1, \ldots, m), \tag{23}$$

where $f^{(j)}(\lambda)$ denotes the jth derivative of $f(\lambda)$. By Leibniz's rule (18), the mapping ω is a *morphism* of the algebra $\mathscr{F}(S_{\mathbf{A}})$ into the algebra $\mathscr{J}(S_{\mathbf{A}})$, in fact, an epimorphism, since for every jet (19) we can find a function $f(\lambda)$ in the algebra $\mathscr{F}(S_{\mathbf{A}})$ (in fact, a polynomial) satisfying (23). But the algebra $\mathscr{J}(S_{\mathbf{A}})$ is in turn isomorphic to the algebra $\mathscr{P}(\mathbf{A})$ of "operator polynomials"

† The open set need not be connected and may depend on $f(\lambda)$.

(see above). It follows that there exists an epimorphism of the algebra $\mathscr{F}(S_{\mathbf{A}})$ into the algebra $\mathscr{P}(\mathbf{A})$. According to the preceding example, this epimorphism is established as follows: With each function $f(\lambda) \in \mathscr{F}(S_{\mathbf{A}})$ we associate the linear operator $f(\mathbf{A})$ whose matrix in the Jordan basis of the operator $\mathbf{A}$ has the same quasi-diagonal structure as the matrix of $\mathbf{A}$ itself, with every Jordan block (20) replaced by a block

$$\left\| \begin{matrix} f(\lambda_k) & f'(\lambda_k) & \frac{1}{2}f''(\lambda_k) & \cdots & \frac{1}{(p-1)!}f^{(p-1)}(\lambda_k) \\ 0 & f(\lambda_k) & f'(\lambda_k) & \cdots & \frac{1}{(p-2)!}f^{(p-2)}(\lambda_k) \\ \cdot & \cdot & \cdot & \cdots & \cdot \\ 0 & 0 & 0 & \cdots & f(\lambda_k) \end{matrix} \right\| \qquad (21')$$

of the same size.

Thus operators like $e^{t\mathbf{A}}$, $\cos t\mathbf{A}$, $\sin t\mathbf{A}$, etc. always make sense. Since the mapping ω carrying $f(\lambda)$ into $f(\mathbf{A})$ is a morphism, the formula $f(\lambda)g(\lambda) = h(\lambda)$, where $f(\lambda), g(\lambda), h(\lambda)$ belong to $\mathscr{F}(S_{\mathbf{A}})$, implies $f(\mathbf{A})g(\mathbf{A}) = h(\mathbf{A})$. For example, the formula

$$e^{(t_1+t_2)\mathbf{A}} = e^{t_1\mathbf{A}}e^{t_2\mathbf{A}} \qquad (t_1, t_2 \in C)$$

always holds.

***g.** A spectrum S with complex $\lambda_1, \ldots, \lambda_m$ (Example 1.19d) is said to be *symmetric* if whenever S contains an imaginary number $\lambda_k = \sigma_k + i\tau_k$, it also contains the conjugate complex number $\bar{\lambda}_k = \sigma_k - i\tau_k$ with the same multiplicity r_k. A jet (19) defined on a symmetric spectrum S is said to be *symmetric* if

$$f_{(j)}(\lambda_k) = \overline{f_{(j)}(\bar{\lambda}_k)} \qquad (j=0, \ldots, r_{k-1}; k=1, \ldots, m).$$

The set of all symmetric jets on a symmetric spectrum S (with the operations indicated in Example 1.19d) forms a real algebra, which we denote by $\mathscr{J}_R(S)$.

***h.** Given a linear operator $\mathbf{A}$ acting in an n-dimensional real space R_n, the set of all real polynomials in the operator $\mathbf{A}$ forms a real algebra which we denote by $\mathscr{P}_R(\mathbf{A})$. It turns out that this algebra is isomorphic to the algebra of symmetric jets on the (always symmetric) spectrum $S_{\mathbf{A}}$ of the operator $\mathbf{A}$, in the complex extension (Lin. Alg., Sec. 6.61) of the real space R_n. The isomorphism is established as follows: To each symmetric jet (19) defined on $S_{\mathbf{A}}$ we assign the operator $f(\mathbf{A})$ whose matrix in the real Jordan basis of the operator **A** has the same quasi-diagonal structure as the real Jordan matrix of **A** itself, with every $p \times p$ Jordan block (20) involving real λ_k replaced by a block of the form (21), and with every $p \times p$ block of the form

$$\begin{Vmatrix} \Lambda_k & E & 0 & \cdots & 0 \\ 0 & \Lambda_k & E & \cdots & 0 \\ \cdot & \cdot & \cdot & \cdots & \cdot \\ 0 & 0 & 0 & \cdots & \Lambda_k \end{Vmatrix} \tag{24}$$

involving the 2×2 matrices

$$\Lambda_k = \begin{Vmatrix} \sigma_k & \tau_k \\ -\tau_k & \sigma_k \end{Vmatrix}, \qquad E = \begin{Vmatrix} 1 & 0 \\ 0 & 1 \end{Vmatrix}, \qquad 0 = \begin{Vmatrix} 0 & 0 \\ 0 & 0 \end{Vmatrix} \qquad (\lambda_k = \sigma_k + i\tau_k)$$

(cf. Sec. 1.17g) replaced by a block of the same form (21), but with Λ_k written for λ_k everywhere; the resulting expression $f_{(j)}(\Lambda_k)$ $(j=0, \ldots, p-1)$ then denotes the 2×2 matrix

$$f_{(j)}(\Lambda_k) = \begin{Vmatrix} \operatorname{Re} f_{(j)}(\lambda_k) & \operatorname{Im} f_{(j)}(\lambda_k) \\ -\operatorname{Im} f_{(j)}(\lambda_k) & \operatorname{Re} f_{(j)}(\lambda_k) \end{Vmatrix}$$

(Lin. Alg., Secs. 6.85–6.86). The operator $f(\mathbf{A})$ is actually of the form $p(\mathbf{A})$, where the polynomial $p(\lambda)$ has real coefficients and satisfies the conditions (22).

***i.** Again let **A** be a linear operator acting in an n-dimensional real space $\mathbf{R}_n$, with spectrum $S_\mathbf{A}$ in the complex extension $\mathbf{C}_n$ of the space $\mathbf{R}_n$. The morphism ω of Example 1.19f associates a symmetric jet (23) with every function $f(\lambda)$ which is *real analytic* on $S_\mathbf{A}$, i.e., which is analytic and satisfies the condition $f(\lambda) = \overline{f(\bar{\lambda})}$ on some open set containing $S_\mathbf{A}$ and symmetric with respect to the real axis (cf. Vol. 1, Sec. 10.39f). But the algebra $\mathcal{J}_R(S_\mathbf{A})$ of all symmetric jets on $S_\mathbf{A}$ is isomorphic to the algebra $\mathcal{P}_R(\mathbf{A})$ of all real polynomials in the operator **A**. It follows that there exists an epimorphism of the algebra $\mathcal{F}_R(S_\mathbf{A})$ of all real analytic functions on $S_\mathbf{A}$ into the algebra $\mathcal{P}_R(\mathbf{A})$. According to the preceding example, this epimorphism is established as follows: To each function $f(\lambda) \in \mathcal{F}_R(S_\mathbf{A})$ we assign the linear operator $f(\mathbf{A})$ whose matrix in the real Jordan basis of the operator **A** has the same quasi-diagonal structure as the real Jordan matrix of **A** itself, with every $p \times p$ Jordan block (20) involving real λ_k replaced by a block of the form (21′), and with every $p \times p$ block of the form (24) replaced by a block of the same form (21′), but with Λ_k written for λ_k everywhere; the resulting expression $f^{(j)}(\Lambda_k)$ $(j=0, \ldots, p-1)$ then denotes the 2×2 matrix

$$f^{(j)}(\Lambda_k) = \begin{Vmatrix} \operatorname{Re} f^{(j)}(\lambda_k) & \operatorname{Im} f^{(j)}(\lambda_k) \\ -\operatorname{Im} f^{(j)}(\lambda_k) & \operatorname{Re} f^{(j)}(\lambda_k) \end{Vmatrix}.$$

1.2. Metric Spaces

1.21. Metric spaces have already played a key role in the first volume of this course (Vol. 1, Chapter 3 ff.). We now continue our study of these spaces, first recalling the definition of a metric space.

a. A set M of elements ("points") $x, y, z, \ldots$ is called a *metric space* if for every pair of points x and y in M there is defined a number $\rho(x,y)$, called the "distance from x to y,"† satisfying the following conditions (axioms):

(a) $\rho(x,y) > 0$ if $x \neq y$, $\rho(x,x) = 0$ for every x;
(b) $\rho(y,x) = \rho(x,y)$ for every x and y (*symmetry of the distance*);
(c) $\rho(x,z) \leqslant \rho(x,y) + \rho(y,z)$ for every x, y, z (*triangle inequality*).

b. A sequence

$$x_1, x_2, \ldots, x_n, \ldots \tag{1}$$

of points in a metric space M is said to *converge* to a point $x_0 \in M$ if

$$\lim_{n\to\infty} (x,x_n) = 0. \tag{2}$$

If (2) holds, we also write $x_n \to x$ (as $n \to \infty$). The sequence (1), with "general term" x_n, is often simply called "the sequence x_n."

c. A point $x_0 \in M$ is said to be a *limit point* of a given subset $E \subset M$ if every "neighborhood" $U_{x_0}(\varepsilon) = \{x \in M: \rho(x_0,x) < \varepsilon\}$ of x_0 contains a point $x \in E$ distinct from x_0 itself; x_0 is said to be a *boundary point* of E if every neighborhood $U_{x_0}(\varepsilon)$ contains both points of E and points of $M - E$ (the complement of E). The set of all boundary points of E is called the *boundary* of E. A point $x_0 \in E$ is said to be *isolated* if x_0 has a neighborhood containing no points of M other than x_0 itself.

1.22. The examples of metric spaces given earlier were sets on the line, in the plane, and in ordinary (Euclidean) space, with the usual concept of distance. However, it is a very important fact that many sets of *functions* can also be regarded as metric spaces when equipped with a suitably defined "metric" $\rho(x,y)$. The particular metric chosen in a given "function space" depends on the requirements of the problem. Once a distance is specified, the "neighboring" elements are just those whose distance apart is small. On the other hand, in analysis we usually reverse this argument, i.e., the conditions of the problem tell us which elements should naturally be regarded as close together, and we then introduce a corresponding definition of the distance.

† Because of the symmetry of the distance, we can also call $\rho(x,y)$ the "distance between x and y."

a. For example, it is often natural to consider continuous functions $x(t)$ and $y(t)$ defined on an interval $a \leqslant t \leqslant b$ as being close together if the quantity

$$\max_{a \leqslant t \leqslant b} |x(t) - y(t)|$$

is small, so that the values of two functions are close together for all values of t. This quantity can be chosen as the definition of the distance between $x(t)$ and $y(t)$, and obviously satisfies Axioms a–c. Therefore any set M of continuous functions defined on an interval $[a,b]$ becomes a metric space when equipped with the distance

$$\rho(x, y) = \max_{a \leqslant t \leqslant b} |x(t) - y(t)|. \tag{3}$$

b. In some cases (for example, in the calculus of variations), when we deal with functions having continuous derivatives up to order m (inclusive), it is natural to regard two functions $x(t)$ and $y(t)$ as being close together if the values not only of the functions themselves but also of their derivatives up to order m are close together for all values of t. This is in keeping with the distance

$$\rho(x, y) = \max_{a \leqslant t \leqslant b} \{|x(t) - y(t)|, |x'(t) - y'(t)|, \ldots, |x^{(m)}(t) - y^{(m)}(t)|\}. \tag{4}$$

Hence any set of functions defined on $[a,b]$ with continuous derivatives up to order m becomes a metric space when equipped with the distance (4).

c. In other cases (for example, in the theory of integral equations), it is natural to regard the functions $x(t)$ and $y(t)$ as close together if they are close in the integral sense, i.e., if the quantity

$$\int_a^b |x(t) - y(t)| \, dt$$

is small. Here the natural choice of the distance is just

$$\rho(x, y) = \int_a^b |x(t) - y(t)| \, dt, \tag{5}$$

which obviously satisfies all the axioms of a metric space.

d. It is sometimes necessary to define closeness of functions by using an integral not of the first power but of some other power, say the pth, of the absolute value of the difference between two functions, with the corresponding distance being defined by the formula

$$\rho(x, y) = \sqrt[p]{\int_a^b |x(t) - y(t)|^p \, dt}. \tag{6}$$

For $p \geqslant 1$ this definition also satisfies the axioms of a metric space. Admittedly, the verification of Axiom c (except for the simple cases $p=1$ and $p=2$) is more complicated than before, and will not be given here (however, see Problem 15).

Thus, finally, we see that the definition of a metric space is flexible enough to satisfy the most varied concrete requirements of mathematical analysis.

1.23. The space of continuous functions on a metric space

a. As just noted, the set of all continuous real functions on an interval $[a,b]$, equipped with the distance (3), is a metric space. This space is denoted by $R^s(a,b)$, just as in Example 1.16b (where the same set appears as a linear space). It is now natural to ask whether the interval $[a,b]$ can be replaced by an arbitrary metric space in this definition. A formula like (3) cannot be the appropriate distance function in such a function space, since continuous functions on an arbitrary metric space M need not be bounded. However, suppose we allow only *bounded* continuous functions on the space M. Then a formula like (3) continues to make sense, provided we replace the symbol max by the symbol sup. Thus we define $R^s(M)$ as the *space of all bounded continuous real functions on a metric space* M, where the distance between two functions $x(t)$ and $y(t)$ is given by

$$\rho(x,y) = \sup_{t \in M} |x(t) - y(t)|. \tag{7}$$

b. More generally, suppose we in turn replace the real line (the range of the given functions) by an arbitrary metric space P. Then we get the space $P^s(M)$ of all bounded continuous functions defined on a metric space M and taking values in another metric space P, where this time the distance between two functions $x(t)$ and $y(t)$ is given by†

$$\rho(x,y) = \sup_{t \in M} \rho\{x(t), y(t)\}. \tag{8}$$

Here ρ denotes the distance in P on the right and the distance in $P^s(M)$ on the left (the meaning of ρ is usually clear from the context).

The rest of this section will be devoted to certain general concepts of the theory of metric spaces, as applied to the space $P^s(M)$ and its special cases.

c. As in Sec. 1.21, a sequence of points $x_1, \ldots, x_n, \ldots$ in the metric space P is said to *converge* to a point $x \in P$ if $\rho(x,x_n) \to 0$ as $n \to \infty$. It follows from (8) that convergence of a sequence $x_n(t)$ to a limit $x(t)$ in the space $P^s(M)$ is equivalent to *uniform* convergence (on M) of the sequence of functions $x_n(t)$ to the limit function $x(t)$.

† Note that (8) reduces to (7) if $P=R$.

d. A set E in a metric space P is said to be *(everywhere) dense relative to another set* $F \subset P$ if every point $x \in F$ is either a point of E or a limit point of E (Vol. 1, Sec. 3.61). If E is dense relative to F and if, in addition, E is a subset of F, we say that E is *dense in* F. A metric space P is said to be *separable* if it has a countable subset $E \subset P$ which is dense in P.

THEOREM. *The space* $R^s(a,b)$ *is separable.*

Proof. A countable dense set $E \subset R^s(a,b)$ can be formed, say, from all piecewise linear functions (polygonal lines) with vertices at the points

$$(a, y_0), (x_1, y_1), \ldots, (x_{n-1}, y_{n-1}), (b, y_n),$$

where $x_j \in (a,b)$ and the numbers x_j, y_j are rational. The countability of E follows from Vol. 1, Sec. 2.35. To show that E is dense in $R^s(a,b)$, we argue as follows: Given an arbitrary function $f(x) \in R^s(a,b)$ and any $\varepsilon > 0$, we choose $\delta > 0$ such that $|x' - x''| < \delta$ implies $|f(x') - f(x'')| < \varepsilon/5$. Let

$$a = x_0 < x_1 < \cdots < x_n = b$$

be a partition of the interval $[a,b]$ by rational points of subdivision x_j $(j = 1, \ldots, n-1)$, and let $\Delta x_j = x_{j+1} - x_j < \delta$. Moreover, let $y_0, y_1, \ldots, y_n$ be rational numbers such that

$$|y_j - f(x_j)| < \frac{\varepsilon}{5} \qquad (j = 1, \ldots, n-1),$$

and consider the piecewise linear function $y(x)$ with consecutive vertices at the points (x_j, y_j). Then we have $\rho(y, f) < \varepsilon$. In fact, given any $x \in [a,b]$, we can find an x_j such that $|x - x_j| < \delta$. But then $|f(x) - f(x_j)| < \varepsilon/5$ and $|f(x_{j\pm 1}) - f(x_j)| < \varepsilon/5$. It follows that

$$|y_j - y_{j\pm 1}| \leqslant |y_j - f(x_j)| + |f(x_j) - f(x_{j\pm 1})| + |f(x_{j\pm 1}) - y_{j\pm 1}| < \frac{3\varepsilon}{5}.$$

Therefore $|y(x) - y_j| < 3\varepsilon/5$ if $x \in (x_{j-1}, x_{j+1})$. Hence, finally,

$$|y(x) - f(x)| \leqslant |y(x) - y_j| + |y_j - f(x_j)| + |f(x_j) - f(x)| < \frac{3\varepsilon}{5} + \frac{\varepsilon}{5} + \frac{\varepsilon}{5} = \varepsilon,$$

$$\rho(y, f) = \max_{a \leqslant x \leqslant b} |y(x) - f(x)| < \varepsilon. \quad \blacksquare$$

It turns out that the space $R^s(0, \infty)$ of all bounded continuous functions on the half-line $0 \leqslant x < \infty$ no longer has a countable dense subset (see Problem 2).

e. A sequence of points $x_1, x_2, \ldots$ in a metric space P is called a *fundamental sequence* if, given any $\varepsilon > 0$, there exists an integer $N > 0$ such that $\rho(x_m, x_n) < \varepsilon$

for all $m, n > N$ (Vol. 1, Sec. 3.71a). A metric space P is said to be *complete* if every fundamental sequence in P has a limit in P (Vol. 1, Sec. 3.71d).

THEOREM. *Let $P^s(M)$ be the space of all bounded continuous functions on a metric space M with distance ρ, taking values in a complete metric space P and equipped with the distance*†

$$\tilde{\rho}(x,y) = \sup_{t \in M} \rho\{x(t),y(t)\}.$$

Then $P^s(M)$ is itself complete.

Proof. Let $x_1(t), x_2(t), \ldots$ be a fundamental sequence of functions in $P^s(M)$. Then, given any $\varepsilon > 0$, there exists an integer $N > 0$ such that

$$\tilde{\rho}(x_m,x_n) = \sup_{t \in M} \rho\{x_m(t),x_n(t)\} < \varepsilon$$

for all $m, n > N$. It follows that every sequence $x_n(t_0) \in P$ obtained for fixed $t = t_0$ is a fundamental sequence. But then

$$x(t_0) = \lim_{n \to \infty} x_n(t_0) \in P$$

exists, because of the completeness of P. Letting $t = t_0$ vary over all of M, we deduce the existence of the limit function

$$x(t) = \lim_{n \to \infty} x_n(t).$$

In the inequality

$$\rho\{x_m(t),x_n(t)\} < \varepsilon$$

valid for all t and $m, n > N$, we now take the limit as $m \to \infty$, holding n fixed. This gives

$$\rho\{x(t),x_n(t)\} \leqslant \varepsilon \tag{9}$$

for all t and $n > N$, by the continuity of the distance between a fixed point of P and a variable point of P (cf. Vol. 1, Sec. 5.12b). Hence the sequence of functions $x_n(t)$ converges to the limit $x(t)$ uniformly on M. But then $x(t)$ is bounded and continuous, being the uniform limit of a sequence of bounded continuous functions (Vol. 1, Secs. 5.94, 5.95b), so that $x(t)$ is an element of the space $P^s(M)$. The inequality (9) can be written in the form

$$\tilde{\rho}(x,x_n) \leqslant \varepsilon,$$

† Here, unlike formula (8), we make the distinction between the two distances ρ and $\tilde{\rho}$ explicit.

implying that $x = x(t) \in P$ is the limit of the sequence of elements $x_n \in P$. ∎

1.24. Arzelà's theorem

a. A metric space P is said to be *compact* if every sequence of points $x_1, x_2, \ldots$ in P contains a convergent subsequence $x_{n_1}, x_{n_2}, \ldots$† (Vol. 1, Secs. 3.42, 3.91a) and *precompact* if every sequence of points $x_1, x_2, \ldots$ in P contains a fundamental subsequence $x_{n_1}, x_{n_2}, \ldots$ (Vol. 1, Sec. 3.93a). A compact metric space is often called a *compactum*, and is necessarily complete (Vol. 1, Sec. 3.92a). Given a compact metric space Q and an arbitrary metric space P, let $P^s(Q)$ be the metric space of all complex functions $x = x(t)$ defined on Q and taking values in P, metrized in accordance with formula (8):

$$\rho(x,y) = \sup_t \rho\{x(t), y(t)\}.$$

The space $P^s(Q)$ will in general not be compact. We now look for conditions guaranteeing the compactness of subsets $E \subset P^s(Q)$. To this end, we introduce the following definitions:

Definition 1. A set E of functions $x(t) \in P^s(Q)$ is said to be *uniformly compact-valued* (*precompact-valued*) if there exists a compact (precompact) set $P_0 \subset P$ containing all values of the functions $x(t)$ for $t \in Q$, $x \in E$.

Definition 2. A set E of functions $x(t) \in P^s(Q)$ is said to be *equicontinuous* if, given any $\varepsilon > 0$, there exists a $\delta > 0$ such that $\rho(t', t'') < \delta$ implies $\rho\{x(t'), x(t'')\} < \varepsilon$ for every $x(t) \in E$.

b. THEOREM (**Arzelà**). *A set $E \subset P^s(Q)$ is precompact if and only if it is uniformly precompact-valued and equicontinuous.*

Proof. Suppose $E \subset P^s(Q)$ is precompact. Then, by Hausdorff's criterion (Vol. 1, Sec. 3.93c), given any $\varepsilon > 0$, there is an $(\varepsilon/3)$-net in E, i.e., a finite set of functions $x_1(t), \ldots, x_m(t)$ such that for each function $x(t) \in E$, the inequality

$$\rho\{x(t), x_k(t)\} \leqslant \frac{\varepsilon}{3} \tag{10}$$

holds for some integer k $(1 \leqslant k \leqslant m)$. For the same ε we now find a $\delta > 0$ such that $\rho(t', t'') < \delta$ implies

$$\rho\{x_k(t'), x_k(t'')\} < \frac{\varepsilon}{3}$$

for all $k = 1, \ldots, m$. Then, given any $x(t) \in E$ with the corresponding $x_k(t)$,

† With limit belonging to P, naturally.

$$\rho\{x(t'),x(t'')\} \leqslant \rho\{x(t'),x_k(t')\} + \rho\{x_k(t'),x_k(t'')\} + \rho\{x_k(t''),x(t'')\} < 3\frac{\varepsilon}{3} = \varepsilon$$

whenever $\rho(t',t'') < \delta$, i.e., the set E is equicontinuous. The set P_k of all values of a function $x_k(t)$ continuous on a compactum Q is itself compact for every $k=1, \ldots, m$ (Vol. 1, Sec. 5.16a). Moreover, the union R of the compact sets $P_1, \ldots, P_m$ is itself obviously compact. The inequality (10) shows that R is an $(\varepsilon/3)$-net for the set $P_0 \subset P$ of all values on Q of all functions $x(t) \in E$. But then P_0 is precompact (Vol. 1, Sec. 3.95), and hence the set E is precompact-valued.

Having just shown that the conditions of Arzelà's theorem are necessary, we now show that they are sufficient. First we note that the space $P^s(Q)$ is isometrically embedded (Vol. 1, Sec. 3.93c) in the space $P(Q)$ of all bounded functions $x(t)$ on Q, both continuous and discontinuous, equipped with the distance

$$\rho(x, y) = \sup_t \{x(t), y(t)\}.$$

By Hausdorff's criterion again, the sufficiency will be proved once we succeed in using the conditions of Arzelà's theorem to construct a finite ε-net (where $\varepsilon > 0$ is arbitrary) for the set E in the space $P(Q)$.

Thus let the set $E \subset P^s(Q)$ be equicontinuous and uniformly precompact-valued. Given any $\varepsilon > 0$, let $\delta > 0$ be a number such that $\rho(t',t'') < \delta$ implies

$$\rho\{x(t'),x(t'')\} < \frac{\varepsilon}{2}$$

for every $x(t) \in E$ (such a δ exists by the equicontinuity of E). We then cover the compactum Q by a finite number of balls of diameter δ. By omitting superfluous points, we can cover Q by a finite number of nonintersecting sets of diameter $\leqslant \delta$, which we denote by $Q_1, \ldots, Q_m$. Next, let $p_1, \ldots, p_k$ be a finite $(\varepsilon/2)$-net in P for the precompact set P_0 containing all values of the functions $x(t)$ for $t \in Q$, $x \in E$. Consider the set G of all functions $x(t) \in P(Q)$ taking constant values from the set $p_1, \ldots, p_k$ on the sets $Q_1, \ldots, Q_m$. There are obviously only a finite number of such functions (no more than k^m), and we now assert that they form an ε-net for the set E. In fact, let $x_0(t)$ be any function in E. This function varies by no more than $\varepsilon/2$ on the set Q_j of diameter $\leqslant \delta$, and hence there exists a point p_j from the set $p_1, \ldots, p_k$ whose distance from all the values of $x_0(t)$ for $t \in Q_j$ is no greater than ε. The function $x(t) \in P(Q)$ taking the corresponding value p_j on each Q_j belongs to the set G, and moreover

$$\rho(x_0,x) = \sup_t \{x_0(t),x(t)\} \leqslant \varepsilon$$

in the space $P(Q)$. Thus G is an ε-net for E, as asserted. ∎

c. Let P be a complete metric space, so that $P^s(Q)$ is also complete (Theorem 1.23e). A subset M of a complete metric space P is compact if and only if M is closed and precompact (Vol. 1, Sec. 3.96e). It follows from Arzelà's theorem that *the compact subsets of $P^s(Q)$ are just the closed subsets of $P^s(Q)$ which are uniformly (pre)compact-valued and equicontinuous* (cf. Vol. 1, Sec. 3.93a).

d. In the case where P is the n-dimensional Euclidean space R_n, the class of precompact sets is just the class of bounded sets (Vol. 1, Secs. 3.93b, 3.94). Hence uniform precompactness of a set E of functions $x(t) \in R_n^s(Q)$ means that there exists a constant $M > 0$ such that $\sup |x(t)| \leqslant M$ for all $t \in Q$ and $x(t) \in E$, in which case the set E is said to be *uniformly bounded.* Thus Arzelà's theorem takes the following form for $P = R_n$: *A set $E \subset R_n^s(Q)$ is precompact if and only if it is uniformly bounded and equicontinuous.*

e. Finally we give a simple sufficient condition for precompactness in the case of numerical functions defined on a closed interval of the real line:

THEOREM. *Given a set E of numerical functions $x(t)$ which are continuous and differentiable on an interval $a \leqslant t \leqslant b$, suppose there exist constants a_0 and a_1 such that*

$$|x(t)| \leqslant a_0, \qquad |x'(t)| \leqslant a_1$$

for all $x(t) \in E$. Then E is precompact in the space $R^s(a,b)$.

Proof. Applying Lagrange's theorem (Vol. 1, Sec. 7.44), we get the inequality

$$|x(t') - x(t'')| \leqslant |t'' - t'| \sup_t |x'(t)| \leqslant a_1 |t'' - t'|,$$

which shows that E is equicontinuous. But E is uniformly bounded, by hypothesis. We now need only apply the result of the preceding section. ∎

1.25. The space of functions with continuous derivatives up to a given order

a. Let $D_m(a,b)$ denote the metric space of all continuous real functions $x(t)$ defined on an interval $a \leqslant t \leqslant b$, with continuous derivatives up to order m, equipped with the distance

$$\rho(x, y) = \max_{a \leqslant t \leqslant b} \{|x(t) - y(t)|, |x'(t) - y'(t)|, \ldots, |x^{(m)}(t) - y^{(m)}(t)|\},$$

as in Sec. 1.22b. Note, in particular, that $D_0(a,b) = R^s(a,b)$. In the space $D_m(a,b)$, convergence of a sequence of functions $x_n(t)$ to a limit function $x(t)$ means uniform convergence of $m+1$ sequences:

$$x_n(t) \to x(t), \ x_n'(t) \to x'(t), \ \ldots, \ x_n^{(m)}(t) \to x^{(m)}(t).$$

b. THEOREM. *The space $D_m(a, b)$ is complete.*

Proof. Let $x_1(t), x_2(t), \ldots$ be a fundamental sequence of functions in the space $D_m(a,b)$. Then it follows from the inequality

$$\max_{a \leqslant t \leqslant b} |x_n^{(k)}(t) - x_p^{(k)}(t)| \leqslant \rho(x_n, x_p)$$

that each of the sequences $x_n(t), x_n'(t), \ldots, x_n^{(m)}(t)$ is itself fundamental in the metric of the space $R^s(a,b)$. But $R^s(a,b)$ is complete, by Theorem 1.23e, and hence as $n \to \infty$ each sequence $x_n^{(k)}(t)$ converges uniformly to a continuous function $y_k(t)$ $(k=0, 1, \ldots, m)$. By the theorem on differentiation of a uniformly convergent sequence of functions (Vol. 1, Sec. 9.106), we have

$$y_1(t) = \lim_{n\to\infty} x_n'(t) = \left\{\lim_{n\to\infty} x_n(t)\right\}' = y_0'(t),$$

$$y_2(t) = y_1'(t) = y_0''(t),$$

$$\ldots$$

$$y_m(t) = y_0^{(m)}(t),$$

so that $y_0(t)$ belongs to the space $D_m(a,b)$. Clearly $y_0(t)$ is the limit of the sequence $x_n(t)$ in the metric of the space $D_m(a,b)$, since as $n \to \infty$ each sequence $x_n^{(k)}(t)$ converges uniformly to $y_k(t) = y_0^{(k)}(t)$. ∎

c. THEOREM. *The space $D_m(a,b)$ is dense in the space $R^s(a,b)$.*†

Proof. Since the property of being dense is transitive (Vol. 1, Sec. 3.62), we need only show that $D_m(a,b)$ is dense relative to the set L of all piecewise linear functions on $[a,b]$.‡ Every piecewise linear function $y(x) \in L$ can be "smoothed out," by replacing it near each "corner" by a function $q(x)$ with continuous derivatives up to order m, whose values at the points where it joins the "linear portions" coincide with the values of the corresponding derivatives of the function $y(x)$ (see Figure 1). For example, $q(x)$ can be constructed in the form of a polynomial of degree $2m+1$. Finally, to complete the proof, we note that by carrying out a similarity transformation with center of similitude at each corner and a sufficiently small ratio of similitude, the deviation of $q(x)$ from $y(x)$ in the metric of $R^s(a,b)$ can be made arbitrarily small. ∎

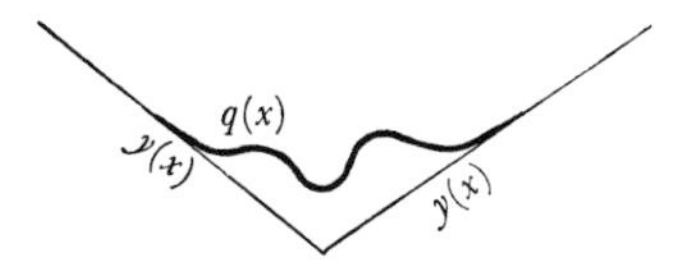

Figure 1

† In the metric of $R^s(a,b)$, naturally.

‡ The fact that L is dense in $R^s(a,b)$ has already been proved in Theorem 1.23d.

1.26. The space of continuous functions with an integral metric

a. Let $L^s(a,b)$ denote the space of all continuous real functions $x(t)$ defined on an interval $[a,b]$, equipped with the metric

$$\rho(x,y)=\int_a^b |x(t)-y(t)|\,dt, \tag{11}$$

as in Sec. 1.22c. The space of the same functions, equipped with the metric

$$\rho(x,y)=\sqrt[p]{\int_a^b |x(t)-y(t)|^p\,dt} \qquad (p\geqslant 1), \tag{12}$$

as in Sec. 1.22d, will be denoted by $L_p^s(a,b)$. The same space has also been equipped with the metric

$$\rho(x,y)=\max_{a\leqslant t\leqslant b} |x(t)-y(t)| \tag{13}$$

in Sec. 1.22a. The metrics (11), (12), and (13) lead to radically different kinds of convergence. Thus the sequence of functions $x_n(t)$ shown in Figure 2 does not converge to zero in the space $R^s(0,1)$, but does converge to zero in any of the spaces $L_p^s(0,1)$, since

$$\int_0^1 x_n^p(t)\,dt=\int_0^{1/n} x_n^p(t)\,dt\leqslant\frac{1}{n}.$$

On the other hand, the sequence of functions $y_n(t)$ shown in Figure 3 converges to zero in every space $L_p^s(0,1)$ with $p<q$, but does not converge to zero in $L_q^s(0,1)$, since

$$\int_0^1 y_n^p(t)\,dt=\frac{1}{p+1}\frac{1}{n}n^{p/q}=\frac{1}{p+1}n^{(p/q)-1}\begin{cases}\to 0 & \text{if } p<q,\\ \leqslant\dfrac{1}{p+1} & \text{if } p=q.\end{cases}$$

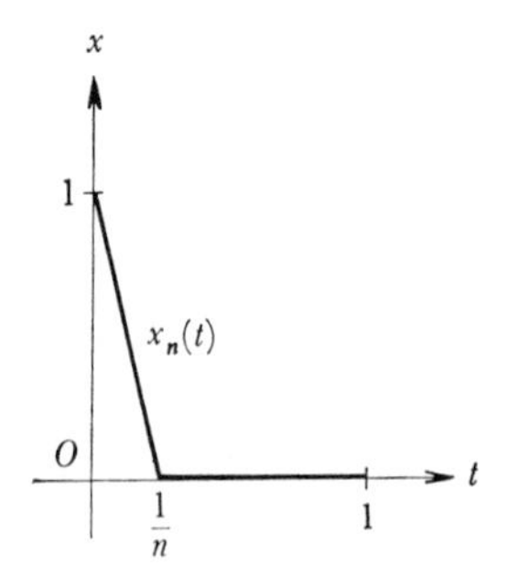

Figure 2

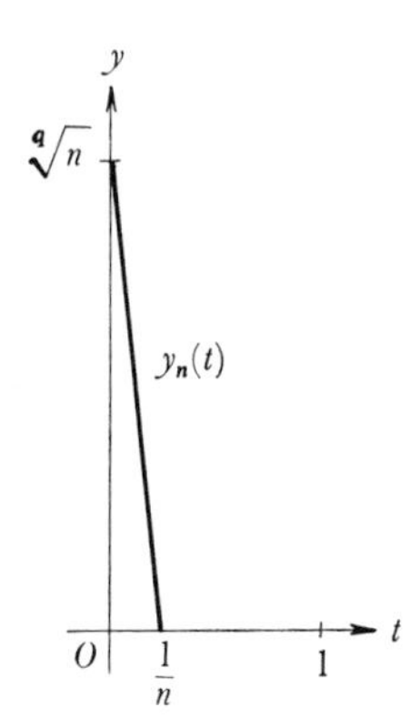

Figure 3

b. THEOREM. *The space $L_p^s(a,b)$ is incomplete for every $p \geqslant 1$.*

Proof. Let $y_n(x)$ be a sequence of continuous functions taking values between 0 and 1, and converging uniformly to 0 (as $m \to \infty$) on every interval $[a,c-\varepsilon]$ and to 1 on every interval $[c+\varepsilon,b]$, where c is a fixed point between a and b. Then $y_n(x)$ is a fundamental sequence in $L_p^s(a,b)$, since, given any $\varepsilon > 0$,

$$\int_a^b |y_m(x)-y_n(x)|^p\,dx = \left\{\int_a^{c-\varepsilon} + \int_{c-\varepsilon}^{c+\varepsilon} + \int_{c+\varepsilon}^b\right\}|y_m(x)-y_n(x)|^p dx$$
$$\leqslant \varepsilon + 2\varepsilon + \varepsilon = 4\varepsilon$$

for all sufficiently large m and n. As we now show, the sequence $y_m(x)$ cannot converge in the metric of $L_p^s(a,b)$ to any continuous function.

To this end, we first note that if a sequence $f_n(x)$ converges in the metric of $L_p^s(a,b)$ to a continuous function $f(x)$ on the interval $\Delta = [a,b]$, while converging uniformly to a function $\varphi(x)$ on some interval $\delta = [c,d]$ contained in Δ, then the identity $f(x) \equiv \varphi(x)$ holds on δ. In fact, in the space $L_p^s(c,d)$ we have

$$\rho^p(f_n,f) = \int_c^d |f_n(x)-f(x)|^p\,dx \leqslant \int_a^b |f_n(x)-f(x)|^p\,dx \to 0,$$

$$\rho^p(f_n,\varphi) = \int_c^d |f_n(x)-\varphi(x)|^p\,dx \leqslant (d-c)\max_{x\in\delta}|f_n(x)-\varphi(x)| \to 0,$$

so that $f(x) \equiv \varphi(x)$ on δ, as asserted, by the uniqueness of the limit (Vol. 1, Sec. 3.33a). Therefore, in particular, if the sequence $y_n(x)$ constructed above converges to a continuous function $f(x)$ in the metric $L_p^s(a,b)$, we must have $f(x) = 0$ for $a \leqslant x < c$ and $f(x) = 1$ for $c < x \leqslant b$. But then, obviously, $f(x)$ cannot be continuous on the interval $[a,b]$, regardless of the value of $f(c)$. ∎

c. By the general completion theorem for metric spaces (Vol. 1, Sec. 3.81), the space $L_p^s(a,b)$ has a completion $\bar{L}_p^s(a,b)$. It is now natural to ask whether the elements of $\bar{L}_p^s(a,b)$, as defined abstractly by the completion theorem, can be assigned a concrete meaning, by interpreting them as functions of some kind. It turns out that this can be done, but not very easily.†

1.3. Normed Linear Spaces

1.31. Suppose we want to furnish a given linear space **K** with a metric. Then it is natural to require that the metric and the linear operations be related to each other in such a way that shifting two points by one and the same vector does not change the distance between the points. It will then only be necessary to define the distance from each point $x \in \mathbf{K}$ to one fixed point, for example, the zero element 0; the number giving the distance from x to 0 is called the *norm* (or *length*) of x.

A real linear space **R** is called a (*real*) *normed linear space*‡ if for each vector $x \in \mathbf{R}$ there is defined a number $|x|$ (sometimes denoted by $\|x\|$), called the *norm* of x, satisfying the following axioms:

(a) $|x| > 0$ if $x \neq 0$, $|0| = 0$;
(b) $|\alpha x| = |\alpha|\ |x|$ for every $x \in \mathbf{R}$ and every real number α;
(c) $|x+y| \leqslant |x| + |y|$ for every $x, y \in \mathbf{R}$ (*triangle inequality*).

Setting

$$\rho(x, y) = |x-y|,$$

by definition, we can easily verify that this distance satisfies the axioms of a metric space (the details are left as an exercise). Thus a normed linear space is a metric space, a fact allowing us to measure distances between vectors and take limits. A complete normed linear space is called a *Banach space.* A linear space is said to be *affine* if it is not equipped with a norm (or metric).

1.32.a. The space $R^s(M)$ of all bounded continuous real functions $x(t)$ on a metric space M, considered as an example of a linear space in Sec. 1.13i and of a metric space in Sec. 1.23a, is also an important starting point for the construction of normed linear spaces. The norm in $R^s(M)$ is given by the formula

$$\|x\| = \sup_t |x(t)|. \tag{1}$$

† For example, see G.E. Shilov and B.L. Gurevich, *Integral, Measure and Derivative: A Unified Approach* (translated by R. A. Silverman), Prentice-Hall, Inc., N. J. (1966), Chapter 2.
‡ The case of complex normed linear spaces will be considered in Sec. 1.39.

Here we denote the norm of the function $x(t)$ by $\|x\|$, rather than by $|x|$, to emphasize the distinction between the norm of $x(t)$, regarded as an element of the space $R^s(M)$, and the absolute value of $x(t)$, regarded as a quantity depending on the point t. In the present case, the validity of Axioms a–c for a norm is almost obvious. In particular, to prove the triangle inequality, we start from

$$|x(t)+y(t)| \leqslant |x(t)|+|y(t)| \leqslant \sup_t |x(t)|+\sup_t |y(t)| = \|x\|+\|y\|,$$

and then take the least upper bound of the left-hand side, obtaining

$$\|x+y\| \leqslant \|x\|+\|y\|.$$

b. Next we generalize the preceding example, replacing the field of real numbers R by an arbitrary real normed linear space $\mathbf{R}$. Thus the elements of our new space $\mathbf{R}^s(M)$ are bounded continuous functions $x(t)$ defined on a metric space M and taking values in a normed linear space $\mathbf{R}$. To specify a norm in the space $\mathbf{R}^s(M)$, we continue to use formula (1), but this time $|x(t)|$ denotes the norm in the space $\mathbf{R}$ (reducing to the absolute value if $\mathbf{R}=R$). Of course, by saying that a function $x(t) \in \mathbf{R}^s(M)$ is bounded, we now mean that the number $\sup_t |x(t)|$ is finite (cf. Vol. 1, Sec. 1.92).

c. We have yet to supply a missing detail, by verifying that linear operations on functions in $\mathbf{R}^s(M)$ lead to functions which again belong to $\mathbf{R}^s(M)$. To this end, suppose the functions $x(t)$ and $y(t)$, defined on a metric space M and taking values in a normed linear space $\mathbf{R}$, are both bounded, so that the numbers

$$\sup_t |x(t)|, \qquad \sup_t |y(t)|$$

are both finite. But

$$|\alpha x(t)+\beta y(t)| \leqslant |\alpha x(t)|+|\beta y(t)| = |\alpha||x(t)|+|\beta||y(t)|$$

for arbitrary real α and β (why?), and hence

$$\sup_t |\alpha x(t)+\beta y(t)| \leqslant |\alpha| \sup_t |x(t)|+|\beta| \sup_t |y(t)|$$

is itself finite, i.e., $\alpha x(t)+\beta y(t)$ is itself a bounded function. Moreover, suppose the functions $x(t)$ and $y(t)$, defined on M and taking values in $\mathbf{R}$, are both continuous at the point $t=t_0$. Then, given any $\varepsilon>0$, we can find a $\delta>0$ such that $\rho(t,t_0)<\delta$ implies

$$|x(t)-x(t_0)|<\frac{\varepsilon}{2|\alpha|}, \qquad |y(t)-y(t_0)|<\frac{\varepsilon}{2|\beta|}$$

for arbitrary real α and β (which can be assumed to be nonzero). It follows that

$$|[\alpha x(t)+\beta y(t)]-[\alpha x(t_0)+\beta y(t_0)]|\leqslant|\alpha||x(t)-x(t_0)|+|\beta||y(t)-y(t_0)| < \frac{\varepsilon}{2}+\frac{\varepsilon}{2}=\varepsilon,$$

which proves the continuity of $\alpha x(t)+\beta y(t)$ at the point $t=t_0$. Thus if $x(t)$ and $y(t)$ both belong to $\mathbf{R}^s(M)$, then so does every linear combination

$\alpha x(t)+\beta y(t)$.

d. Finally, we note that if the space $\mathbf{R}$ is complete, then, by Theorem 1.23e, so is the space $\mathbf{R}^s(M)$.

1.33. Other examples of normed linear spaces

a. The metric space $D_m(a,b)$ of Sec. 1.25a can be made into a normed linear space by introducing the norm

$$\|x\|=\max\{|x(t)|,\ |x'(t)|,\ \ldots,\ |x^{(m)}(t)|\}.$$

The fact that this norm satisfies Axioms a–c of Sec. 1.31 is easily verified.

b. The metric space $L_p^s(a,b)$ of Sec. 1.26a can be made into a normed linear space by introducing the norm

$$\|x\|_p=\sqrt[p]{\int_a^b |x(t)|^p\,dt} \qquad (p\geqslant 1).$$

Here the norm axioms are easily verified for $p=1$, but it takes a little work to verify the triangle inequality in the general case $p>1$ (see Problem 15).

c. The normed linear spaces $L_p^s(a,b)$ and $R^s(a,b)$ have interesting finite-dimensional analogues. Suppose we introduce the following three norms in the n-dimensional space R_n of vectors $x=(\xi_1,\ \ldots,\ \xi_n)$:

$$|x|_1=\sum_{k=1}^{n}|\xi_k|, \tag{2}$$

$$|x|_p=\sqrt[p]{\sum_{k=1}^{n}|\xi_k|^p} \qquad (p>1), \tag{3}$$

$$|x|_\infty=\max_{1\leqslant k\leqslant n}|\xi_k|. \tag{4}$$

The *Euclidean norm*

$$|x|=\sqrt{\sum_{k=1}^{n}\xi_k^2} \tag{5}$$

(Vol. 1, Sec. 3.14a) is just the special case of the norm $|x|_p$ corresponding to $p=2$. The norms (2) and (3) are analogous to the integral norms in the spaces $L_1^s(a,b)$ and $L_p^s(a,b)$, while the norm (4) is analogous to the norm (1) in the space $R^s(a,b)$, the notation $|x|_\infty$ being motivated by the limiting relation

$$\max_{1\leqslant k\leqslant n} |\xi_k| = \lim_{p\to\infty} \sqrt[p]{\sum_{k=1}^{n} |\xi_k|^p}$$

(see Vol. 1, Chapter 4, Problem 9).

The fact that (5) satisfies Axioms a–c for a norm has already been verified in Vol. 1, Sec. 3.14a. The verification of these axioms is easy for the norms (2) and (4), but not so easy for the norm (3) (see Problem 17).

Unlike the corresponding norms in a function space, *the norms* (2)–(5) *all lead to exactly the same kind of convergence*; moreover, a sequence of vectors

$$x^{(m)} = (\xi_1^{(m)}, \ldots, \xi_n^{(m)}) \qquad (m=1, 2, \ldots)$$

converges to a vector $x=(\xi_1, \ldots, \xi_n)$ as $m\to\infty$ in any of the norms (2)–(5) if and only if the n numerical sequences $\xi_1^{(m)}, \ldots, \xi_n^{(m)}$ converge to the limits $\xi_1, \ldots, \xi_n$, respectively:

$$\lim_{m\to\infty} \xi_1^{(m)} = \xi_1, \ldots, \lim_{m\to\infty} \xi_n^{(m)} = \xi_n$$

(see Theorem 1.35e and its corollary).

d. Replacing n by ∞ in the preceding examples, we get an interesting family of infinite-dimensional spaces. More exactly, let l_p $(1\leqslant p<\infty)$ denote the set of all numerical sequences $x=(\xi_1, \ldots, \xi_k, \ldots)$† such that

$$\sum_{k=1}^{\infty} |\xi_k|^p < \infty,$$

and let

$$\|x\|_p = \sqrt[p]{\sum_{k=1}^{\infty} |\xi_k|^p}.$$

It follows from the triangle inequality for the norm $|x|_p$ in the space R_n, with $x=(\xi_1, \ldots, \xi_k, \ldots)\in l_p$, $y=(\eta_1, \ldots, \eta_k, \ldots)\in l_p$, that

$$\begin{aligned}\sqrt[p]{\sum_{k=1}^{n} |\xi_k+\eta_k|^p} &\leqslant \sqrt[p]{\sum_{k=1}^{n} |\xi_k|^p} + \sqrt[p]{\sum_{k=1}^{n} |\eta_k|^p} \\ &\leqslant \sqrt[p]{\sum_{k=1}^{\infty} |\xi_k|^p} + \sqrt[p]{\sum_{k=1}^{\infty} |\eta_k|^p} = \|x\|_p + \|y\|_p.\end{aligned}$$

† Here we write the infinite sequence $\xi_1, \ldots, \xi_k, \ldots$ in "vector notation" as $x=(\xi_1, \ldots, \xi_k, \ldots)$, i.e., as an "ordered ∞-tuple," generalizing the notion of an ordered n-tuple in an obvious way.

Taking the limit as $n\to\infty$ in the left-hand side, we find that the series

$$\sum_{k=1}^{\infty} |\xi_k+\eta_k|^p$$

converges and that the triangle inequality

$$\|x+y\|_p=\sqrt[p]{\sum_{k=1}^{\infty} |\xi_k+\eta_k|^p} \leqslant \|x\|_p+\|y\|_p$$

holds in the space l_p. It is obvious that if $x=(\xi_1, \ldots, \xi_k, \ldots)$ belongs to l_p, then so does $\alpha x=\{\alpha\xi_1, \alpha\xi_2, \ldots\}$ for every real α, and moreover that

$$\|\alpha x\|_p=|\alpha|\,\|x\|_p.$$

Thus *the set l_p is a normed linear space for every $p\geqslant 1$*. It can be shown that the space l_p is complete for every $p\geqslant 1$ (see Problem 19).

1.34. Naturally, all definitions and theorems involving affine linear spaces (without assuming the presence of a metric) and metric spaces (without assuming the presence of a linear structure) continue to hold in normed linear spaces. In particular, from the theory of affine linear spaces we can introduce the concept of a centrally symmetric set and that of a convex set:

a. A set E in a linear space X is said to be *centrally symmetric* if, whenever E contains a point x, it also contains the point $-x$.

b. A set E in a linear space X is said to be *convex* if, whenever E contains two points x and y, it also contains all points of the form $z=\alpha x+\beta y$, where $\alpha\geqslant 0$, $\beta\geqslant 0$, $\alpha+\beta=1$ (or, in geometric language, if, whenever E contains two points x and y, it also contains the whole segment with end points x and y).

c. The following theorem involves both a linear structure and a norm, and hence its natural context is that of a normed linear space:

THEOREM. *The ball $S=\{x:|x|\leqslant\rho\}$ in a normed linear space* $\mathbf{R}$ *is a centrally symmetric closed convex set.*

Proof. If $|x|\leqslant\rho$, then $|-x|=|x|\leqslant\rho$ as well, and hence S is centrally symmetric. The fact that S is closed has already been noted (Vol. 1, Sec. 3.51). The convexity of S follows from the triangle inequality, since if $|x|\leqslant\rho$, $|y|\leqslant\rho$, then

$$|\alpha x+\beta y|\leqslant|\alpha||x|+|\beta||y|\leqslant(\alpha+\beta)\rho=\rho$$

for $\alpha\geqslant 0$, $\beta\geqslant 0$, $\alpha+\beta=1$.

d. The convexity of the unit ball in a normed linear space is such an essential property that it can be substituted for the triangle inequality itself in our axiomatics. More exactly, we can replace Axiom c of Sec. 1.31 by the following new axiom:

(c′) The unit ball $\{x \in \mathbf{X}: |x| \leqslant 1\}$ is a convex set.

The legitimacy of this replacement is shown by the following

THEOREM. *Axioms* a, b, *and* c′ *together imply Axiom* c.

Proof. Given any nonzero vectors $x, y \in \mathbf{X}$, the vectors

$$\frac{x}{|x|}, \frac{y}{|y|}$$

(of unit norm) belong to the unit ball. Hence, by Axiom c′, the vector

$$\frac{\alpha x}{|x|}+\frac{\beta y}{|\beta|}$$

also lies in the unit ball for arbitrary $\alpha \geqslant 0$, $\beta \geqslant 0$, $\alpha+\beta=1$, i.e.,

$$\left|\alpha\frac{x}{|x|}+\beta\frac{y}{|y|}\right| \leqslant 1.$$

Choosing

$$\alpha=\frac{|x|}{|x|+|y|}, \qquad \beta=\frac{|y|}{|x|+|y|},$$

we get

$$\left|\frac{|x|}{|x|+|y|}\frac{x}{|x|}+\frac{|y|}{|x|+|y|}\frac{y}{|y|}\right|=\frac{1}{|x|+|y|}|x+y| \leqslant 1,$$

or equivalently,

$$|x+y| \leqslant |x|+|y|, \tag{6}$$

as required. The triangle inequality (6) is trivial if one or both of the vectors $x, y \in \mathbf{X}$ is zero. ∎

1.35. Equivalent norms

a. Two norms $|x|_1$ and $|x|_2$ in the same linear space $\mathbf{X}$ are said to be *equivalent* (or *homeomorphic*) if they generate homeomorphic metrics (Vol. 1, Sec. 3.34), i.e., if convergence of x_n to x in one metric implies convergence of x_n to x in the other metric, and conversely. Therefore every set in X which is closed (open) relative to one metric is closed (open) relative to the other metric (why?).

b. Next we show how equivalence of two norms $|x|_1$ and $|x|_2$ expresses itself in terms of the geometric properties of the balls

$$S_1(\rho)=\{x\in\mathbf{X}: |x|_1\leqslant\rho\},\qquad S_2(\rho)=\{x\in\mathbf{X}: |x|_2\leqslant\rho\}.$$

According to Theorem 1.34c, these balls are centrally symmetric convex sets, which are closed with respect to the appropriate norms.

LEMMA. *Two norms $|x|_1$ and $|x|_2$ are equivalent if and only if there exist positive constants c_1 and c_2 such that every ball $S_1(\rho)$ contains the ball $S_2(c_1\rho)$ while every ball $S_2(\rho)$ contains the ball $S_1(c_2\rho)$.*

Proof. Let c_1 and c_2 be constants with the indicated properties, and suppose $|x-x_n|_1=\varepsilon_n\to 0$. Since the sphere $S_2(\varepsilon_n/c_2)$ contains the sphere $S_1(\varepsilon_n)$, it also contains the element $x-x_n$, so that $|x-x_n|_2\leqslant\varepsilon_n/c_2$. But then $|x-x_n|_2\to 0$. Similarly, we find that $|x-x_n|_2\to 0$ implies $|x-x_n|_1\to 0$. Hence the norms $|x|_1$ and $|x|_2$ are equivalent.

Conversely, suppose the norms $|x|_1$ and $|x|_2$ are equivalent, but the required constant c_1 does not exist. Then for any $n=1,2,\ldots$ we can find balls $S_1(\rho_n)$, $S_2(\rho_n/n)$ such that the first does not contain the second, i.e., there exists a point x_n such that $|x_n|_1>\rho_n$, $|x_n|_2\leqslant\rho_n/n$. Setting $y_n=x_n/\rho_n$, we have $|y_n|_1>1$, $|y_n|_2\leqslant 1/n$. Hence the sequence y_n converges to 0 in the second norm, but not in the first norm. This contradicts the assumed equivalence of the two norms, thereby proving the existence of c_1. The existence of c_2 is proved similarly. ∎

c. COROLLARY. *Two norms $|x|_1$ and $|x|_2$ are equivalent if and only if there exist positive constants c_1 and c_2 such that*

$$c_1|x|_1\leqslant|x|_2\leqslant\frac{|x|_1}{c_2}\tag{7}$$

for every $x\in\mathbf{X}$.

Proof. If (7) holds, then $|x-x_n|_1\to 0$ implies

$$|x-x_n|_2\leqslant\frac{1}{c_2}|x-x_n|_1\to 0,$$

and similarly $|x-x_n|_2\to 0$ implies

$$|x-x_n|_1\leqslant\frac{1}{c_1}|x-x_n|_2\to 0,$$

so that the norms $|x|_1$ and $|x|_2$ are equivalent. Conversely, if $|x|_1$ and $|x|_2$ are equivalent, then, by the lemma, there exists a constant c_2 such that every ball $S_2(\rho)$ contains the ball $S_1(c_2\rho)$. Let $|x|_1=a$, so that $x\in S_1(a)$. The ball

$S_2(a/c_2)$ contains the ball $S_1(a)$, and hence $|x|_2 \leqslant a/c_2 = |x|_1/c_2$. The other inequality in (7) is proved similarly. ∎

d. We now give a geometric description of every norm $|x|_2$ equivalent to a given norm $|x|_1$:

THEOREM. *Given a normed linear space* $\mathbf{R}$ *with norm* $|x|_1$, *let* S *be a centrally symmetric convex set containing some ball* $S_1(\rho)$ *and contained in some ball* $S_1(r)$. *Then there exists a norm* $|x|_2$ *equivalent to the norm* $|x|_1$ *such that* $S_2(1) = S$.

Proof. Given any nonzero vector $x \in \mathbf{R}$, consider the ray $\{x/t: 0<t<\infty\}$. By our assumptions, the points of this ray with sufficiently large t belong to the set S, while the points with sufficiently small t do not belong to S. Let

$$|x|_2 = \inf\left\{t: \frac{x}{t} \in S\right\}, \qquad |0|_2 = 0.$$

Then, as we now show, $|x|_2$ satisfies the norm axioms of Sec. 1.31 and the required condition $\{x: |x|_2 \leqslant 1\} = S$.

Since $x \neq 0$, we have $0 < |x|_2 < \infty$, so that Axiom a is satisfied. Moreover, if $\alpha > 0$, then

$$|\alpha x|_2 = \inf\left\{t: \frac{\alpha x}{t} \in S\right\} = \inf\left\{\alpha \frac{t}{\alpha}: \frac{\alpha x}{t} \in S\right\} = \alpha \inf\left\{\tau: \frac{x}{\tau} \in S\right\} = \alpha |x|_2.$$

Moreover, it is obvious from the central symmetry of the set S that $|-x|_2 = |x|_2$, and hence $|\alpha x|_2 = |\alpha||x|_2$ for arbitrary real α, so that Axiom b is satisfied.

Next we verify that $S = \{x: |x|_2 \leqslant 1\}$. If $x \in S$, then obviously $|x|_2 = \inf\{t: x/t \in S\} \leqslant 1$. Moreover, given any $x \in \mathbf{R}$, the convexity of S implies the convexity of the set S_x of points of the ray $\{x/t: 0<t<\infty\}$ which belong to S. Hence S_x contains *all* points x/t with $t > \inf\{\tau: x/\tau \in S\}$. But S is closed, and hence the point x/t with $t = \inf\{\tau: x/\tau \in S\}$ also belongs to S. Therefore if $|x|_2 = \inf\{\tau: x/\tau \in S\} \leqslant 1$, then $x = x/1$ belongs to S, as required. As for the triangle inequality (Axiom c), it follows from Theorem 1.34d, since the ball $\{x: |x|_2 \leqslant 1\}$ coincides with the set S, as just shown, and since S is convex, by hypothesis.

Finally, to verify that the norm $|x|_2$ is equivalent to the norm $|x|_1$, we apply the lemma, noting that

$$S_1(\rho) \subset S = S_2(1) \subset S_1(r)$$

implies

$$S_1(\rho p) \subset S_2(p) \subset S_1(rp)$$

for arbitrary $p > 0$ (why?), so that the conditions of the lemma are satisfied for $c_1 = 1/r$ and $c_2 = \rho$. ∎

e. Norms in finite-dimensional spaces. We now prove a result anticipated in Sec. 1.33c.

THEOREM: *All norms are equivalent in a finite-dimensional linear space* $\mathbf{R}_n$.

Proof. Since equivalence of norms is obviously a transitive relation (two norms equivalent to a third are necessarily equivalent to each other), we need only show that an arbitrary norm $|x|_1$ is equivalent to the Euclidean norm

$$|x|_2 = \sqrt{\sum_{k=1}^{n} \xi_k^2},$$

where $\xi_1, \ldots, \xi_n$ are the components of the vector in some basis $e_1, \ldots, e_n$. By the triangle inequality for the norm $|x|_1$, we have

$$|x|_1 = \left|\sum_{k=1}^{n} \xi_k e_k\right|_1 \leqslant \sum_{k=1}^{n} |\xi_k| |e_k|_1 \leqslant |x|_2 \sum_{k=1}^{n} |e_k|_1 \leqslant c_1 |x|_2 \tag{8}$$

for every $|x|_2$, where

$$c_1 = \sum_{k=1}^{n} |e_k|_1.$$

Moreover, there exists a constant c_2 such that

$$|x|_1 \geqslant c_2 |x|_2. \tag{9}$$

To prove this, suppose to the contrary that there is no such c_2. Then there exists a sequence of vectors $x^{(m)}$ $(m=1, 2, \ldots)$ such that

$$|x^{(m)}|_1 < \frac{1}{m} |x^{(m)}|_2. \tag{10}$$

Writing

$$y^{(m)} = \frac{x^{(m)}}{|x^{(m)}|_2} = (\eta_1^{(m)}, \ldots, \eta_n^{(m)}) \qquad (m=1, \ldots, n), \tag{11}$$

we get

$$|y^{(m)}|_2^2 = \sum_{k=1}^{n} (\eta_k^{(m)})^2 = 1, \tag{12}$$

so that $|\eta_k^{(m)}| \leqslant 1$ for all k and m. Since the unit Euclidean ball is compact (Vol. 1, Sec. 3.96e), the sequence (11) has a convergent subsequence. By dropping superfluous vectors and changing the numbering, we can assume that the sequence (11) itself converges to some vector $y = (\eta_1, \ldots, \eta_n)$, where in fact

$$\eta_1 = \lim_{m\to\infty} \eta_1^{(m)}, \ldots, \eta_n = \lim_{m\to\infty} \eta_n^{(m)}$$

(Vol. 1, Sec. 3.32f). Then, taking the limit as $m\to\infty$ in (12), we get

$$|y|_2^2 = \sum_{k=1}^{n} \eta_k^2 = 1,$$

so that $y\neq 0$. On the one hand, using (8), we have

$$|y-y^{(m)}|_1 \leqslant c_1 |y-y^{(m)}|_2 \to 0,$$

so that $y^{(m)}\to y\neq 0$ in the norm $|x|_1$; on the other hand,

$$|y^{(m)}|_1 = \frac{|x^{(m)}|_1}{|x^{(m)}|_2} \to 0,$$

because of (10), so that $y^{(m)}\to 0$ in the same norm. But this is impossible, by the uniqueness of the limit (Vol. 1, Sec. 3.33a). It follows by contradiction that (9) holds, as well as (8). To complete the proof, we now need only apply Corollary 1.35c. ∎

In particular, since the space $\mathbf{R}_n$ is complete in the Euclidean norm $|x|_2$ (Vol. 1, Sec. 3.72d), it is also complete in every other norm $|x|_1$.

f. COROLLARY.† *A sequence $x^{(m)} = (\xi_1^{(m)}, \ldots, \xi_n^{(m)})$ in a finite-dimensional space R_n converges to a limit $x = (\xi_1, \ldots, \xi_n)$ in some (and hence every) norm if and only if*

$$\lim_{m\to\infty} \xi_1^{(m)} = \xi_1, \ldots, \lim_{m\to\infty} \xi_n^{(m)} = \xi_n.$$

Proof. Merely note that these conditions are equivalent to convergence in the norm

$$|x|_\infty = \max_{1\leqslant k\leqslant n} |\xi_k|. \quad ∎$$

Figures 4–8 show the unit spheres of the norms $|x|_\infty$, $|x|_1$, $|x|_p$ of Sec. 1.33 for the case $n=2$. Figure 9 shows a norm of a type different from the others. Concerning the case $p<1$, see Problem 20.

1.36. The ball $|x|\leqslant 1$ in n-dimensional Euclidean space is compact (Vol. 1, Sec. 3.96e). The ball $\|x\|\leqslant 1$ in an arbitrary n-dimensional normed linear space is also compact, since every norm is equivalent to the Euclidean norm. Does there exist an infinite-dimensional normed linear space with a compact unit ball $\|x\|\leqslant 1$? It turns out that there does not, so that compactness of

† This result has also been anticipated in Sec. 1.33c.

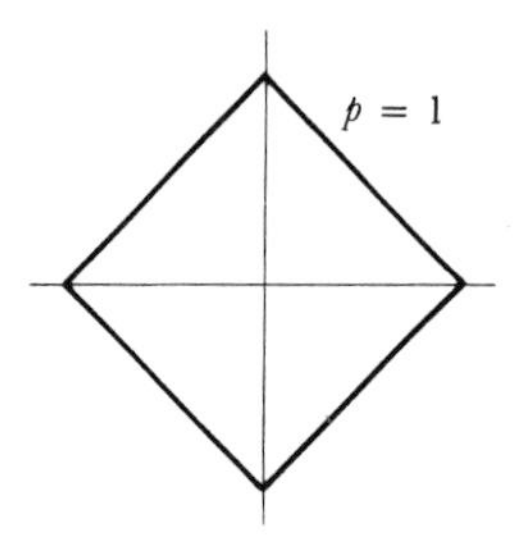

Figure 4

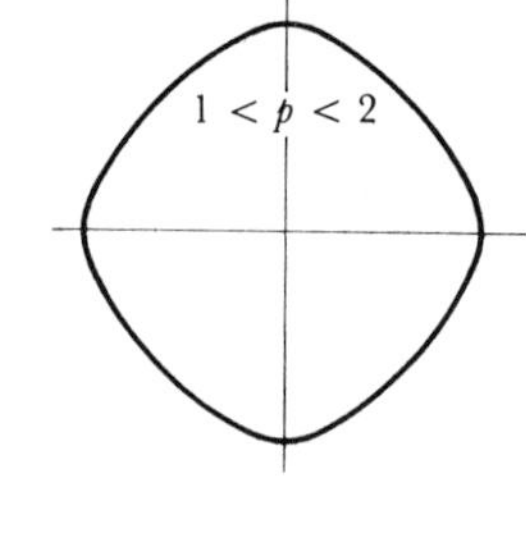

Figure 5

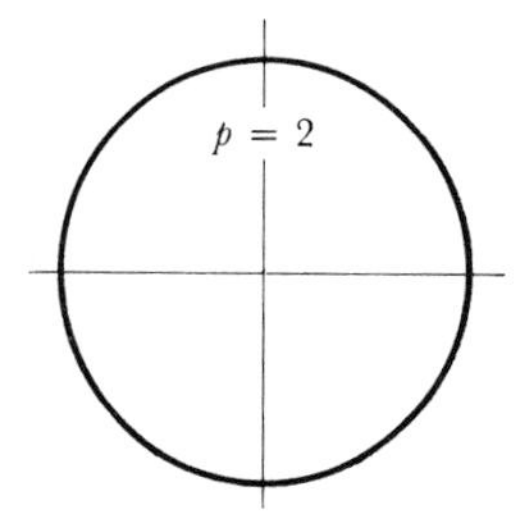

Figure 6

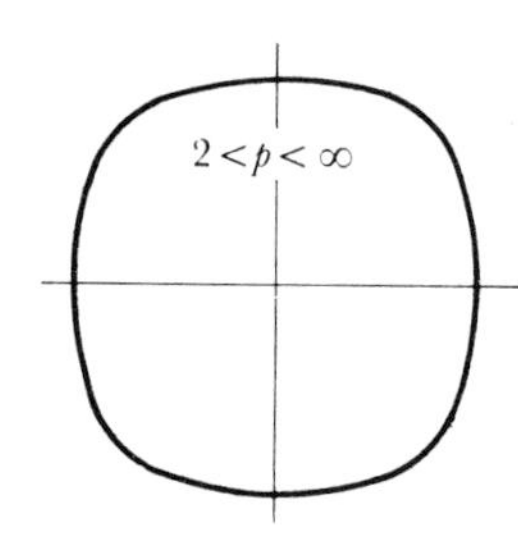

Figure 7

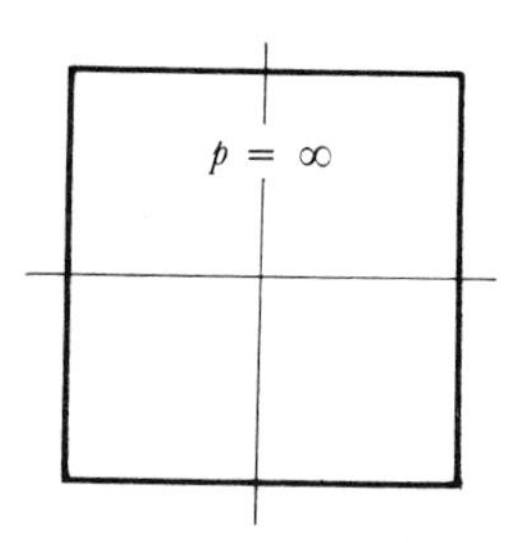

Figure 8

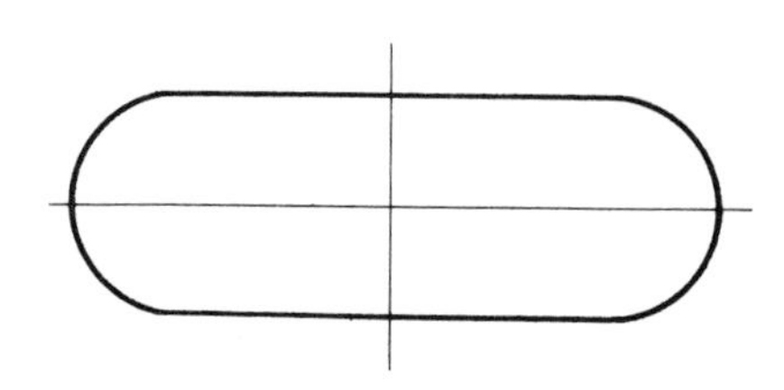

Figure 9

the unit ball is a property distinguishing finite-dimensional from infinite-dimensional spaces.

a. LEMMA. *Given a normed linear space* $\mathbf{R}$, *let* E *be a closed subset of* $\mathbf{R}$ *distinct from the whole space* $\mathbf{R}$. *Then there exists a vector* $y \in \mathbf{R}$ *such that* $|y| = 1$ *and* $|y - x| > \frac{1}{2}$ *for all* $x \in E$.

Proof. Let $y_0 \in \mathbf{R} - E$, and let

$$d = \inf_{x \in E} |y_0 - x|.$$

If $\inf |y_0 - x| = 0$, there exists a sequence $x_n \in E$ converging to y_0. But then, since E is closed,

$$y_0 = \lim_{n \to \infty} x_n \in E,$$

contrary to hypothesis. Therefore $d > 0$. Let $x_0 \in E$ be any vector such that $|y_0 - x_0| < 2d$. If

$$y = \frac{y_0 - x_0}{|y_0 - x_0|},$$

then $|y| = 1$. Moreover, if x belongs to E, so does $x_0 + x|y_0 - x_0|$, and

$$|y - x| = \left| \frac{y_0 - x_0}{|y_0 - x_0|} - x \right| = \left| \frac{y_0 - x_0 - x|y_0 - x_0|}{|y_0 - x_0|} \right| > \frac{d}{2d} = \frac{1}{2}$$

for all $x \in E$. ∎

b. THEOREM (**F. Riesz**). *The unit ball in an infinite-dimensional normed linear space* $\mathbf{R}$ *is not precompact.*†

Proof. We will construct a sequence of points $x_1, x_2, \ldots$ in the unit ball of the space $\mathbf{R}$ such that the distances $|x_m - x_n|$ between points of the sequence all exceed $\frac{1}{2}$. This sequence obviously cannot contain a fundamental sequence, and hence the unit ball $S = \{x : |x| \leqslant 1\}$ cannot be precompact, as asserted.

To construct the sequence in question, we first choose as x_1 any vector $x_1 \in S$, $|x_1| = 1$. The multiples λx_1 (λ real) form a closed subspace $E_1 \subset \mathbf{R}$. Hence, by the lemma, there exists a vector $x_2 \in S$, $|x_2| = 1$, such that $|x_2 - x| > \frac{1}{2}$ for all $x \in E_1$; in particular, $|x_2 - x_1| > \frac{1}{2}$. The linear combinations $\lambda_1 x_1 + \lambda_2 x_2$ form a closed subspace $E_2 \subset \mathbf{R}$. By the lemma again, there exists a vector $x_3 \in S$, $|x_3| = 1$, such that $|x_3 - x| > \frac{1}{2}$ for all $x \in E_2$; in particular, $|x_3 - x_1| > \frac{1}{2}$, $|x_3 - x_2| > \frac{1}{2}$. Continuing this process, we construct an

† And hence not compact (Vol. 1, Sec. 3.93a).

"expanding chain" of finite-dimensional subspaces $E_1 \subset E_2 \subset \cdots$, each of which is a proper subset of **R** (since **R** is infinite-dimensional), and a sequence of points $x_1, x_2, \ldots$ in S such that $|x_m - x_n| > \frac{1}{2}$ for all distinct pairs of integers $m, n = 1, 2, \ldots$ ∎

1.37. Series of vectors in a normed linear space. In a metric space we can talk about convergent *sequences*, but the concept of a convergent *series* makes no sense. In a normed linear space, however, the concept of a convergent series (of vectors) does make sense.

a. Given a series

$$x_1 + x_2 + \cdots + x_n + \cdots, \tag{13}$$

whose terms are elements of a normed linear space **R**, suppose the sequence of partial sums

$$s_1 = x_1,\ s_2 = x_1 + x_2,\ \ldots,\ s_n = x_1 + x_2 + \cdots + x_n,\ \ldots \tag{13'}$$

converges to a limit

$$s = \lim_{n \to \infty} s_n$$

in **R**. Then the series (13) is said to be *convergent* or to *converge* (in **R**), with *sum* s. If, however, the sequence (13′) diverges, then the series (13) is said to be *divergent* or to *diverge* (in **R**), with no sum at all.†

b. THEOREM (**Cauchy convergence criterion**). *Let the space* **R** *be complete. Then the series* (13) *converges if and only if, given any* $\varepsilon > 0$, *there exists an integer* $N > 0$ *such that*

$$|s_n - s_m| = |x_{m+1} + \cdots + x_n| < \varepsilon$$

for all $n > m \geqslant N$.

Proof. The exact analogue of the proof for series of vectors in R_m (Vol. 1, Sec. 6.43). ∎

c. THEOREM. *If the numerical series*

$$|x_1| + |x_2| + \cdots + |x_n| + \cdots \tag{14}$$

converges, then so does the series (13) *itself, provided* **R** *is complete.*

Proof. An immediate consequence of the Cauchy convergence criterion, since

$$|x_{m+1} + \cdots + x_n| \leqslant |x_{m+1}| + \cdots + |x_n|,$$

by the triangle inequality. ∎

† A detailed treatment of the case $\mathbf{R} = R_m$ is given in Vol. 1, Sec. 6.4.

d. THEOREM (**Weierstrass' test**). *If* $|x_n| \leqslant \alpha_n$ *and if the numerical series*

$$\alpha_1 + \alpha_2 + \cdots + \alpha_n + \cdots \tag{15}$$

converges, then the series (13) *also converges, provided* **R** *is complete.*

Proof. If (15) converges, then so does the series (14), by the comparison test for numerical series. But then (13) converges, by the preceding theorem. ∎

e. THEOREM (**Cauchy's test**). *The series* (13) *converges if*

$$\overline{\lim_{n\to\infty}} \sqrt[n]{|x_n|} < 1$$

and diverges if

$$\overline{\lim_{n\to\infty}} \sqrt[n]{|x_n|} > 1.$$

Proof. The exact analogue of the proof for numerical series (Vol. 1, Sec. 6.14c). ∎

f. THEOREM (**Abel-Dirichlet test**). *Given a numerical sequence* α_n *and a vector sequence* $x_n \in \mathbf{R}$, *suppose* α_n *decreases monotonically to zero (written* $\alpha_n \searrow 0$), *while the sequence of partial sums* $s_n = x_1 + \cdots + x_n$ *is bounded (in norm), i.e.,*

$$|s_n| \leqslant C \qquad (n = 1, 2, \ldots)$$

for some fixed constant C. *Then the series*

$$\alpha_1 x_1 + \alpha_2 x_2 + \cdots + \alpha_n x_n + \cdots$$

is convergent.

Proof. The exact analogue of the proof for series of vectors in R_m (Vol. 1, Sec. 6.47c), based on the use of Abel's transformation (Vol. 1, Sec. 6.47a). ∎

g. Example. Consider the series

$$\sum_{n=0}^{\infty} \alpha_n \cos nt, \tag{16}$$

$$\sum_{n=0}^{\infty} \alpha_n \sin nt \tag{17}$$

in the space $R^s(a,b)$, where it will be recalled that $R^s(a,b)$ is complete (Sec. 1.23e) and that convergence in the norm of $R^s(a,b)$ means uniform convergence on the interval $[a,b]$. The norms of the functions $\cos nt$ and $\sin nt$ in the space $R^s(a,b)$ do not exceed unity. Therefore if the series

$$\sum_{n=1}^{\infty} |\alpha_n| \tag{18}$$

converges, the series (16) and (17) converge in $R^s(a,b)$, by Weierstrass' test, i.e., the series converge uniformly on $[a,b]$ for arbitrary a and b $(a<b)$.

In the case where the series (18) diverges but $\alpha_n \searrow 0$, we can use the Abel-Dirichlet test. First we note that

$$\left|\sum_{m=0}^{n} \cos mt\right| \leqslant \sqrt{\frac{2}{1-\cos t}}$$

(Vol. 1, Sec. 6.47e), and similarly with cos replaced by sin. Therefore, if $0<\varepsilon<\pi$, we have

$$\max_{\varepsilon \leqslant t \leqslant 2\pi-\varepsilon} \left|\sum_{m=0}^{n} \cos mt\right| \leqslant \sqrt{\frac{2}{1-\cos \varepsilon}},$$

and similarly with cos replaced by sin, so that the Abel-Dirichlet test is applicable in the space $R^s(\varepsilon,2\pi-\varepsilon)$. Thus the series (16) and (17) converge uniformly on every interval $[\varepsilon,2\pi-\varepsilon]$ if $\alpha_n \searrow 0$. However, the series may fail to converge uniformly on the whole interval $[0,2\pi]$, even though the sine series converges at every point of $[0,2\pi]$! We shall see later (Sec. 4.44b) that

$$\sum_{n=1}^{\infty} \frac{1}{n} \cos nt = -\ln 2\left|\sin \frac{t}{2}\right| \qquad (0<t<2\pi), \tag{19}$$

$$\sum_{n=1}^{\infty} \frac{1}{n} \sin nt = \begin{cases} 0 & \text{if } \ t=0 \ \text{ or } \ t=2\pi, \\ \dfrac{\pi-t}{2} & \text{if } \ 0<t<2\pi. \end{cases} \tag{20}$$

If the series (20) converges uniformly on $[0,2\pi]$, i.e., if it converges in the norm of the space $R^s(0,2\pi)$, then its sum $s(t)$ must belong to $R^s(0,2\pi)$, i.e., $s(t)$ must be a continuous function on $[0,2\pi]$. But it is clear from (20) that $s(t)$ is discontinuous at the points 0 and 2π. Thus the series (20) cannot converge uniformly on the whole interval $[0,2\pi]$. As for the series (19), it is unbounded on $[0,2\pi]$, and hence certainly cannot converge uniformly on $[0,2\pi]$, or, for that matter, even on the open interval $(0,2\pi)$ (cf. Vol. 1, Sec. 5.94). It can be shown that the series (20) also fails to converge uniformly on $(0,2\pi)$.†

1.38. Completion of a normed linear space. Just like any metric space, a normed linear space **R** can be either complete or incomplete. In the latter case, we can complete the space **R** by embedding it in a larger complete metric space $\overline{\mathbf{R}}$, as described in Vol. 1, Sec. 3.8. It turns out that the completion of a normed linear space is not only a metric space, but also a normed linear space. This assertion will be proved below, after we have introduced linear operations and a norm in $\overline{\mathbf{R}}$.

† See e.g., R. Courant and D. Hilbert, *Methods of Mathematical Physics*, Vol. 1, John Wiley-Interscience, New York (1953), p. 107.

It will be recalled that each element X of the completion $\overline{\mathbf{R}}$ of a metric space $\mathbf{R}$ is defined as a class of cofinal fundamental sequences of the space $\mathbf{R}$ (Vol. 1, Sec. 3.82). Suppose now that $\mathbf{R}$ is a normed linear space. Then, adding two fundamental sequences $x_1, x_2, \ldots, x_n, \ldots$ and $y_1, y_2, \ldots, y_n, \ldots$ term by term, we get a new sequence

$$x_1+y_1, x_2+y_2, \ldots, x_n+y_n, \ldots$$

which is fundamental, since

$$|(x_n+y_n)-(x_m+y_m)| \leqslant |x_n-x_m|+|y_n-y_m|,$$

by the triangle inequality for the norm $|\cdots|$. Moreover, replacing the sequence x_n by a cofinal sequence x_n' and the sequence y_n by a cofinal sequence y_n', we get a sum sequence $x_n'+y_n'$ which is cofinal with the original sequence x_n+y_n, since

$$|(x_n'+y_n')-(x_n+y_n)| \leqslant |x_n'-x_n|+|y_n'-y_n|.$$

We now define addition of elements of the space $\overline{\mathbf{R}}$. Given two classes X, $Y \in \overline{\mathbf{R}}$, let x_n be any fundamental sequence in X and y_n any fundamental sequence in Y. Then by the *sum* of X and Y, denoted by $X+Y$, we mean the class containing the sequence x_n+y_n. This definition makes sense, since, as just shown, the sequence x_n+y_n is fundamental, and the class $X+Y$ does not depend on the particular sequences x_n and y_n chosen from the classes X and Y. The *product* of a class X and a number λ is defined similarly: Thus, choosing any fundamental sequence $x_n \in X$, by the class λX we mean the class containing the sequence λx_n (the reader should verify that this definition also makes sense, i.e., that λx_n is fundamental and λX is unique).

The space $\overline{\mathbf{R}}$ equipped with these operations clearly satisfies the axioms of a linear space (Sec. 1.11). In fact, by their very definition, linear operations on the classes reduce to the corresponding operations on the elements of the original space $\mathbf{R}$. In particular, the class 0 consists of all sequences in the space $\mathbf{R}$ which converge to zero.

We must still introduce a norm in the space $\overline{\mathbf{R}}$, and verify that this norm satisfies Axioms a–c of Sec. 1.31. To this end, let ρ denote the distance in the completion $\overline{\mathbf{R}}$ of the metric space $\mathbf{R}$ (Vol. 1, Sec. 3.82), and then define the norm of the class $X \in \overline{\mathbf{R}}$ by the formula

$$\|X\| = \rho(X,0).$$

This means that

$$\|X\| = \lim_{n\to\infty} \rho(x_n,0),$$

where x_n is any fundamental sequence in the class X and ρ now denotes the distance in $\mathbf{R}$ itself, or equivalently,

$$\|X\| = \lim_{n\to\infty} |x_n|,$$

where $|x|$ is the norm in $\mathbf{R}$. If $\|X\| = 0$, then

$$\lim_{n\to\infty} |x_n| = 0,$$

so that the sequence x_n is cofinal with the sequence $0, 0, \ldots, 0, \ldots$ defining the class 0. Therefore $X = 0$, so that Axiom a holds. Moreover, if $x_n \in X$, $y_n \in Y$, then

$$|x_n + y_n| \leqslant |x_n| + |y_n|,$$

and hence

$$\|X+Y\| = \lim_{n\to\infty} |x_n + y_n| \leqslant \lim_{n\to\infty} |x_n| + \lim_{n\to\infty} |y_n| = \|X\| + \|Y\|,$$

which proves the triangle inequality (Axiom c) in $\overline{\mathbf{R}}$. Similarly,

$$\|\lambda X\| = \lim_{n\to\infty} |\lambda x_n| = |\lambda| \lim_{n\to\infty} |x_n| = |\lambda| \|X\|,$$

which proves Axiom b. This completes the proof of our assertion that the completion $\overline{\mathbf{R}}$ of a normed linear space is itself a normed linear space.

1.39. Complex normed linear spaces

a. So far we have considered only real normed linear spaces, but the above considerations are easily extended to the case of a normed linear space over the field of complex numbers.† Thus a complex linear space $\mathbf{C}$ is called a (*complex*) *normed linear space* if for each vector $x \in \mathbf{C}$ there is defined a number $|x|$ (sometimes denoted by $\|x\|$), called the *norm* of x, satisfying the following axioms:

(a) $|x| > 0$ if $x \neq 0$, $|0| = 0$;
(b) $|\alpha x| = |\alpha| |x|$ for every $x \in \mathbf{C}$ and every complex number α;
(c) $|x+y| \leqslant |x| + |y|$ for every $x, y \in \mathbf{C}$ (*triangle inequality*).

Obviously, every complex normed linear space $\mathbf{C}$ is simultaneously a real normed linear space, since we can multiply vectors of $\mathbf{C}$ by arbitrary complex numbers, in particular, by real numbers. Thus everything we know about real normed linear spaces can be carried over to the case of complex normed linear spaces, either directly or with slight modifications. In particular, a complex normed linear space, just like a real normed linear space,

† We cannot define a normed linear space over an *arbitrary* field K, since the absolute value $|\alpha|$ is not defined for an element α of an arbitrary field K.

becomes a metric space, when equipped with the distance

$$\rho(x, y) = |x - y|.$$

b. The space of all bounded continuous complex (-valued) functions $x(t)$ on a metric space M, equipped with the norm

$$\|x\| = \sup_t |x(t)|,$$

is a complex normed linear space, denoted by $C^s(M)$. This space is complete, by Theorem 1.23e.

c. The space of all continuous complex functions $x(t)$ on the interval $[a,b]$, equipped with the norm

$$\|x\|_p = \sqrt[p]{\int_a^b |x(t)|^p \, dt},$$

is a complex normed linear space, denoted by $CL_p^s(a,b)$, or simply by $L_p^s(a,b)$ as in the case of the corresponding space of continuous real functions (provided this briefer notation causes no confusion).

d. The space of all bounded continuous functions $x(t)$, defined on a metric space M and taking values in a complex normed linear space $\mathbf{C}$, equipped with the norm

$$\|x\| = \sup_t |x(t)|$$

(where $|x(t)|$ is the norm in the space $\mathbf{C}$), is a complex normed linear space, denoted by $\mathbf{C}^s(M)$. If $\mathbf{C}$ is complete, then so is $\mathbf{C}^s(M)$, by Theorem 1.23e.

e. The examples of finite-dimensional real normed linear spaces given in Sec. 1.33c lead, after slight changes, to analogous examples of finite-dimensional complex normed linear spaces. We need only change the real vector $x = (\xi_1, \ldots, \xi_n)$ to a complex vector, by regarding the components $\xi_1, \ldots, \xi_n$ as complex numbers, and write $|\xi_k|^2$ instead of ξ_k^2 in formula (5), p. 38. In the same way, we get complex analogues of the infinite-dimensional spaces l_p of Sec. 1.33d.

f. A set E in a complex normed linear space $\mathbf{C}$ is said to be *absolutely convex* if, whenever E contains two points x and y, it also contains all points of the form $z = \alpha x + \beta y$, where α and β are arbitrary complex numbers such that $|\alpha| + |\beta| \leqslant 1$. Every ball $\{x \in \mathbf{C}: |x - x_0| \leqslant \rho\}$ in a complex normed linear space is an absolutely convex set.

g. The conditions for equivalence of norms in a real normed linear space

(Sec. 1.35) continue to hold in the complex case. In particular, all norms are equivalent in a finite-dimensional complex linear space, and they are all equivalent to convergence in the norm

$$\|x\|_\infty = \max_{1 \leqslant k \leqslant n} |\xi_k|$$

(component-by-component convergence). All finite-dimensional complex normed linear spaces are complete, just as in the real case.

h. Riesz's theorem on the noncompactness of the unit ball in an infinite-dimensional normed linear space, proved for real spaces in Sec. 1.36b, is equally valid for complex spaces.

i. The whole theory of convergence of series of vectors in a normed linear space, described in Sec. 1.37 for the real case, carries over without change to the case of complex spaces. Consider, for example, the power series

$$\sum_{n=0}^{\infty} a_n(z-z_0)^n, \tag{21}$$

where z and z_0 are complex numbers, and the coefficients a_n are elements of a complete complex normed linear space **C**. Then (21) converges inside the disk of radius

$$R = \frac{1}{\overline{\lim\limits_{n\to\infty}} \sqrt[n]{\|a_n\|}}$$

centered at the point z_0, and diverges outside this disk. The proof is the same as that of the Cauchy-Hadamard theorem (Vol. 1, Sec. 6.62), with the help of Cauchy's test (Theorem 1.37e).

j. The completion of a complex normed linear space **C** is carried out in just the same way as in the real case (Sec. 1.38), and leads to a complete complex normed linear space $\overline{\mathbf{C}}$.

1.4. Hilbert Spaces

1.41. In a normed linear space we can measure distances, but we still cannot measure angles, and this prevents us from making full use of geometrical intuition. In a Hilbert space, however, every pair of vectors is equipped (by definition) with a "scalar product," in terms of which we can express both the length of vectors and angles between vectors. More exactly, a real linear space **H** is said to be a *Hilbert space* if for every pair of vectors $x, y \in \mathbf{H}$ there is defined a real number (x, y), called the *scalar product* of x and y, satisfying the following conditions (axioms):†

(a) $(x,x)>0$ if $x\neq 0$, $(0,0)=0$;
(b) $(y,x)=(x,y)$ for every $x,y\in\mathbf{H}$;
(c) $(\alpha x,y)=\alpha(x,y)$ for every $x,y\in\mathbf{H}$ and every real number α;
(d) $(x+y,z)=(x,z)+(y,z)$ for every $x,y,z\in\mathbf{H}$.

Axioms b–d imply the more general formula

$$\left(\sum_{j=1}^{m}\alpha_j x_j,\ \sum_{k=1}^{n}\beta_k y_k\right)=\sum_{j=1}^{m}\sum_{k=1}^{n}\alpha_j\beta_k(x_j,y_k), \tag{1}$$

with the help of induction.

1.42. Examples

a. Suppose we equip the n-dimensional real space R_n (Sec. 1.14d) of vectors $x=(\xi_1,\ldots,\xi_n), y=(\eta_1,\ldots,\eta_n),\ldots$ with the scalar product

$$(x,y)=\sum_{k=1}^{n}\xi_k\eta_k. \tag{2}$$

Then R_n becomes a Hilbert space, called *n-dimensional Euclidean space* (Vol. 1, Sec. 3.14a).

b. Formula (2) is not the only way of equipping the n-dimensional space R_n with a scalar product. In fact, we can easily describe all possible scalar products in R_n. If (x,y) is any scalar product in R_n and if the vectors $x,y\in R_n$ have expansions

$$x=\sum_{j=1}^{n}\xi_j e_j,\qquad y=\sum_{k=1}^{n}\eta_k e_k$$

with respect to the basis $e_1=(1,\ldots,0),\ldots,e_n=(0,\ldots,1)$, then (1) implies

$$(x,y)=\left(\sum_{j=1}^{n}\xi_j e_j,\ \sum_{k=1}^{n}\eta_k e_k\right)=\sum_{j=1}^{n}\sum_{k=1}^{n}\xi_j\eta_k(e_j,e_k).$$

Thus the scalar products

$$\omega_{jk}=(e_j,e_k)\qquad (j,k=1,\ldots,n)$$

of the basis vectors $e_1,\ldots,e_n$ are sufficient to uniquely determine the scalar product (x,y) of an arbitrary pair of vectors x and y. The numbers ω_{jk} must

† These axioms apply to a real Hilbert space. The corresponding axioms for a complex Hilbert space will be given in Sec. 1.44b. What we call a Hilbert space is often called a *Euclidean space*, with the term "Hilbert space" being reserved for a Euclidean space which is complete, separable, and infinite-dimensional.

satisfy the symmetry condition

$$\omega_{jk} = (e_j, e_k) = (e_k, e_j) = \omega_{kj}$$

and the inequality

$$(x,x) = \sum_{j=1}^{n} \sum_{k=1}^{n} \xi_j \eta_k \omega_{jk} > 0$$

for arbitrary $x \neq 0$. This means that the matrix $\|\omega_{jk}\|$ must be symmetric and positive definite. It can be shown (see Lin. Alg., Sec. 7.96) that a necessary and sufficient condition for a given symmetric matrix $\|\omega_{jk}\|$ to be positive definite is that the following inequalities (called *Sylvester's conditions*) be satisfied:

$$\omega_{11} > 0, \qquad \begin{vmatrix} \omega_{11} & \omega_{12} \\ \omega_{21} & \omega_{22} \end{vmatrix} > 0, \qquad \ldots, \qquad \begin{vmatrix} \omega_{11} & \cdots & \omega_{1n} \\ \cdot & \cdots & \cdot \\ \omega_{n1} & \cdots & \omega_{nn} \end{vmatrix} > 0.$$

Conversely, every symmetric positive definite matrix $\|\omega_{jk}\|$ leads to a scalar product

$$(x,y) = \sum_{j=1}^{n} \sum_{k=1}^{n} \xi_j \eta_k \omega_{jk}$$

in the space R_n, satisfying Axioms a–d (why?).

c. One way of equipping the space $R^s(a,b)$ of continuous real functions on the interval $[a,b]$ with a scalar product is to use the continuous analogue of formula (2):

$$(x(t), y(t)) = \int_a^b x(t) y(t)\, dt. \tag{3}$$

The axioms of a Hilbert space then follow at once from elementary properties of the integral. There are also other ways of equipping $R^s(a,b)$ with a scalar product.

d. As in Sec. 1.33d, let l_2 be the linear space consisting of all numerical sequences $x = (\xi_1, \ldots, \xi_n, \ldots)$ such that

$$\sum_{n=1}^{\infty} \xi_n^2 < \infty,$$

and let the scalar product of two vectors $x = (\xi_1, \ldots, \xi_n, \ldots)$, $y = (\eta_1, \ldots, \eta_n, \ldots) \in l_2$ be defined by the formula

$$(x,y) = \sum_{n=1}^{\infty} \xi_n \eta_n. \tag{4}$$

The convergence of the series on the right (in fact, its absolute convergence) follows from the inequality

$$|ab| \leqslant \tfrac{1}{2}(a^2+b^2),$$

valid for arbitrary real numbers a and b. The validity of Axioms a–d for the scalar product (3) is obvious. Thus the space l_2, equipped with the norm (4), is a Hilbert space.†

1.43. Geometry in Hilbert space

a. THEOREM. *Any two vectors x and y in a Hilbert space* **H** *satisfy* **Schwarz's inequality**‡

$$|(x,y)| \leqslant \sqrt{(x,x)(y,y)}. \tag{5}$$

Proof. Consider the expression

$$\varphi(\lambda) = (x-\lambda y,\ x-\lambda y),$$

where λ is an arbitrary real number. By Axioms b–d of Sec. 1.41,

$$\varphi(\lambda) = (x,x) - 2\lambda(x,y) + \lambda^2(y,y) = A - 2B\lambda + C\lambda^2,$$

where

$$A = (x,x), \qquad B = (x,y), \qquad C = (y,y).$$

Moreover

$$\varphi(\lambda) \geqslant 0$$

for all real λ, by Axiom a. Hence the polynomial $\varphi(\lambda)$ cannot have distinct real roots, since otherwise it would take both positive and negative values. It follows that the quantity

$$B^2 - AC = (x,y)^2 - (x,x)(y,y)$$

cannot be positive, and hence

$$(x,y)^2 \leqslant (x,x)(y,y),$$

which is equivalent to (5). ∎

b. Next we equip the Hilbert space **H** with the norm

$$|x| = \sqrt{(x,x)}. \tag{6}$$

It is easy to see that this choice of $|x|$ satisfies all the axioms for a normed linear space (see Sec. 1.31), namely

† Formulas (4) and (6) lead to the same norm in l_2 as that introduced in Sec. 1.33d.
‡ As always, the radical denotes the *positive* square root.

(a′) $|x| > 0$ if $x \neq 0$, $|0| = 0$;
(b′) $|\alpha x| = |\alpha||x|$ for every $x \in \mathbf{H}$ and every real number α;
(c′) $|x+y| \leqslant |x| + |y|$ for every $x, y \in \mathbf{H}$.

In fact, Axiom a′ follows from Axiom a of Sec. 1.41, while Axiom b′ follows from Axiom c. As for Axiom c′ (the triangle inequality), we need only note that

$$|x+y|^2 = (x+y,\, x+y) = (x,x) + 2(x,y) + (y,y)$$
$$\leqslant |x|^2 + 2|x||y| + |y|^2 = (|x| + |y|)^2$$

(by Schwarz's inequality), which is equivalent to $|x+y| \leqslant |x| + |y|$.

Thus all the concepts and properties associated with the existence of a norm obtain in a Hilbert space. But since a Hilbert space is a very special kind of normed linear space, there is reason to expect that the norm in a Hilbert space has properties which are valid only for such a space. One of these properties is given by the following

THEOREM. *Any two vectors x and y in a Hilbert space* $\mathbf{H}$ *satisfy the* **parallelogram equality**

$$|x+y|^2 + |x-y|^2 = 2|x|^2 + 2|y|^2 \tag{7}$$

(*"the sum of the squares of the diagonals of a parallelogram equals the sum of the squares of its sides"*).

Proof. We need only observe that

$$|x+y|^2 + |x-y|^2 = (x+y,\, x+y) + (x-y,\, x-y)$$
$$= (x,x) + 2(x,y) + (y,y) + (x,x) - 2(x,y) + (y,y)$$
$$= 2(x,x) + 2(y,y) = 2|x|^2 + 2|y|^2. \quad \blacksquare$$

Conversely, it can be shown that if a normed linear space has a norm satisfying the condition (7), then the norm is "generated" by some scalar product (see Problem 4).

c. It should be noted that if the norm $|x|$ in the n-dimensional space R_n is obtained from a scalar product (x,y) via formula (6), then the surface $|x| = 1$ is an ellipsoid. In fact, $|x| = 1$ implies

$$1 = (x,x) = \left(\sum_{j=1}^{n} \xi_j e_j, \sum_{k=1}^{n} \xi_k e_k \right) = \sum_{j=1}^{n} \sum_{k=1}^{n} \xi_j \xi_k \omega_{jk},$$

where $(e_j, e_k) = \omega_{jk}$. This is the equation of a central quadric surface in R_n, and being bounded, the surface must be an ellipsoid (see Lin. Alg., Chapter

10). In the case of the "ordinary" scalar product (4), we have

$$\omega_{jk} = \begin{cases} 1 & \text{if } j=k, \\ 0 & \text{if } j \neq k, \end{cases}$$

and the ellipsoid reduces to a sphere, in keeping with the norm

$$|x| = \sqrt{\sum_{k=1}^{n} \xi_k^2}\,.$$

d. THEOREM. *The scalar product in a Hilbert space* **H** *is continuous, in the sense that if* x_n, y_n *are two convergent sequences in* **H** *such that* $x_n \to x, y_n \to y$, *then* $(x_n, y_n) \to (x, y)$.

Proof. Clearly

$$(x,y) - (x_n, y_n) = (x, y - y_n) + (x - x_n, y_n),$$

and hence, by Schwarz's inequality,

$$|(x,y) - (x_n, y_n)| \leqslant |x||y - y_n| + |x - x_n||y_n|.$$

But the right-hand side approaches 0 as $n \to \infty$, since a convergent sequence is bounded (Vol. 1, Sec. 3.33b) and $|y - y_n| \to 0$, $|x - x_n| \to 0$. ∎

e. In a Hilbert space **H** we can measure not only lengths (norms) of vectors, but also angles between vectors. Let x and y be any two vectors in **H**. Then, by Schwarz's inequality, there is a unique number θ in the interval $0 \leqslant \theta \leqslant 2\pi$ such that

$$\cos\theta = \frac{(x,y)}{|x||y|}, \tag{8}$$

and this number is called the *angle between* x *and* y.

f. Two vectors x and y in a Hilbert space **H** are said to be *orthogonal* if $(x, y) = 0$. It follows from (8) that the angle between two nonzero orthogonal vectors x and y equals $\pi/2$. Note that the zero vector is orthogonal to every vector.

In the Euclidean space R_n, with scalar product (4), the condition for two vectors $x = (\xi_1, \ldots, \xi_n), y = (\eta_1, \ldots, \eta_n)$ to be orthogonal takes the form

$$\sum_{k=1}^{n} \xi_k \eta_k = 0.$$

In the space $R^s(a,b)$, with scalar product (3), the condition for two vectors $x = x(t), y = y(t)$ to be orthogonal takes the form

$$\int_a^b x(t)y(t)\,dt = 0.$$

g. Suppose a vector x in a Hilbert space $\mathbf{H}$ is orthogonal to the vectors $y_1,\dots,y_k \in \mathbf{H}$. Then x is orthogonal to every linear combination $\alpha_1 y_1 + \cdots + \alpha_k y_k$ of these vectors, since

$$(x,\alpha_1 y_1 + \cdots + \alpha_k y_k) = 0.$$

It follows that the set of *all* vectors orthogonal to any given vector $x \in \mathbf{H}$ (or to every vector in a given *set* $X \subset \mathbf{H}$) is a subspace of $\mathbf{H}$. This subspace is called the *orthogonal complement* of the vector x (or of the set X).

h. The Pythagorean theorem. Let x and y be be orthogonal vectors in a Hilbert space. Then, by analogy with elementary geometry, we can call the vector $x+y$ the *hypotenuse* of the right triangle constructed on the vectors x and y. Taking the scalar product of $x+y$ with itself and using the orthogonality of the vectors x and y, we get

$$|x+y|^2 = (x+y,\, x+y) = (x,x) + 2(x,y) + (y,y) = |x|^2 + |y|^2.$$

This is the *Pythagorean theorem* in a general Hilbert space: "The square of the hypotenuse equals the sum of the squares of the sides." The theorem can immediately be generalized to the case of any finite number of vectors. In fact, if the vectors $x_1,\dots,x_k$ are (pairwise) orthogonal and if $y = x_1 + \cdots + x_k$, then

$$|y|^2 = (x_1 + \cdots + x_k,\, x_1 + \cdots + x_k) = |x_1|^2 + \cdots + |x_k|^2.$$

i. Orthogonalization. A common method of obtaining an orthogonal set of vectors consists of "orthogonalizing" a given nonorthogonal set of vectors:

THEOREM (**Orthogonalization theorem**). *Let $x_1, x_2, \dots, x_n, \dots$ be a finite or infinite system of vectors in a Hilbert space $\mathbf{H}$ such that the first n vectors $x_1, x_2, \dots, x_n$ are linearly independent for every n. Then there exist unique coefficients a_{jk} such that the vectors*

$$\begin{aligned}
&y_1 = x_1,\\
&y_2 = a_{21}x_1 + x_2,\\
&y_3 = a_{31}x_1 + a_{32}x_2 + x_3,\\
&\cdots\\
&y_n = a_{n1}x_1 + a_{n2}x_2 + \cdots + a_{n,n-1}x_{n-1} + x_n,\\
&\cdots
\end{aligned} \tag{9}$$

are nonzero and (pairwise) orthogonal.

Proof. We will prove the theorem inductively, i.e., assuming that $n-1$ nonzero orthogonal vectors $y_1,\dots,y_{n-1}$ satisfying the first $n-1$ equations (9) have been found, we will construct a vector $y_n \neq 0$ satisfying the nth equation

(9) and orthogonal to the vectors $y_1,\ldots,y_{n-1}$. To this end, we represent y_n in the form

$$y_n=b_{n1}y_1+\cdots+b_{n,n-1}y_{n-1}+x_n, \tag{10}$$

where $y_1,\ldots,y_{n-1}$ are the vectors already found and $b_{n1},\ldots,b_{n,n-1}$ are coefficients to be suitably determined. Taking the scalar product of (10) with y_k $(k<n)$ and using the assumed orthogonality of y_k to $y_1,\ldots,y_{k-1},y_{k+1},\ldots,y_{n-1}$, we get

$$(y_n,y_k)=b_{nk}(y_k,y_k)+(x_n,y_k). \tag{11}$$

Equating the right-hand side of (11) to zero and noting that $(y_k,y_k)\neq 0$, by hypothesis, we find that

$$b_{nk}=-\frac{(x_n,y_k)}{(y_k,y_k)} \qquad (k=1,\ldots,n-1). \tag{12}$$

Substituting (12) and the first $n-1$ equations (9) into (10), we get

$$y_n=a_{n1}x_1+a_{n2}x_2+\cdots+a_{n,n-1}x_{n-1}+x_n,$$

as required, after collecting coefficients of each vector x_k $(k<n)$. Obviously y_n is orthogonal to each of the vectors $y_1,\ldots,y_{n-1}$, by construction. Moreover, $y_n\neq 0$, since otherwise we would have a vanishing linear combination of the vectors $x_1, x_2,\ldots, x_n$, with coefficients that are not all zero (the coefficient of a_n is always 1). But this is impossible, since the vectors $x_1, x_2,\ldots, x_n$ are linearly independent, by hypothesis. Finally we note that the coefficients a_{jk} are unique, by their very construction. ∎

The resulting system of vectors $y_1, y_2,\ldots, y_n,\ldots$ can be made even nicer by "normalizing" each vector y_n, i.e., by dividing each y_n by its norm. This gives a system of vectors

$$e_n=\frac{y_n}{|y_n|} \qquad (n=1,2,\ldots),$$

which are not only orthogonal but also all of unit length (of norm 1). Such systems of vectors, which are both orthogonal and normalized, are simply said to be *orthonormal*.

j. In accordance with the general definition of an isomorphism of two mathematical structures (Vol. 1, Sec. 2.52), two Hilbert spaces **H** and **H**′ are said to be *isomorphic* if they are isomorphic as linear spaces (Sec. 1.14j) and if, in addition, $x\leftrightarrow x'$, $y\leftrightarrow y'$ (where the symbol $\leftrightarrow$ denotes a one-to-one correspondence) implies $(x,y)\leftrightarrow(x',y')$.

For a finite-dimensional Hilbert space, we have the following analogue of Theorem 1.14j:

THEOREM. *Every n-dimensional Hilbert space* $\mathbf{H}_n$ *is isomorphic to the n-dimensional Euclidean space* R_n.

Proof. Starting from any system $x_1, \ldots, x_n$ of n linearly independent vectors in $\mathbf{H}_n$, we use the preceding theorem to "orthogonalize" $x_1, \ldots, x_n$, obtaining an orthonormal basis in $\mathbf{H}_n$. Then every vector $x \in \mathbf{H}_n$ has a (unique) expansion $x = \xi_1 e_1 + \cdots + \xi_n e_n$. Assigning the vector $x' = (\xi_1, \ldots, \xi_n) \in R_n$ to the vector $x \in \mathbf{H}_n$, we get an isomorphism $\omega: \mathbf{H}_n \to R_n$, regarding $\mathbf{H}_n$ and R_n as linear spaces. Moreover, if x and y are any two vectors in $\mathbf{H}_n$, with expansions

$$x = \sum_{j=1}^{n} \xi_j e_j, \qquad y = \sum_{k=1}^{n} \eta_k e_k,$$

then, by the orthonormality of the vectors $e_1, \ldots, e_n$, we have

$$(x, y) = \left(\sum_{j=1}^{n} \xi_j e_j, \sum_{k=1}^{n} \eta_k e_k \right) = \sum_{j=1}^{n} \sum_{k=1}^{n} \xi_j \eta_k (e_j, e_k) = \sum_{k=1}^{n} \xi_k \eta_k.$$

But this is precisely the definition (2) of the scalar product in the n-dimensional Euclidean space R_n. Hence ω also establishes an isomorphism between $\mathbf{H}_n$ and R_n, regarded as Hilbert spaces. ∎

COROLLARY. *Any two n-dimensional Hilbert spaces* $\mathbf{H}_n$ *and* $\mathbf{H}_n'$ *are isomorphic.*

Proof. Both $\mathbf{H}_n$ and $\mathbf{H}_n'$ are isomorphic to R_n, and hence isomorphic to each other. ∎

1.44. Complex Hilbert spaces

a. Since complex functions play an important role in analysis, we must enlarge the concept of a Hilbert space, going from the real Hilbert space of Sec. 1.41 to complex Hilbert spaces. In equipping a complex linear space with a scalar product, we want to allow the scalar product to take complex values. But it is then impossible to preserve Axioms a–c of Sec. 1.41 unaltered, since the expression (ix, ix) must be positive, by Axiom a, at the same time that

$$(ix, ix) = i(x, ix) = i(ix, x) = i^2(x, x) < 0,$$

by Axioms b and c. This difficulty can be overcome by a slight modification of Axiom b, as we now show.

b. Definition. A complex linear space $\mathbf{H}$ (i.e., a linear space $\mathbf{H}$ over the field of complex numbers) is said to be a *Hilbert space* if for every pair of vectors $x, y \in \mathbf{H}$ there is defined a complex number (x,y), called the *scalar product* of x and y, satisfying the following conditions (axioms):

(a) $(x,x) > 0$ if $x \neq 0$, $(0,0) = 0$;
(b) $(y,x) = \overline{(x,y)}$ for every $x, y \in \mathbf{H}$;†
(c) $(\alpha x, y) = \alpha(x,y)$ for every $x, y \in \mathbf{H}$ and every complex number α;
(d) $(x+y,z) = (x,z) + (y,z)$ for every $x, y, z \in \mathbf{H}$.

It follows from Axioms b and c that if the second entry in the scalar product has a numerical factor, then this factor is replaced by its complex conjugate when brought out in front of the scalar product:

(b′) $(x,\alpha y) = \overline{(\alpha y, x)} = \bar{\alpha}\overline{(y,x)} = \bar{\alpha}(x,y)$.

This, together with Axioms b–d, immediately leads to the more general formula

$$\left(\sum_{j=1}^{m} \alpha_j x_j, \sum_{k=1}^{n} \beta_k y_k \right) = \sum_{j=1}^{m} \sum_{k=1}^{n} \alpha_j \bar{\beta}_k (x_j, y_k). \tag{1′}$$

1.45. Examples

a. A simple example of a complex Hilbert space is given by the n-dimensional complex space C_n of vectors $x = (\xi_1, \ldots, \xi_n), y = (\eta_1, \ldots, \eta_n), \ldots$ with complex components, equipped with the usual linear operations (see Sec. 1.13f) and the scalar product

$$(x,y) = \sum_{k=1}^{n} \xi_k \bar{\eta}_k \tag{2′}$$

(note that the second factor in the sum is a complex conjugate). The validity of Axioms a–d in this case is obvious.‡

b. Another example of a complex Hilbert space is given by the space $C^s(a,b)$ of continuous complex-valued functions $x(t)$ defined on the interval $[a,b]$, equipped with the scalar product

$$(x(t), y(t)) = \int_a^b x(t)\overline{y(t)}\, dt. \tag{3′}$$

Here Axioms a–d follow at once from elementary properties of the integral.

c. The real space l_2 (Example 1.42d) has as its complex analogue the space of all complex numerical sequences $x = (\xi_1, \ldots, \xi_n, \ldots)$ such that

$$\sum_{n=1}^{\infty} |\xi_n|^2 < \infty.$$

† As usual, the overbar is used to denote the complex conjugate.
‡ The space C_n can also be equipped with other scalar products (cf. Lin. Alg., Sec. 9.21).

The scalar product of two vectors $x=(\xi_1,\ldots,\xi_n,\ldots)$, $y=(\eta_1,\ldots,\eta_n,\ldots)$ in this space is defined by the formula

$$(x,y)=\sum_{n=1}^{\infty}\xi_n\bar{\eta}_n, \tag{4'}$$

and it is again an easy matter to verify the validity of Axioms a–d.

1.46.a. Schwarz's inequality

$$|(x,y)|\leqslant\sqrt{(x,x)(y,y)} \tag{13}$$

(Theorem 1.43a) continues to hold in a complex Hilbert space $\mathbf{H}$, but the proof is slightly different. We now have

$$\varphi(\lambda)=(x-\lambda y,\ x-\lambda y)=(x,x)-\lambda(x,y)-\bar{\lambda}\overline{(x,y)}+\lambda\bar{\lambda}(y,y)\geqslant 0 \tag{14}$$

for all complex λ, by Axiom a. Let

$$\lambda=te^{-i\arg(x,y)},$$

where t is real. Then $\lambda(x,y)=t|(x,y)|$ and (14) implies

$$\psi(t)\equiv\varphi(te^{-i\arg(x,y)})=A-2Bt+Ct^2,$$

where this time

$$A=(x,x),\qquad B=|(x,y)|,\qquad C=(y,y).$$

Therefore

$$B^2-AC=|(x,y)|^2-(x,x)(y,y)\leqslant 0$$

for the same reason as in the real case, and hence

$$|(x,y)|^2\leqslant(x,x)(y,y),$$

which is equivalent to (13).

b. Next we equip the complex Hilbert space $\mathbf{H}$ with the norm

$$|x|=\sqrt{(x,x)},$$

just as in the real case (Sec. 1.43b). Schwarz's inequality again leads to the triangle inequality. In fact,

$$\begin{aligned}|x+y|^2&=(x+y,\ x+y)=(x,x)+(x,y)+(y,x)+(y,y)\\&=|x|^2+(x,y)+(y,x)+|y|^2=|x|^2+2\ \mathrm{Re}\ (x,y)+|y|^2\\&\leqslant|x|^2+2|(x,y)|+|y|^2\leqslant|x|^2+2|x||y|+|y|^2=(|x|+|y|)^2,\end{aligned}$$

and hence

$$|x+y|\leqslant|x|+|y|.$$

c. Just as in the real case, two vectors x and y in a complex Hilbert space $\mathbf{H}$ are said to be *orthogonal* if $(x,y)=0$. Let $x_1,x_2,\ldots, x_n,\ldots$ be a system of vectors in $\mathbf{H}$ such that the first n vectors $x_1,x_2,\ldots, x_n$ are linearly independent for every n. Then we can again "orthogonalize" $x_1,x_2,\ldots,x_n,\ldots$, i.e., we can find a system of nonzero orthogonal vectors $y_1,y_2,\ldots,y_n,\ldots$ related to $x_1,x_2,\ldots,x_n,\ldots$ by the formulas (9). In particular, every n-dimensional complex Hilbert space $\mathbf{H}_n$ has an orthonormal basis $e_1,e_2,\ldots,e_n$. Given any two vectors $x, y \in \mathbf{H}_n$, let

$$x=\sum_{j=1}^{n} \xi_j e_j, \quad y=\sum_{k=1}^{n} \eta_k e_k$$

be the expansions of x and y with respect to this basis. Then the scalar product of x and y is just

$$(x,y)=\left(\sum_{j=1}^{n} \xi_j e_j, \sum_{k=1}^{n} \eta_k e_k\right)=\sum_{j=1}^{n}\sum_{k=1}^{n} \xi_j \bar{\eta}_k (e_j,e_k)=\sum_{k=1}^{n} \xi_k \bar{\eta}_k. \tag{15}$$

It follows from (15), just as in the real case, that every n-dimensional complex Hilbert space $\mathbf{H}_n$ is isomorphic to the n-dimensional complex space C_n of Example 1.45a. Therefore any two n-dimensional complex Hilbert spaces $\mathbf{H}_n$ and $\mathbf{H}_n'$ are isomorphic.

1.47. Completion of a Hilbert space. Like any other normed linear space, a (real or complex) Hilbert space may be either complete or incomplete. For example, finite-dimensional Hilbert spaces, both real and complex (Examples 1.42a, 1.42b, 1.45a) are always complete (see Secs. 1.35e, 1.39g). The spaces of functions with an integral scalar product (Examples 1.42c, 1.45b) are incomplete (cf. Theorem 1.26b), while both the real and complex space l_2 (Examples 1.42d, 1.45c) are complete (see Problem 18).

If a Hilbert space $\mathbf{H}$ is incomplete, then it can be completed by embedding it in a larger normed linear space $\overline{\mathbf{H}}$, just as was done in Sec. 1.38. We now show that the completion of a Hilbert space is not only a normed linear space, *but also a Hilbert space.* To prove this, we must equip $\overline{\mathbf{H}}$ with a scalar product satisfying either the axioms of Sec. 1.41 in the real case or the axioms of Sec. 1.44b in the complex case.

Let $\overline{\mathbf{H}}$ be the completion of $\mathbf{H}$, regarded as a normed linear space. Then it will be recalled from Sec. 1.38 that each element X of $\overline{\mathbf{H}}$ is defined as a class of cofinal fundamental sequences of $\mathbf{H}$. Given any two classes $X, Y \in \overline{\mathbf{H}}$, let x_n be any fundamental sequence in X and y_n any fundamental sequence in Y. Then the sequence of scalar products (x_n,y_n) approaches a limit as $n\to\infty$. To see this, we observe that

$$\begin{aligned}|(x_m,y_m)-(x_n,y_n)| &= |(x_m-x_n,y_m)+(x_n,y_m-y_n)|\\ &\leqslant |x_m-x_n||y_m|+|x_n||y_m-y_n|.\end{aligned} \tag{16}$$

Since the fundamental sequences x_n and y_n are bounded (Vol. 1, Sec. 3.71c), the right-hand side of (16) can be made arbitrarily small for m and n exceeding a sufficiently large integer N. But then the numerical sequence (x_n, y_n) is itself fundamental, and hence has a limit. This limit is independent of the particular sequences x_n and y_n chosen from the classes X and Y. In fact, if x_n' and y_n' are other sequences from the same classes, then

$$|(x_n, y_n) - (x_n', y_n')| = |(x_n - x_n', y_n) + (x_n', y_n - y_n')|$$
$$\leqslant |x_n - x_n'||y_n| + |x_n'||y_n - y_n'| \to 0$$

as $n \to \infty$, so that the numerical sequences (x_n, y_n) and (x_n', y_n') have the same limit.

We now equip $\overline{\mathbf{H}}$ with the scalar product

$$(X, Y) = \lim_{n\to\infty} (x_n, y_n),$$

where, as just shown, the number (X, Y) depends only on the classes X and Y, and not on the choice of the sequences x_n and y_n from these classes. In particular, the number

$$\sqrt{(X,X)} = \lim_{n\to\infty} \sqrt{(x_n, x_n)} = \lim_{n\to\infty} |x_n|$$

is just the norm of the class X in the space $\overline{\mathbf{H}}$, regarded as a normed linear space (see Sec. 1.38). This proves Axiom a of Sec. 1.41 in the space $\overline{\mathbf{H}}$. The validity of the remaining axioms of Sec. 1.41 (or of Sec. 1.44b in the complex case) in the space $\overline{\mathbf{H}}$ is proved by starting from the same axioms in the space $\mathbf{H}$ and then taking the limit. For example,

$$(Y, X) = \lim_{n\to\infty} (y_n, x_n) = \lim_{n\to\infty} (x_n, y_n) = (X, Y)$$

in the real case, and the validity of all the other axioms is proved similarly.

1.48. Pre-Hilbert spaces

a. Let $\mathbf{L}$ be a linear space, which, for the time being, we assume to be real. Suppose $\mathbf{L}$ can be equipped with a function (x, y) which satisfies Axioms b–d of Sec. 1.41, but not Axiom a. More exactly, suppose Axiom a is replaced by the weaker axiom

(a′) $(x,x) \geqslant 0$ for every $x \in \mathbf{L}$, $(0,0) = 0$, (z,z) may vanish for certain nonzero elements of $\mathbf{L}$.

Then $\mathbf{L}$ is said to be a *pre-Hilbert space*. As it turns out (see below), the space $\mathbf{L}$, which is not a Hilbert space itself, has a certain factor space $\mathbf{L}/E$ (Sec. 1.14i) which is a Hilbert space.

b. For the set E in question we choose the set of all elements $z \in \mathbf{L}$ such that $(z,z)=0$. If $(z,z)=0$, then, by Schwarz's inequality,†

$$|(y,z)| \leqslant \sqrt{(y,y)(z,z)}$$

for every $y \in \mathbf{L}$, so that $(y,z)=0$ for every $y \in \mathbf{L}$. The set E is clearly a subspace of $\mathbf{L}$. In fact, if z_1 and z_2 belong to $\mathbf{L}$, then so does every linear combination $\alpha_1 z_1 + \alpha_2 z_2$, since

$$(\alpha_1 z_1 + \alpha_2 z_2, \alpha_1 z_1 + \alpha_2 z_2) = \alpha_1^2 (z_1, z_1) + 2\alpha_1 \alpha_2 (z_1, z_2) + \alpha_2^2 (z_2, z_2) = 0.$$

We now form the factor space $\mathbf{H} = \mathbf{L}/E$ and equip $\mathbf{H}$ with the scalar product

$$(X,Y) = (x,y), \tag{17}$$

where x and y are arbitrary elements of the classes X and Y, respectively. The quantity (17) does not depend on the choice of the elements $x \in X, y \in Y$. In fact, given any other elements $x' \in X$, $y' \in Y$, we have $x' = x + z$, $y' = y + \zeta$ where $z \in E$, $\zeta \in E$, and hence

$$(x', y') = (x,y) + (y,z) + (x,\zeta) + (z,\zeta) = (x,y),$$

since $(y,z)=0$ and $(y,\zeta)=0$ for every $y \in \mathbf{L}$. To verify that (17) is actually a scalar product for $\mathbf{H}$, we note that $(X,X)=0$ implies $(x,x)=0$ for every $x \in X$, so that X coincides with the class E which is the zero element of the space $\mathbf{H}=\mathbf{L}/E$ (Sec. 1.14i). This proves Axiom a, and the validity of Axioms b–d in $\mathbf{H}$ is an immediate consequence of (17) and the validity of the axioms in $\mathbf{L}$ (give the details).

c. A completely analogous construction can be carried out for a complex pre-Hilbert space $\mathbf{L}$: If E is the set of all elements $z \in \mathbf{L}$ such that $(z,z)=0$, then the factor space $\mathbf{H}=\mathbf{L}/E$ is a complex Hilbert space.

d. As an example, consider the real linear space $\mathbf{G}_R = \mathbf{G}_R(a,b)$ of all piecewise continuous functions on the interval $[a,b]$, with the scalar product

$$(x(t), y(t)) = \int_a^b x(t) y(t)\, dt. \tag{18}$$

Here Axioms b–d of Sec. 1.41 are satisfied, but not Axiom a, since

$$(z(t), z(t)) = \int_a^b z^2(t)\, dt = 0 \tag{19}$$

for every function $x(t) \in \mathbf{G}_R$ which differs from zero at only a finite number

† Note that Axiom a can be replaced by Axiom a′ in the proof of Schwarz's inequality.

of points (cf. Vol. 1, Sec. 9.16c). However, Axiom a obviously holds, and hence $\mathbf{G}_R$ is a pre-Hilbert space rather than a Hilbert space. To get a Hilbert space, we go from $\mathbf{G}_R$ to the factor space $\mathbf{G}_R/E$, where E is the set of all functions $z(t) \in \mathbf{G}_R$ satisfying (19). These are just the functions differing from zero at only a finite number of points (Vol. 1, Sec. 9.16e). The factor space $\mathbf{G}_R/E$ consists of classes of functions $x(t) \in \mathbf{G}_R$, where two functions belong to the same class if they differ at only a finite number of points. We denote this real Hilbert space by $\mathbf{H}_R(a,b)$.

e. In just the same way we can go from the complex pre-Hilbert space $\mathbf{G}_C = \mathbf{G}_C(a,b)$ of all piecewise continuous complex functions on the interval $[a,b]$, with the scalar product

$$(x(t), y(t)) = \int_a^b x(t)\overline{y(t)}\, dt, \tag{18'}$$

to the Hilbert space $\mathbf{G}_C/E$, where $\mathbf{G}_C/E$ is the factor space of $\mathbf{G}_C$ relative to the subspace E of all functions in $\mathbf{G}_C$ differing from zero at only a finite number of points. We denote this complex Hilbert space by $\mathbf{H}_C(a,b)$.

Hilbert spaces will be considered further in Chapter 4, in connection with certain applications of Hilbert spaces to analysis.

1.5. Approximation on a Compactum

1.51. As shown above, the space $R^s(Q)$ (or $C^s(Q)$) of all continuous real (or complex) functions on a compactum Q is a complete normed linear space.† We now consider various linear manifolds $B(Q) \subset R^s(Q)$ (or $B(Q) \subset C^s(Q)$),‡ with the objective of finding conditions on $B(Q)$ guaranteeing that $\overline{B(Q)}$, the closure of $B(Q)$ with respect to uniform convergence on Q,§ contain *all* continuous functions on Q, i.e., that $\overline{B(Q)}$ *coincide* with $R^s(Q)$ (or $C^s(Q)$).

a. A set of functions $B(Q)$ defined on Q is said to *separate the points y and z* ($y \neq z$) of the set Q if $B(Q)$ contains a function $\varphi(x)$ such that $\varphi(y) \neq \varphi(z)$. (By the same token, the function $\varphi(x)$ itself is then said to separate the points y and z.) Suppose a set $B(Q)$, defined on a compactum Q, does not separate the points y and z, which means that

$$f(y) = f(z) \tag{1}$$

† See Secs. 1.13i, 1.13k, 1.23e, 1.32a, and 1.39b; recall that a continuous real (or complex) function on a compactum (Sec. 1.24a) is automatically bounded (Vol. 1, Secs. 5.16a and 3.96e).

‡ A subset E of a linear space $\mathbf{K}$ over a field K is called a *linear manifold* if it is *closed under linear operations*, i.e., if whenever E contains two elements $x, y \in K$, it also contains all linear combinations $z = \alpha x + \beta y$ with coefficients $\alpha, \beta \in K$.

§ I.e., with respect to the norm of the space $R^s(Q)$ (or $C^s(Q)$).

for all functions $f(x) \in B(Q)$. Then, since (1) continues to hold after taking the closure of $B(Q)$ with respect to uniform convergence, the closure of $B(Q)$ certainly does not contain all continuous functions on Q; for example, it cannot contain the function $\rho(x,y)$ equal to 0 if $x=y$ and different from zero if $x=z$. Thus, if the closure of the set $B(Q)$ is to contain all continuous functions on Q, it must be assumed that $B(Q)$ separates every pair of points $x,y \in Q$.

b. A linear manifold $B(Q)$ consisting of real functions defined on a set Q is called a *linear net* if, whenever $B(Q)$ contains the function $f(x)$, it also contains the function $|f(x)|$. Let α and β be arbitrary real numbers. Then

$$\max\{\alpha,\beta\}+\min\{\alpha,\beta\}=\alpha+\beta,$$
$$\max\{\alpha,\beta\}-\min\{\alpha,\beta\}=|\alpha-\beta|.$$

Therefore

$$\max\{f(x),g(x)\}+\min\{f(x),g(x)\}=f(x)+g(x),$$
$$\max\{f(x),g(x)\}-\min\{f(x),g(x)\}=|f(x)-g(x)|$$

for arbitrary real functions $f(x)$ and $g(x)$. Solving these equations for $\max\{f(x),g(x)\}$ and $\min\{f(x),g(x)\}$, we see that if a linear net contains two functions $f(x)$ and $g(x)$, then it also contains the functions $\max\{f(x),g(x)\}$ and $\min\{f(x),g(x)\}$. Moreover, it follows by induction that if a linear net contains the functions $f_1(x),\ldots, f_n(x)$, then it also contains the functions $\max\{f_1(x),\ldots, f_n(x)\}$ and $\min\{f_1(x),\ldots, f_n(x)\}$.

c. THEOREM. *Let $B(Q)$ be a linear net on a compactum Q, separating every pair of points of Q. Then $B(Q)$ is dense in the space $R^s(Q)$ of all continuous functions on Q, provided $B(Q)$ contains the function $e(x)\equiv 1$.*

Proof. A linear net $B(Q)$, containing 1 and separating the points y and z, contains the function taking arbitrary preassigned values p and q at the points y and z. In fact, we can find such a function in the form $a\varphi(x)+b\cdot 1$ for suitable constants a and b, where $\varphi(x)$ is a function in $B(Q)$ separating the points y and z.

Now, given any $\varepsilon>0$, let $f(x)$ be any element of $R^s(Q)$, i.e., any continuous function on Q, and let y and z be any two points of Q (not necessarily distinct). Then, as just shown, we can find a function $\varphi_{yz}(x) \in B(Q)$ such that $\varphi_{yz}(y)=f(y)$, $\varphi_{yz}(z)=f(z)$. Let

$$U_{yz}=\{x \in Q:\varphi_{yz}(x)< f(x)+\varepsilon\}.$$

The set U_{yz} is open and contains the points y and z. Let y be fixed. Then the open sets U_{yz}, where z varies over all points of Q, form a covering of Q. By the finite covering theorem (Vol. 1, Sec. 3.97), this covering contains a finite subcovering $U_{yz_1},\ldots, U_{yz_m}$. Consider the function

$$\varphi_y(x) = \min\{\varphi_{yz_1}(x),\ldots, \varphi_{yz_m}(x)\},$$

which belongs to the linear net $B(Q)$. For fixed y, at least one of the inequalities defining the sets U_{yz_k} holds at every point $x \in Q$, and hence

$$\varphi_y(x) = \min_k \varphi_{yz_k}(x) < f(x) + \varepsilon$$

for all $x \in Q$, while, at the same time,

$$\varphi_y(y) = \min_k \varphi_{yz_k}(y) = f(y).$$

Next let

$$V_y = \{x \in Q : \varphi_y(x) > f(x) - \varepsilon\}.$$

The set V_y is open and contains the point y. Moreover, the open sets V_y, where y varies over all points of Q, form another covering of Q. By the open covering theorem again, this covering contains a finite subcovering $V_{y_1},\ldots, V_{y_n}$. Let

$$\varphi(x) = \max\{\varphi_{y_1}(x),\ldots, \varphi_{y_n}(x)\}.$$

Then $\varphi(x)$ also belongs to the linear net $B(Q)$, and

$$\varphi(x) = \max_j \varphi_{y_j}(x) < f(x) + \varepsilon,$$

by construction. On the other hand, at least one of the inequalities defining the sets V_{y_j} holds at every point $x \in Q$, and hence

$$\varphi(x) = \max_j \varphi_{y_j}(x) > f(x) - \varepsilon$$

for all $x \in Q$. It follows that

$$f(x) - \varepsilon < \varphi(x) < f(x) + \varepsilon$$

for all $x \in Q$. ∎

The theorem is false without the assumption that $e(x) \equiv 1 \in B(Q)$. For example, the linear net of all continuous functions $f(x)$ satisfying the condition $f(y) = 2f(z)$ for two given points y and z is not dense in the space $R^s(Q)$.

1.52. Stone's theorem

a. According to the general definition of an algebra (Sec. 1.18a), a linear manifold $B(Q)$, consisting of (real) functions on a compactum Q, is said to be an *algebra* if, whenever $B(Q)$ contains two functions $f(x)$ and $g(x)$, it also contains the function $f(x)g(x)$.

b. LEMMA. *An algebra $B(Q)$ of continuous real functions on a compactum Q is a linear net, provided it is closed under uniform convergence*† *and contains the unit function $e(x) \equiv 1$.*

Proof. We wish to show that $B(Q)$ contains $|f(x)|$ whenever it contains $f(x)$. There is no loss of generality in assuming that $\max_x |f(x)| = 1$. Consider the Taylor series

$$(1-\xi)^{1/2} = 1 - \tfrac{1}{2}\xi + \frac{\frac{1}{2}(\frac{1}{2}-1)}{1\cdot 2}\xi^2 + \cdots + \frac{\frac{1}{2}(\frac{1}{2}-1)\cdots(\frac{1}{2}-n+1)}{1\cdot 2\cdots n}(-\xi)^n + \cdots, \quad (2)$$

which, as we know, is uniformly convergent for $0 \leqslant \xi \leqslant 1$ (the convergence of (2) for $\xi = 1$ follows from Vol. 1, Sec. 9.52d, with $\alpha = \frac{1}{2}$, and the uniform convergence then follows from Vol. 1, Sec. 6.66). But $0 \leqslant f^2(x) \leqslant 1$ on the compactum Q, and hence

$$|f(x)| = \sqrt{1-[1-f^2(x)]} = 1 - \tfrac{1}{2}[1-f^2(x)] - \tfrac{1}{8}[1-f^2(x)]^2 + \cdots,$$

where the series on the right is uniformly convergent on Q. Since the algebra $B(Q)$ is closed under uniform convergence, it follows that $|f(x)| \in B(Q)$. ▌

c. THEOREM (**Stone's theorem for a real algebra**). *Let $B(Q)$ be an algebra of real functions on a compactum Q, such that $B(Q)$ contains the unit function $e(x) \equiv 1$ and separates every pair of points $x, y \in Q$. Then $B(Q)$ is dense in the space $R^s(Q)$.*

Proof. Let $\overline{(BQ)}$ denote the closure of $B(Q)$ under uniform convergence. Then $\overline{(BQ)}$ is obviously an algebra. In fact, if $f_n(x) \to f(x)$, $g_n(x) \to g(x)$ uniformly on Q, then $f_n(x)g_n(x) \to f(x)g(x)$ uniformly on Q (why?), so that $f(x) \in \overline{(BQ)}$, $g(x) \in \overline{(BQ)}$ implies $f(x)g(x) \in \overline{(BQ)}$. It follows from the lemma that $\overline{(BQ)}$ is a linear net, and then from Theorem 1.51c that $\overline{B(Q)}$ is dense in $R^s(Q)$. But the algebra $\overline{(BQ)}$ is closed, and hence $\overline{(BQ)} = R^s(Q)$. ▌

1.53.a. Now let $B(Q)$ be an algebra of *complex* functions defined on a compactum Q. One might expect that $B(Q)$ will be dense in the space $C^s(Q)$ of all continuous complex functions defined on Q, provided that $B(Q)$ contains the unit function and separates every pair of points in Q. However, it turns out that this form of Stone's theorem is in fact false (see Problem 5).

b. However, Stone's theorem can be extended to algebras of complex functions, provided we impose a further condition on $B(Q)$. To this end, we

† We say that $B(Q)$ is *closed under uniform convergence* if, whenever $B(Q)$ contains a uniformly convergent sequence of functions $f_1(x), \ldots, f_n(x), \ldots$, it also contains the function $\lim_n f_n(x)$.

introduce the following condition: A complex algebra $B(Q)$ is said to be *symmetric* if, whenever $B(Q)$ contains a function $\varphi(x)=u(x)+iv(x)$, it also contains the conjugate complex function $\bar{\varphi}(x)=u(x)-iv(x)$.

THEOREM (**Stone's theorem for a complex algebra**). *Let $B(Q)$ be an algebra of complex functions on a compactum Q, such that $B(Q)$ is symmetric, contains the unit function $e(x)\equiv 1$, and separates every pair of points $x, y \in Q$. Then $B(Q)$ is dense in the space $C^s(Q)$.*

Proof. Under these conditions, $B(Q)$ contains the real functions

$$u(x)=\frac{1}{2}[\varphi(x)+\bar{\varphi}(x)], \qquad v(x)=\frac{1}{2i}[\varphi(x)-\bar{\varphi}(x)]$$

whenever it contains the function $\varphi(x)=u(x)+iv(x)$. Let $B_R(Q)$ denote the subalgebra of real functions $h(x)\in B(Q)$. This subalgebra $B_R(Q)$ contains the unit function and separates every pair of points $x, y \in Q$ (if $\varphi(y)\neq\varphi(z)$, then either $u(y)\neq u(z)$ or $v(y)\neq v(z)$). But $\overline{B_R(Q)}=R^s(Q)$, by Stone's theorem for a real algebra, and hence $\overline{B(Q)}=C^s(Q)$. ∎

1.54. Some implications of Stone's theorem

a. First let the compactum Q be a closed bounded set in R_n. Then we have the following

THEOREM (**Weierstrass**). *Every continuous real function $f(x)$ on a closed bounded set $Q \in R_n$ is the limit of a uniformly convergent sequence of polynomials $p_m(x_1,\ldots,x_n)$ in the variables $x_1,\ldots,x_n$.*

Proof. The algebra $B(Q)$ of all real polynomials $p(x_1,\ldots,x_n)$ on Q obviously satisfies the conditions of Stone's theorem for a real algebra (Theorem 1.52c). ∎

b. Similarly, it is an immediate consequence of Stone's theorem for a complex algebra (Theorem 1.53b) that *every continuous complex-valued function on a closed bounded set $Q\subset R_n$ is the limit of a uniformly convergent sequence of (complex-valued) polynomials $p_m(x_1,\ldots,x_n)$ in the variables $x_1,\ldots,x_n$.* In particular, *every continuous (real or complex) function on the interval $a\leqslant x\leqslant b$ is the limit of a uniformly convergent sequence of (real or complex) polynomials $p_m(x)$.*

c. Next let the compactum Q be the unit circle $x^2+y^2=1$ in the xy-plane. The position of a point on this circle is determined by the polar angle θ. Let the algebra $B(Q)$ be the set of all trigonometric polynomials

$$p(\theta)=\sum_{k=0}^{n}(a_k\cos k\theta+b_k\sin k\theta) \tag{3}$$

with real coefficients. It follows from the formulas

$$2 \cos k\theta \cos m\theta = \cos (k-m)\theta + \cos (k+m)\theta,$$
$$2 \cos k\theta \sin m\theta = \sin (m-k)\,\theta + \sin (m+k)\theta,$$
$$2 \sin k\theta \sin m\theta = \cos (k-m)\theta - \cos (k+m)\theta$$

(cf. Vol. 1, Secs. 5.61, 5.63) that the product of two functions of the form (3) is itself a function of the same form, so that the set $B(Q)$ of all functions of this form is in fact an algebra. This leads to the following variant of Theorem 1.54a:

THEOREM (**Weierstrass**). *Every continuous real function $f(\theta)$ on the circle Q is the limit of a uniformly convergent sequence of trigonometric polynomials* (3) *with real coefficients.*

Proof. Apply Theorem 1.52c, noting that any two distinct points θ_1, θ_2 of the circle Q are separated by either $\cos\theta$ or $\sin\theta$, both functions in the algebra $B(Q)$. ▮

d. Let $g(t)$ be any continuous real function defined on the whole real line, which is periodic with period 2π. Then the formula

$$f(\theta) = g(\theta + 2k\pi) \qquad (k \text{ any integer})$$

defines a continuous real function on the circle Q. Conversely, given any continuous real function $f(\theta)$ on the circle Q, the formula

$$g(t+2k\pi) = f(t) \qquad (k \text{ any integer})$$

defines a continuous real function $g(t)$ on the whole real line, which is periodic with period 2π.

THEOREM. *Every periodic continuous real function on the whole real line, with period 2π, is the limit of a uniformly convergent sequence of trigonometric polynomials* (*converging on the whole real line*).

Proof. Apply the preceding theorem, noting that every polynomial (3) is periodic with period 2π. ▮

e. The complex versions of the preceding two theorems are, in a certain sense, even simpler. Suppose we use Euler's formulas

$$\cos k\theta = \frac{1}{2}(e^{ik\theta} + e^{-ik\theta}),$$

$$\sin k\theta = \frac{1}{2i}(e^{ik\theta} - e^{-ik\theta})$$

(Vol. 1, Sec. 8.63) to replace the real trigonometric polynomials (3) by the complex trigonometric polynomials

$$q(\theta) = \sum_{k=-n}^{n} c_k e^{ik\theta} \tag{3'}$$

(with complex coefficients). Then we have the following

THEOREM. *Every continuous complex function on the circle Q (or, equivalently, every periodic continuous complex function on the whole real line, with period 2π) is the limit of a uniformly convergent sequence of complex trigonometric polynomials* (3′).

Proof. The fact that the polynomials (3′) form an algebra $B(Q)$ is obvious from the rules for multiplying exponentials. The symmetry of $B(Q)$ follows from the formula

$$\overline{\sum c_k e^{ik\theta}} = \sum \bar{c}_k e^{-ik\theta}.$$

Moreover, any two distinct points θ_1, θ_2 of the circle Q are separated by the function $e^{i\theta}$. The proof is now an immediate consequence of Theorem 1.53b. ▮

1.55. Delta-like sequences. Stone's theorem establishes the theoretical possibility of approximating any continuous function by a function from the algebra $B(Q)$, but it does not tell us how to actually construct the approximating function. We now consider ways of obtaining concrete approximations. Since integration will play a role in what follows, we assume that the compactum Q is either a closed interval of the real line or a circle of unit radius (the interval $[-\pi,\pi]$ with identification of the end points).

a. Let $U_\rho(y)$ denote the open interval of length 2ρ centered at the point y. Suppose that for a given point $y \in Q$, there is a sequence of nonnegative functions $D_n(x;y)$ $(n=1,2,\ldots)$ such that, for every $\rho > 0$,

$$\lim_{n\to\infty} \int_{U_\rho(y)} D_n(x;y)\, dx = 1, \tag{4}$$

$$\lim_{n\to\infty} \int_{Q-U_\rho(y)} D_n(x;y)\, dx = 0. \tag{5}$$

Then $D_n(x;y)$ is called a *delta-like sequence* (for the point y). The origin of this term will be explained in Sec. 1.55h.

b. THEOREM. *If $D_n(x;y)$ is a delta-like sequence for the point y and if $f(x)$ is a piecewise continuous function which is continuous at the point y, then*

$$\lim_{n\to\infty} \int_Q D_n(x;y) f(x)\, dx = f(y). \tag{6}$$

Proof. Given any $\varepsilon>0$, let $\delta>0$ be such that $|f(x)-f(y)|<\varepsilon$ whenever $|x-y|<\delta$, $x\in Q$. Then

$$\begin{aligned}
&\left|\int_Q D_n(x;y)f(x)\,dx-f(y)\right| \\
&\quad=\left|\int_Q D_n(x;y)[f(x)-f(y)]\,dx+f(y)\left\{\int_Q D_n(x;y)\,dx-1\right\}\right| \\
&\quad\leqslant \int_{U_\delta(y)} D_n(x;y)|f(x)-f(y)|\,dx+\int_{Q-U_\delta(y)} D_n(x;y)|f(x)-f(y)|\,dx \\
&\qquad+|f(y)|\left|\int_Q D_n(x;y)\,dx-1\right| \\
&\quad<\varepsilon\int_{U_\delta(y)} D_n(x;y)\,dx+2M\int_{Q-U_\delta(y)} D_n(x;y)\,dx \\
&\qquad+M\left|\int_Q D_n(x;y)\,dx-1\right|, \qquad (7)
\end{aligned}$$

where $M=\sup|f(x)|$. But the right-hand side of (7) can be made less than 2ε for all sufficiently large n, because of (4) and (5). ∎

Somewhat more generally, suppose $D_n(x;y)$ is a delta-like sequence for every point y in some subset $E\subset Q$, while $f(x)$ is a piecewise continuous function which is continuous at every point $y\in E$. Then obviously (6) holds for every $y\in E$.

c. We note in passing that if $D_n(x;y)$ is continuous in both variables $x,y\in Q$ and if $f(x)$ is piecewise continuous, as before, then every function

$$f_n(y)=\int_Q D_n(x;y)f(x)\,dx \qquad (n=1,2,\ldots) \qquad (8)$$

is continuous on Q. In fact, given any $\varepsilon>0$, let $\delta>0$ be such that $|y'-y''|<\delta$ implies

$$|D_n(x;y')-D_n(x;y'')|<\frac{\varepsilon}{M(b-a)}$$

for all $x\in Q=[a,b]$, where $M=\sup|f(x)|$. Then $|y'-y''|<\delta$ implies

$$\begin{aligned}
|f_n(y')-f_n(y'')|&=\left|\int_Q [D_n(x;y')-D_n(x;y'')]f(x)\,dx\right| \\
&\leqslant M\int_Q |D_n(x;y')-D_n(x;y'')|\,dx<M\frac{\varepsilon}{M(b-a)}(b-a)=\varepsilon.
\end{aligned}$$

d. Next we further generalize the preceding theorem, first introducing two new definitions. We say that formulas (4) and (5) *hold uniformly on a set* $E\subset Q$

if, given any $\varepsilon>0$, there exists an integer $N>0$ such that

$$\left|\int_{U_\rho(y)} D_n(x;y)\,dx-1\right|<\varepsilon, \tag{4'}$$

$$\int_{Q-U_\rho(y)} D_n(x;y)\,dx<\varepsilon \tag{5'}$$

for all $n>N$ and all $y\in E$. Similarly, we say that a function $f(x)$ is *uniformly continuous on E relative to Q* if, given any $\varepsilon>0$, there exists a $\delta>0$ such that $|f(x)-f(y)|<\varepsilon$ whenever $|x-y|<\delta$, $x\in Q$, $y\in E$.

THEOREM. *If formulas (4) and (5) hold uniformly on E for every $\rho>0$ and if $f(x)$ is piecewise continuous on Q and uniformly continuous on E relative to Q, then the sequence (8) converges uniformly on E to its limit function $f(y)$.*

Proof. This time (7) holds for all $y\in E$. But, because of (4′) and (5′), the right-hand side of (7) can now be made less than 2ε for all sufficiently large n and all $y\in E$. ∎

e. To facilitate the application of the theorem just proved, we now give a simple test for uniform continuity of $f(x)$ on E relative to Q:

LEMMA. *Let $f(x)$ be piecewise continuous on Q, and let $E\subset Q$ be a closed set of continuity points of $f(x)$. Then $f(x)$ is uniformly continuous on E relative to Q.*

Proof. By Heine's theorem (Vol. 1, Sec. 5.17b), $f(x)$ is uniformly continuous on E. Hence, given any $\varepsilon>0$, we can find a $\delta_0>0$ such that $|f(x)-f(y)|<\varepsilon/2$ whenever $|x-y|<\delta_0$, $x\in E$, $y\in E$. With every $y\in E$ we now associate an interval $|x-y|<\delta(y)\leqslant\delta_0/2$ in which $|f(x)-f(y)|<\varepsilon/2$, thereby obtaining a covering of E. By the finite covering theorem (Vol. 1, Sec. 3.97), this covering contains a finite subcovering by intervals $|x-y_1|<\delta_1,\ldots,|x-y_n|<\delta_n$. Let $\delta=\min\{\delta_1,\ldots,\delta_n\}$. Then, given any points $x\in Q$, $y\in E$ with $|x-y|<\delta$, having found a point y_k with $|x-y_k|<\delta_k$, we have

$$|y_k-y|\leqslant|y_k-x|+|x-y|<\delta_k+\delta\leqslant\frac{\delta_0}{2}+\frac{\delta_0}{2}=\delta_0,$$

and hence

$$|f(x)-f(y)|\leqslant|f(x)-f(y_k)|+|f(y_k)-f(y)|<\frac{\varepsilon}{2}+\frac{\varepsilon}{2}=\varepsilon. \quad ∎$$

f. There is a somewhat stronger version of Theorem 1.55b, which goes as follows: *If $D_n(x;y)$ is a delta-like sequence for the point y and if $f(x)$ is a piecewise continuous function which is continuous at the point y, then*

$$\lim_{n\to\infty}\int_Q D_n(x;y_n)f(x)\,dx=f(y) \tag{6'}$$

for every sequence y_n *converging to* y. The proof of (6′) is left as an exercise for the reader.

g. Next we consider the case involving a continuous parameter t, rather than a discrete parameter n. Let $D(t,x,y)$ be a function of three variables, where x and y vary over the compactum Q and t varies over the half-open interval $0<t\leqslant b$, and suppose that, for every $\rho>0$,

$$\lim_{t\to 0}\int_{|x-y|\leqslant\rho} D(t,x,y)\,dx=1, \tag{4″}$$

$$\lim_{t\to 0}\int_{|x-y|\geqslant\rho} D(t,x,y)\,dx=0. \tag{5″}$$

Suppose further that $y(t)$ is a function such that

$$\lim_{t\to 0} y(t)=y.$$

Then

$$\lim_{t\to 0}\int_Q D(t,x,y(t))f(x)\,dx=f(y). \tag{6″}$$

The proof of (6″) entails a calculation like that made in the proof of Theorem 1.55b (give the details).† The function

$$F(t,y)=\int_Q D(t,x,y)f(x)\,dx \qquad (0<t\leqslant b,\, y\in Q),$$

defined for $t=0$ by the condition

$$F(0,y)=f(y),$$

can be regarded as continuous in the closed region $0\leqslant t\leqslant b, y\in Q$.

h. Later (Sec. 4.53), we will need to relax the condition that $D_n(x;y)$ be nonnegative in Theorem 1.55b (and its generalization Theorem 1.55d). It is easy to see (give the details) that Theorem 1.55b remains true (for continuous $f(x)$, say) if

$$\int_Q |D_n(x;y)|\,dx\leqslant C, \tag{9}$$

where C is a constant independent of n, and

$$\lim_{n\to\infty}\int_{Q-U_\rho(y)} D_n(x;y)\varphi(x)\,dx=0, \tag{10}$$

where $\varphi(x)$ is an arbitrary continuous function. On the other hand, the con-

† See also the discussion concerning (6′).

dition (9) is absolutely essential, since without it, the theorem becomes false, as will be shown in Sec. 4.51.

i. Remark. The term "delta-like sequence" stems from Dirac's "delta function." In his classic book on quantum mechanics,† Dirac defined the delta function $\delta(x)$ as a function on the real line $-\infty<x<\infty$ which equals zero everywhere except at the point $x=0$ and has the property

$$\int_{-\infty}^{\infty} \delta(x)\, dx=1. \tag{11}$$

He then "proved" the following theorem: If $f(x)$ is continuous at $x=\xi$, then

$$\int_{-\infty}^{\infty} \delta(x-\xi)f(\xi)\, d\xi=f(x).$$

The "proof" is simple enough: The function $\delta(x-\xi)$ vanishes for $\xi\neq x$, and hence the values of $f(\xi)$ for $\xi\neq x$ play no role here. Thus we can replace $f(\xi)$ by a constant equal to $f(x)$, and the theorem then follows from (11). There is no function in classical analysis with the properties prescribed by Dirac, and the real content of his theorem is essentially that of Theorem 1.55b. It was only some years later, in the work of S. L. Sobolev (1935) and L. Schwartz (1947), that the delta function received a proper mathematical formulation, as a *generalized function*‡ rather than as an ordinary function. Dirac's delta function is a typical example of how a physicist's unerring mathematical intuition can go beyond the level of the mathematics of his time.

1.56. The use of delta-like sequences to construct approximating functions. The problem of approximating a given function $f(y)$ by a function $f_n(y)$ belonging to the algebra $B(Q)$ will be solved once we manage to find a delta-like sequence $D_n(x;y)$ such that

$$f_n(y)=\int_Q D_n(x;y)f(x)\, dx\in B(Q).$$

a. Choosing $Q=[0,1]$, let $B(Q)$ be the algebra of all real polynomials defined on $[0,1]$. Let

$$D_n(x;y)=C_n[1-(x-y)^2]^n \qquad (n=1,2,\ldots), \tag{12}$$

† P. A. M. Dirac, *The Principles of Quantum Mechanics*, Oxford University Press, New York (1930).

‡ Concerning generalized functions, see, for example, A. N. Kolmogorov and S. V. Fomin, *Introductory Real Analysis* (translated by R. A. Silverman), Prentice-Hall, Inc., Englewood Cliffs, N. J. (1970), Sec. 21.

where

$$C_n = \frac{1}{\displaystyle\int_{-1}^{1} (1-t^2)^n \, dt}.$$

As we will see in a moment, $D_n(x;y)$ is a delta-like sequence. Then, since the function

$$f_n(y) = C_n \int_0^1 [1-(x-y)^2]^n f(x) \, dx \qquad (n=1,2,\ldots) \tag{13}$$

is obviously a polynomial in y, of degree $\leqslant 2n$, (13) gives an explicit formula for a sequence of polynomials approximating the function $f(y)$.

b. LEMMA. *For every $\rho \in (0,1)$,*

$$\lim_{n\to\infty} \frac{\displaystyle\int_\rho^1 (1-t^2)^n \, dt}{\displaystyle\int_0^1 (1-t^2)^n \, dt} = 0.$$

Proof. An immediate consequence of the simple estimates

$$\int_\rho^1 (1-t^2)^n \, dt < (1-\rho^2)^n (1-\rho) < (1-\rho^2)^n,$$

$$\int_0^1 (1-t^2)^n \, dt > \int_0^1 (1-t)^n \, dt = \frac{1}{n+1},$$

and the formula†

$$\lim_{n\to\infty} (n+1)(1-\rho^2)^n = 0. \quad ∎$$

As a corollary, we find that

$$\lim_{n\to\infty} \frac{\displaystyle\int_0^\rho (1-t^2)^n \, dt}{\displaystyle\int_0^1 (1-t^2)^n \, dt} = 1. \tag{14}$$

c. We now confirm that the functions $D_n(x;y)$ defined by (12) actually form a delta-like sequence. By the lemma, for every $\rho \in (0,1)$,

† This follows from Vol. 1, Sec. 5.57, formula (24), together with Vol. 1, Sec. 4.37a.

$$\begin{aligned}\int_{\substack{|x-y|\geqslant\rho\\0\leqslant x\leqslant 1}} D_n(x;y)\,dx &= C_n\int_{\substack{|x-y|\geqslant\rho\\0\leqslant x\leqslant 1}}[1-(x-y)^2]^n\,dx\\ &= C_n\int_{\substack{|t|\geqslant\rho\\-y\leqslant t\leqslant 1-y}}(1-t^2)^n\,dt\leqslant 2C_n\int_\rho^1(1-t^2)^n\,dt\\ &= \frac{\displaystyle\int_\rho^1(1-t^2)^n\,dt}{\displaystyle\int_0^1(1-t^2)^n\,dt}\to 0\end{aligned}$$

as $n\to\infty$, so that formula (5) holds uniformly on the set $0\leqslant y\leqslant 1$. Similarly, by (14), for every $y\in[\rho_0, 1-\rho_0]$, $0<\rho<\rho_0$,

$$\begin{aligned}\int_{\substack{|x-y|\leqslant\rho\\0\leqslant x\leqslant 1}} D_n(x;y)\,dx &= C_n\int_{\substack{|x-y|\leqslant\rho\\0\leqslant x\leqslant 1}}[1-(x-y)^2]^n\,dx\\ &= C_n\int_{-\rho}^{\rho}(1-t^2)^n\,dt=\frac{\displaystyle\int_0^\rho(1-t^2)^n\,dt}{\displaystyle\int_0^1(1-t^2)^n\,dt}\to 1\end{aligned}$$

as $n\to\infty$, so that formula (4) holds uniformly on the set $\rho_0\leqslant y\leqslant 1-\rho_0$. Suppose $f(y)$ is continuous on $[0,1]$. Then it follows from Theorems 1.55b and 1.55d that the sequence of polynomials (13) converges to $f(y)$ for every $y\in[0,1]$, and that the convergence is uniform on every interval $[\rho_0, 1-\rho_0]$, $0<\rho_0<\frac{1}{2}$. Note that this gives a direct proof of Weierstrass' theorem (Sec. 1.54a) for the interval $[\rho_0, 1-\rho_0]$ and hence for an arbitrary interval $[a,b]$, since $[\rho_0, 1-\rho_0]$ can be transformed into $[a,b]$ by making a suitable expansion and shift.

1.57.a. We can get approximating trigonometric polynomials by carrying out a completely analogous construction. Let Q be the unit circle $x^2+y^2=1$, let θ be the polar angle determining the position of a point on Q, and let $B(Q)$ be the algebra of all real trigonometric polynomials. Moreover, let

$$D_n(\theta;\varphi)=C_n\cos^{2n}\frac{\theta-\varphi}{2}\qquad(n=1,2,\ldots),\tag{15}$$

where this time

$$C_n=\frac{1}{\displaystyle\int_0^{2\pi}\cos^{2n}t\,dt}.$$

As we will see in a moment, $D_n(\theta;\varphi)$ is a delta-like sequence (for any given

φ). Then, since the function

$$f_n(\varphi)=C_n\int_0^{2\pi}\cos^{2n}\frac{\theta-\varphi}{2}f(\theta)\,d\theta \qquad (n=1,2,\ldots) \tag{16}$$

is obviously a trigonometric polynomial in φ, of order $\leqslant 2n$, (16) gives an explicit formula for a sequence of trigonometric polynomials approximating the function $f(\varphi)$.

b. LEMMA. *For every* $\rho \in (0,\pi/2)$,

$$\lim_{n\to\infty}\frac{\displaystyle\int_\rho^{\pi/2}\cos^{2n}t\,dt}{\displaystyle\int_0^{\pi/2}\cos^{2n}t\,dt}=0.$$

Proof. The function $\cos t$ is decreasing for $0\leqslant t\leqslant\pi/2$, and hence

$$\int_\rho^{\pi/2}\cos^{2n}t\,dt\leqslant\left(\frac{\pi}{2}-\rho\right)\cos^{2n}\rho\leqslant\frac{\pi}{2}\cos^{2n}\rho.$$

Moreover, $\cos t$ is concave upward in the same interval (cf. Vol. 1, Sec. 7.53), so that

$$\cos t\geqslant 1-\frac{2t}{\pi},$$

and hence

$$\int_0^{\pi/2}\cos^{2n}t\,dt\geqslant\int_0^{\pi/2}\left(1-\frac{2t}{\pi}\right)^{2n}dt=\frac{\pi}{2(2n+1)}.$$

It follows that

$$\frac{\displaystyle\int_\rho^{\pi/2}\cos^{2n}t\,dt}{\displaystyle\int_0^{\pi/2}\cos^{2n}t\,dt}\leqslant\frac{2(2n+1)}{\pi}\frac{\pi}{2}\cos^{2n}\rho\to 0$$

as $n\to\infty$ (cf. Vol. 1, Sec. 5.55). ∎

As a corollary, we find that

$$\lim_{n\to\infty}\frac{\displaystyle\int_0^{\rho}\cos^{2n}t\,dt}{\displaystyle\int_0^{\pi/2}\cos^{2n}t\,dt}=1. \tag{17}$$

c. We now confirm that the functions $D_n(\theta;\varphi)$ defined by (15) actually form a delta-like sequence. By the lemma, for every $\rho \in (0,\pi)$,

$$\int_{\substack{|\theta-\varphi|\geqslant\rho \\ -\pi\leqslant\theta\leqslant\pi}} D_n(\theta;\varphi)\, d\theta = C_n \int_{\substack{|\theta-\varphi|\geqslant\rho \\ -\pi\leqslant\theta\leqslant\pi}} \cos^{2n}\frac{\theta-\varphi}{2}\, d\theta$$

$$= 2C_n \int_{\substack{|t|\geqslant\rho/2 \\ (-\pi-\varphi)/2\leqslant t\leqslant(\pi-\varphi)/2}} \cos^{2n} t\, dt \leqslant 4C_n \int_{\rho/2}^{\pi/2} \cos^{2n} t\, dt$$

$$= \frac{\int_{\rho/2}^{\pi/2} \cos^{2n} t\, dt}{\int_{0}^{\pi/2} \cos^{2n} t\, dt} \to 0$$

as $n\to\infty$, so that formula (5), written in the variables θ and φ, holds uniformly on the set $-\pi\leqslant\varphi\leqslant\pi$. Similarly, by (17), for every φ in the interval $[-\pi+\rho_0, \pi-\rho_0]$, $0<\rho<\rho_0$,

$$\int_{\substack{|\theta-\varphi|\leqslant\rho \\ -\pi\leqslant\theta\leqslant\pi}} D_n(\theta;\varphi)\, d\theta = C_n \int_{\substack{|\theta-\varphi|\leqslant\rho \\ -\pi\leqslant\theta\leqslant\pi}} \cos^{2n}\frac{\theta-\varphi}{2}\, d\theta$$

$$= 2C_n \int_{-\rho/2}^{\rho/2} \cos^{2n} t\, dt = \frac{\int_{0}^{\rho/2} \cos^{2n} t\, dt}{\int_{0}^{\pi/2} \cos^{2n} t\, dt} \to 1$$

as $n\to\infty$, so that formula (4) holds uniformly on the set $-\pi+\rho_0\leqslant\varphi\leqslant\pi-\rho_0$. Suppose $f(\varphi)$ is piecewise continuous on $[-\pi,\pi]$. Then it follows from Theorem 1.55d that the sequence of polynomials (16) converges uniformly to $f(\varphi)$ on every set $E\subset Q$ on which the function $f(\varphi)$ is uniformly continuous relative to Q, in particular on every closed set on which it is continuous (see Lemma 1.55c).

d. Remark. In both of the cases just considered, we can estimate the degree of the algebraic polynomial (13) or trigonometric polynomial (16) approximating the function $f(y)$ or $f(\varphi)$ to within a given accuracy ε. Although the polynomials (13) and (16) have a very simple structure, they are in general not the best approximations among all polynomials of a given degree. It can be shown that among all polynomials of degree n, there is a polynomial which differs from a given continuous function $f(x)$ on the interval $[a,b]$ by no more than

$$6\omega\left(\frac{b-a}{2n}\right), \qquad (18)$$

where

$$\omega(\delta) = \max_{|x-y| \leqslant \delta} |f(x) - f(y)|$$

is the *modulus of continuity* of $f(x)$ on $[a,b]$ (Vol. 1, Sec. 5.17c). For trigonometric polynomials (on the circle Q), the estimate (18) is replaced by†

$$6\omega\left(\frac{1}{n}\right). \tag{18'}$$

1.6. Differentiation and Integration in a Normed Linear Space

1.61. Differentiation

a. Let $x(t)$ be a function defined on an interval $a \leqslant t \leqslant b$, taking values in a real or complex normed linear space $\mathbf{X}$. Then $x(t)$ is said to be *differentiable at a point* $t_0 \in [a,b]$ if the limit

$$x'(t_0) = \lim_{\Delta t \to 0} \frac{x(t_0 + \Delta t) - x(t_0)}{\Delta t}, \tag{1}$$

called the *derivative* of $x(t)$ at $t = t_0$, exists in $\mathbf{X}$.

b. Suppose $x(t)$ has a derivative $x'(t)$ at every point of an interval $[a,b]$. Then $x(t)$ is said to be *differentiable on* $[a,b]$, and the derivative $x'(t)$ is itself a function defined on $[a,b]$ with values in the space $\mathbf{X}$.

c. Suppose $x(t)$ is differentiable at a point t_0. Then, according to (1),

$$x(t) - x(t_0) = x'(t_0)(t - t_0) + \varepsilon(t,t_0)(t - t_0), \tag{2}$$

where $\varepsilon(t,t_0)$ approaches zero in the space $\mathbf{X}$ as $t \to t_0$. In particular, if $x(t)$ is differentiable at a point t_0, then $x(t)$ is continuous at t_0, while if $x(t)$ is differentiable on an interval $[a,b]$, then $x(t)$ is continuous on $[a,b]$.

d. The following basic rules for calculating derivatives are the exact analogues of those for numerical functions, and are easily verified:

(a) If $x(t) = x_0$ is a constant element of the space $\mathbf{X}$, then $x'(t) \equiv 0$.
(b) If $x(t)$ and $y(t)$ are differentiable functions with values in $\mathbf{X}$, then so is the sum $x(t) + y(t)$, and

$$[x(t) + y(t)]' = x'(t) + y'(t).$$

† For the proof of the estimates (18) and (18′), called *Jackson's theorems*, see e.g., T. J. Rivlin, *An Introduction to the Approximation of Functions*, Blaisdell Publ. Co., Waltham, Mass. (1969), pp. 21–22.

(c) If $x(t)$ is a differentiable function with values in $\mathbf{X}$ and if $\gamma(t)$ is a differentiable numerical function, then the product $\gamma(t)x(t)$ is a differentiable function with values in $\mathbf{X}$, and

$$[\gamma(t)x(t)]'=\gamma'(t)x(t)+\gamma(t)x'(t);$$

in particular,

$$[\alpha x(t)]'=\alpha x'(t)$$

for every constant α.

(d) If $x(t)$ is a differentiable function on $[a,b]$ with values in $\mathbf{X}$, while $t(\tau)$ is a numerical function with values in $[a,b]$ which is differentiable at τ_0, then the composite function $y(\tau)=x(t(\tau))$ is differentiable at τ_0 and

$$y'(\tau_0)=x'(t_0)t'(\tau_0), \tag{3}$$

where $t_0=t(\tau_0)$.

e. Let $x(t)$ be a function with values in a normed linear space $\mathbf{X}$, and suppose the independent variable t changes from t_0 to t_0+dt. Then by the *differential* of $x(t)$ we mean the vector

$$dx=x'(t_0)\,dt\in\mathbf{X}.$$

Consulting (2), we see that the differential dx is just the principal linear part of the increment of $x(t)$ corresponding to the increment dt of the independent variable t.

Just as in the case of numerical functions (Vol. 1, Sec. 7.34), the differential of a function $x(t)$ with values in $\mathbf{X}$ does not depend on whether its argument t is the independent variable or a function $t=t(\tau)$ of some other independent variable τ.† In fact, let

$$y=g(\tau)\equiv x(t(\tau)).$$

Then, by (3),

$$g'(\tau_0)=x'(t_0)t'(\tau_0), \tag{4}$$

where $t_0=t(\tau_0)$. Multiplying both sides of (4) by $d\tau$, we get

$$dy=g'(\tau_0)\,d\tau=x'(t_0)t'(\tau_0)\,d\tau,$$

and hence

$$dy=x'(t_0)\,dt, \tag{5}$$

since $dt=t'(\tau_0)\,d\tau$. But (5) is precisely the expression for the differential of

† This property might be called the "invariance of the differential under the change of variables $t=t(\tau)$."

the function y in the case where t is the independent variable rather than a function of τ.

f. A function $x(t)$ with values in a normed linear space $\mathbf{X}$ is said to be *piecewise continuous* on the interval $a \leqslant t \leqslant b$ if there exists a partition

$$a=t_0<t_1<t_2<\cdots<t_n=b$$

such that $x(t)$ is continuous on every open interval $t_k<t<t_{k+1}$ and has finite limits†

$$x(t_0+0),\ x(t_1-0),\ x(t_1+0),\ x(t_2-0),\ x(t_2+0),\ldots,\ x(t_n-0).$$

A function $x(t)$ with values in $\mathbf{X}$ is said to be *piecewise smooth* on $[a,b]$ if it is continuous on $[a,b]$ and has a derivative $x'(t)$ at all but a finite number of points of $[a,b]$, and if $x'(t)$ is piecewise continuous on $[a,b]$.

We now prove a proposition which is essentially the converse of the first differentiation rule listed in Sec. 1.61d:

THEOREM. *If $x(t)$, $t \in [a,b]$, is a piecewise smooth function with values in a normed linear space $\mathbf{X}$ and if the derivative $x'(t)$ vanishes at every point where it exists, then $x(t) \equiv x_0$, i.e., $x(t)$ is a constant element of $\mathbf{X}$.*

Proof. Assuming first that $x'(t)=0$ at every interior point of $[a,b]$, we fix a point $c \in (a,b)$ and a number $\varepsilon>0$. Then, since $x'(c)=0$, there exists a neighborhood of the point c in which the inequality

$$|x(t)-x(c)| \leqslant \varepsilon |t-c| \tag{6}$$

holds. Let $T_\varepsilon(c)$ denote the set of all $t>b$ together with all those $t \in [a,b]$ for which (6) does not hold, and let $t_0=\inf T_\varepsilon(c)$. Suppose $t_0<b$. Then, since $x(t)$ is continuous, the inequality (6), being valid near t_0, remains valid at the point t_0 itself. Moreover, since $x'(t_0)=0$, there exists a neighborhood of the point t_0 in which

$$|x(t)-x(t_0)| < \frac{\varepsilon}{2} |t-t_0|. \tag{7}$$

Let $t>t_0$ be a point satisfying (7). Then (6) and (7) imply

$$\begin{aligned} |x(t)-x(c)| &\leqslant |x(t)-x(t_0)|+|x(t_0)-x(c)| \\ &\leqslant \frac{\varepsilon}{2}(t-t_0)+\varepsilon(t_0-c)=\varepsilon\frac{t-t_0}{2}+t_0-c<\varepsilon(t-c), \end{aligned}$$

† By definition,

$$x(t_k-0)=\lim_{t \nearrow t_k} x(t), \qquad x(t_k+0)=\lim_{t \searrow t_k} x(t)$$

(cf. Vol. 1, Sec. 5.32). The values of $x(t)$ at the points t_k can be arbitrary or even undefined.

so that the point t does not belong to $T_\varepsilon(c)$. But this contradicts the definition $t_0 = \inf T_\varepsilon(c)$. It follows that $t_0 = b$, and hence that

$$|x(t) - x(c)| \leqslant \varepsilon(t - c)$$

for all $t \in [c,b]$. But then, since $\varepsilon > 0$ can be made arbitrarily small, we have

$$x(t) - x(c) = 0$$

for all $t \in [c,b]$, so that $x(t)$ has the constant value $x(c)$ on the interval $[c,b]$. But c can be chosen arbitrarily near the point a, and hence $x(t)$ is constant on the whole interval $[a,b]$.

Next we consider the general case, where the interval $[a,b]$ contains a finite number of points c_k, where $a = c_0 < c_1 < c_2 < \cdots < c_n = b$, such that the derivative $x'(t)$ exists and equals zero in each of the intervals

$$(c_k, c_{k+1}) \qquad (k = 0, 1, \ldots, n-1), \tag{8}$$

but fails to exist at the points c_k themselves. The argument just given shows that the function $x(t)$ is constant on each of the intervals (8). But since $x(t)$ is continuous on $[a,b]$, its values on neighboring intervals (c_{k-1}, c_k), (c_k, c_{k+1}) must coincide, i.e., $x(t)$ must be constant on the whole interval $[a,b]$. ∎

1.62. Integration

a. Let $\mathbf{X}$ be a Banach space, i.e., a complete normed linear space (real or complex), and let $x(t)$ be a function defined on an interval $[a,b]$ and taking values in $\mathbf{X}$. Given any partition

$$\Pi = \{a = t_0 \leqslant \tau_0 \leqslant t_1 \leqslant \tau_1 \leqslant t_2 \leqslant \cdots \leqslant t_{n-1} \leqslant \tau_{n-1} \leqslant t_n = b\}$$

of $[a,b]$, with marked points $\tau_0, \tau_1, \ldots, \tau_{n-1}$ and parameter

$$d(\Pi) = \max\{\Delta t_0, \Delta t_1, \ldots, \Delta t_{n-1}\} \qquad (\Delta t_k = t_{k+1} - t_k)$$

(see Vol. 1, p. 274), we can form the Riemann sum

$$S_\Pi(x) = \sum_{k=0}^{n-1} x(\tau_k)\Delta t_k, \tag{9}$$

which is obviously itself an element of the space $\mathbf{X}$. Suppose the sum (9) approaches a limit in the space $\mathbf{X}$ under arbitrary refinement of the partition Π, i.e., as $d(\Pi) \to 0$. Then this limit is called the *integral of* $x(t)$ *over the interval* $[a,b]$, denoted by

$$\int_a^b x(t)\,dt,$$

and the function $x(t)$ is said to be *integrable on* $[a,b]$.

b. THEOREM. *If $x(t)$ is continuous on $[a,b]$, then $x(t)$ is integrable on $[a,b]$.*

Proof. We need only reproduce the steps of the analogous proof for a continuous *numerical* function (Vol. 1, Sec. 9.14). Since $x(t)$ is (uniformly) continuous on $[a,b]$, we have

$$\lim_{\delta \to 0} \omega_x(\delta) = 0, \tag{10}$$

where

$$\omega_x(\delta) = \sup_{|t'-t''| \leqslant \delta} \|x(t') - x(t'')\|$$

is the *modulus of continuity* of $x(t)$ on $[a,b]$ (cf. Vol. 1, Sec. 5.17c), written in the form appropriate to a function taking values in a normed linear space. Moreover,

$$\|S_{\Pi_1}(x) - S_{\Pi_2}(x)\| \leqslant 2\omega_x(\delta)(b-a) \tag{11}$$

for any two partitions Π_1 and Π_2 of $[a,b]$ such that $d(\Pi_1) < \delta$, $d(\Pi_2) < \delta$, just as in the case of a numerical function (Vol. 1, Sec. 9.14d). According to (10), given any $\varepsilon > 0$, we can find a $\delta > 0$ such that

$$\omega_x(\delta) < \frac{\varepsilon}{2(b-a)}.$$

But then, by (11),

$$\|S_{\Pi_1}(x) - S_{\Pi_2}(x)\| > 2\frac{\varepsilon}{2(b-a)}(b-a) = \varepsilon$$

for any two partitions Π_1 and Π_2 such that $d(\Pi_1) < \delta$, $d(\Pi_2) < \delta$. The proof is now an immediate consequence of the Cauchy convergence criterion (cf. Vol. 1, Sec. 4.19). ∎

c. THEOREM. *If $x(t)$ is piecewise continuous on $[a,b]$, then $x(t)$ is integrable on $[a,b]$.*

Proof. The exact analogue of the proof for a piecewise continuous numerical function (Vol. 1, Sec. 9.16c). ∎

d. THEOREM. *Every integrable function $x(t)$ is bounded (in norm).*†

Proof. The exact analogue of the proof for an integrable numerical function (Vol. 1, Sec. 9.15c). ∎

† I.e., there is a constant $M > 0$ such that $\|x(t)\| \leqslant M$ for all $t \in [a,b]$.

e. The following key properties of integrals of integrable functions are easily verified:

(a) $\int_a^b \alpha x(t)\, dt = \alpha \int_a^b x(t)\, dt$ (for any number α);

(b) $\int_a^b [x(t)+y(t)]\, dt = \int_a^b x(t)\, dt + \int_a^b y(t)\, dt$;

(c) $\int_a^c x(t)\, dt + \int_c^b x(t)\, dt = \int_a^b x(t)\, dt$ $(a<c<b)$;

(d) $\left\| \int_a^b x(t)\, dt \right\| \leqslant (b-a) \max_{a \leqslant t \leqslant b} \|x(t)\|$;

(e) $\left\| \int_a^b x(t)\, dt \right\| \leqslant \int_a^b \|x(t)\|\, dt.$

To obtain each formula, we need only pass to the limit in the (obviously true) analogous formula for Riemann sums.

f. The mean value of a function. Given an integrable function $x(t)$ with values in a Banach space $\mathbf{X}$, the quantity

$$\frac{1}{b-a} \int_a^b x(t)\, dt$$

is called the *(integral) mean value* of $f(x)$ on the interval $[a,b]$, just as in the case of a numerical function (Vol. 1, Sec. 9.15f). The mean value of a real function $x(t)$ lies between its smallest and largest values on $[a,b]$ and coincides with some value $x(t_0)$ if $x(t)$ is continuous. The mean value of a continuous function with values in a Banach space (or even of a continuous complex function) may fail to coincide with its value at any point of the interval $[a,b]$. For example,

$$\frac{1}{2\pi} \int_0^{2\pi} ie^{it}\, dt = e^{it} \Big|_0^{2\pi} = 0,$$

even though the function ie^{it} does not vanish at any point of the interval of integration.

g. Given a subset E of a linear space $\mathbf{L}$, by the *convex hull* of E, we mean the set $V(E)$ of all vectors of the form

$$x = \sum_{j=1}^{m} \alpha_j x_j \qquad (x_j \in E), \tag{12}$$

where $m = 1,2,\ldots$ is any positive integer and

$$\alpha_j \geqslant 0, \qquad \sum_{j=1}^{m} \alpha_j = 1. \tag{13}$$

The set $V(E)$ is convex (Sec. 1.34b). In fact, if

$$x = \sum_{j=1}^{m} \alpha_j x_j \in V(E), \qquad y = \sum_{k=1}^{n} \beta_k y_k \in V(E) \qquad (x_j, y_k \in E),$$

and if $\alpha \geqslant 0$, $\beta \geqslant 0$, $\alpha + \beta = 1$, then

$$\alpha x + \beta y = \alpha \sum_{j=1}^{m} \alpha_j x_j + \beta \sum_{k=1}^{n} \beta_k y_k = \sum_{j=1}^{m} \alpha\alpha_j x_j + \sum_{k=1}^{n} \beta\beta_k y_k$$

also belongs to $V(E)$, since $\alpha\alpha_j \geqslant 0$, $\beta\beta_k \geqslant 0$, and

$$\sum_{j=1}^{m} \alpha\alpha_j + \sum_{k=1}^{n} \beta\beta_k = \alpha \sum_{j=1}^{m} \alpha_j + \beta \sum_{k=1}^{n} \beta_k = \alpha + \beta = 1.$$

On the other hand, *every convex set P containing a given set E also contains the set $V(E)$ of all vectors of the form* (12). For $m = 2$ this follows from the very definition of a convex set. To prove the assertion for $m > 2$, we use mathematical induction: Suppose the assertion is true for any set of $m - 1$ vectors in E. Let $x_1, \ldots, x_m$ be any set of m vectors in E, and let $\alpha_1, \ldots, \alpha_m$ be numbers satisfying (13). Then

$$\begin{aligned} z &= \alpha_1 x_1 + \cdots + \alpha_{m-1} x_{m-1} + \alpha_m x_m \\ &= (\alpha_1 + \cdots + \alpha_{m-1}) \left(\frac{\alpha_1}{\alpha_1 + \cdots + \alpha_{m-1}} x_1 + \cdots + \frac{\alpha_{m-1}}{\alpha_1 + \cdots + \alpha_{m-1}} x_{m-1} \right) + \alpha_m x_m \\ &= (\alpha_1 + \cdots + \alpha_{m-1}) z_1 + \alpha_m x_m, \end{aligned}$$

where z_1 belongs to P, by the induction hypothesis, since obviously

$$\frac{\alpha_j}{\alpha_1 + \cdots + \alpha_{m-1}} \geqslant 0, \qquad \sum_{j=0}^{m-1} \frac{\alpha_j}{\alpha_1 + \cdots + \alpha_{m-1}} = 1.$$

Therefore z belongs to P, being a point of the segment joining z_1 and x_m (why?). Hence the assertion is true for any set of m vectors in E, and the proof for arbitrary $m \geqslant 2$ follows by induction.

h. The above considerations show that the set $V(E)$ is the smallest convex set containing the set E. If E is itself convex, then obviously $V(E) = E$.

i. A convex set in a Banach space $\mathbf{X}$ need not be closed (consider, for example, an open interval on the real line). Starting from a set $E \subset \mathbf{X}$, we can form first the convex hull $V(E)$ and then take the closure of $V(E)$, obtaining a set $\overline{V(E)}$ called the *closed convex hull* of E. The set $\overline{V(E)}$ is itself convex. More generally, the closure $\bar{P}$ of any convex set $P \subset \mathbf{X}$ is itself convex, since if

$x, y \in \bar{P}$, then

$$x = \lim_{n \to \infty} x_n, \qquad y = \lim_{n \to \infty} y_n$$

where $x_n, y_n \in P$, and hence

$$\alpha x + \beta y = \lim_{n \to \infty} (\alpha x_n + \beta y_n) \in \bar{P}.$$

The set $\overline{V(E)}$ is the smallest closed convex set containing the given set E.

THEOREM. *Let $x(t)$ be an integrable function on an interval $[a,b]$, with values in a Banach space $\mathbf{X}$. Then the mean value of $x(t)$ on $[a,b]$ belongs to the closed convex hull of the set of values of $x(t)$ on $[a,b]$.*

Proof. An immediate consequence of the definition

$$\frac{1}{b-a} \int_a^b x(t)\, dt = \frac{1}{b-a} \lim_{d(\Pi) \to 0} \sum_{k=0}^{n-1} x(\tau_k) \Delta t_k,$$

since the Riemann sum on the right belongs to the convex hull of the set of values of $x(t)$.† ∎

The mean value of the function ie^{it} on $[0,2\pi]$, found to be 0 in Sec. 1.62f, belongs to the (closed) convex hull of all values of ie^{it} on $[0,2\pi]$. In fact, these values fill up a circle Q of radius 1 centered at 0, whose convex hull is just the disk bounded by Q.

j. Improper integrals. Next we construct a theory of improper integrals of functions with values in a Banach space, patterned after the theory of improper integrals of *numerical* functions (Vol. 1, Chapter 11). Given a function $x(t)$ with values in a Banach space $\mathbf{X}$, suppose $x(t)$ is integrable (being piecewise continuous, say) on every finite interval $a \leqslant t \leqslant b < \infty$, where a is fixed and b variable. Consider the function

$$I(T) = \int_0^T x(t)\, dt,$$

defined for all $T \geqslant a$, and suppose $I(T)$ approaches a limit $I \in \mathbf{X}$ (in the norm of $\mathbf{X}$) as $T \to \infty$. Then the expression

$$\int_a^\infty x(t)\, dt, \tag{14}$$

called an *improper integral of the first kind*, is said to be *convergent*, with *value I*.

† Note that $\frac{1}{b-a} \sum_{k=0}^{n-1} \Delta t_k = 1$.

Improper integrals of the second and third kinds are defined similarly, by analogy with the case of numerical functions (cf. Vol. 1, Sec. 11.3).

An improper integral (14) is said to be *absolutely convergent* if the ordinary improper integral

$$\int_a^\infty \|x(t)\| \, dt$$

is convergent (cf. Vol. 1, Sec. 11.21b).

THEOREM. *If the integral* (14) *is absolutely convergent, then it is also convergent and*

$$\left\| \int_a^\infty x(t) \, dt \right\| \leqslant \int_a^\infty \|x(t)\| \, dt.$$

Proof. Virtually the same as the corresponding proof for numerical functions (Vol. 1, Sec. 11.21a), based on the following *Cauchy convergence criterion for improper integrals*: The integral (14) is convergent if and only if, given any $\varepsilon > 0$, there exists a number $T_0 > a$ such that

$$|I(T') - I(T'')| = \left| \int_{T'}^{T''} x(t) \, dt \right| < \varepsilon$$

for all $T' \geqslant T_0$, $T'' \geqslant T_0$ (cf. Vol. 1, Sec. 11.12). ∎

k. The integral as a function of its upper limit

THEOREM. *Let $x(t)$ be a piecewise continuous function on $[a,b]$, with values in a Banach space $\mathbf{X}$, and suppose $x(t)$ is continuous at a point $t_0 \in [a,b]$. Then the function*

$$F(t) = \int_a^t x(\tau) \, d\tau \tag{15}$$

is differentiable at t_0, with derivative $x(t_0)$.

Proof. Clearly

$$\frac{F(t) - F(t_0)}{t - t_0} = \frac{1}{t - t_0} \int_{t_0}^t x(\tau) \, d\tau = \frac{1}{t - t_0} \int_{t_0}^t x(t_0) \, d\tau + \frac{1}{t - t_0} \int_{t_0}^t [x(\tau) - x(t_0)] \, d\tau$$

$$= x(t_0) + \frac{1}{t - t_0} \int_{t_0}^t [x(\tau) - x(t_0)] \, d\tau,$$

where

$$\left\| \frac{1}{t - t_0} \int_{t_0}^t [x(\tau) - x(t_0)] \, d\tau \right\| \leqslant \max_{t_0 \leqslant \tau \leqslant t} \|x(\tau) - x(t_0)\| \to 0 \tag{16}$$

as $t \to t_0$, since $x(t)$ is continuous at t_0.† ∎

† Note that the inequality (16) immediately implies the continuity of $F(t)$ at *every* point $t_0 \in [a,b]$, whether or not t_0 is a continuity point of $x(t)$.

1.63. Antiderivatives

a. A continuous function $G(t)$ with values in a Banach space $\mathbf{X}$ is called an *antiderivative* of a piecewise continuous function $x(t)$ if $G'(t)=x(t)$ at every continuity point of $x(t)$. If $x(t)$ has two antiderivatives, say $G(t)$ and $H(t)$, then

$$[H(t)-G(t)]'=H'(t)-G'(t)=x(t)-x(t)=0$$

at every continuity point of $x(t)$, and hence $H(t)-G(t)$ equals a constant, by Theorem 1.61f. Therefore *two antiderivatives of $x(t)$ can only differ by a constant element of the space* $\mathbf{X}$. Since, as just shown, one antiderivative of $x(t)$ is the function (15), it follows that every antiderivative of $x(t)$ is of the form

$$G(t)=\int_a^t x(\tau)\,d\tau+x_0,$$

where x_0 is a fixed element of $\mathbf{X}$. In particular, every antiderivative $G(t)$ satisfies the following formula, often called the "fundamental theorem of calculus":

$$\int_a^b x(\tau)\,d\tau=G(b)-G(a). \tag{17}$$

b. Having just shown how a function is the derivative of its own integral (Theorem 1.62k), we now give conditions under which a function is the integral of its own derivative:

THEOREM. *Let $G(t)$ be a function with values in a Banach space $\mathbf{X}$, and suppose $G(t)$ is differentiable on the interval $[a,b]$, with a piecewise continuous derivative. Then*

$$G(t)=G(a)+\int_a^t G'(\tau)\,d\tau \tag{18}$$

for every $t\in[a,b]$.

Proof. Let $G^*(t)$ denote the right-hand side of (18). Then, by Theorem 1.62k, $G^*(t)$ has the derivative $G'(t)$ at every continuity point of $G'(t)$. But $G(t)$ has the same property, by hypothesis, and hence both $G(t)$ and $G^*(t)$, being continuous, are antiderivatives of $G'(t)$. Therefore $G^*(t)-G(t)$ equals a constant element $x_0\in\mathbf{X}$. But $x_0=0$, since $G^*(a)=G(a)$, and hence $G(t)=G^*(t)$. ∎

c. For a differentiable numerical function $G(t)$, we have Lagrange's theorem

$$G(b)-G(a)=(b-a)A$$

(Vol. 1, Sec. 7.44), where A is the value of $G'(t)$ at a suitable point of $[a,b]$.

For a differentiable function $G(t)$ with values in a Banach space $\mathbf{X}$, the same formula continues to hold, except that now A is a suitable point of the closed convex hull of the set of values of $G'(t)$ on $[a,b]$. This is an immediate consequence of formula (17) and Theorem 1.62i.

d. Let $u(t)$ and $v(t)$ be two piecewise smooth functions on $[a,b]$, one a numerical function and the other a vector function (with values in $\mathbf{X}$). Then (17) implies the usual formula

$$\int_a^b u(t)\,dv(t) = u(t)v(t)\Big|_a^b - \int_a^b v(t)\,du(t) \tag{19}$$

for integration by parts, just as in the case where both $u(t)$ and $v(t)$ are numerical functions (Vol. 1, Sec. 9.51a).

e. In the same way, we get the usual formula

$$\int_a^b x(t)\,dt = \int_\alpha^\beta x(t(\tau))t'(\tau)\,d\tau$$

for integration by substitution, under the same conditions on the vector function $x(t)$, the numerical function $t(\tau)$, and the numbers α, β, $a=t(\alpha)$, $b=t(\beta)$ as in the case where $x(t)$ is itself a numerical function (Vol. 1, Sec. 9.53).

1.64. Taylor's formula

a. The higher derivatives of a function $x(t)$ with values in a Banach space $\mathbf{X}$ are defined inductively, just as in the case of a numerical function. Thus we define the *derivative of order n* as the (first) derivative of the derivative of order $n-1$, assuming that the latter is a differentiable function on $[a,b]$. The higher derivatives of the vector function $x(t)$ are written in the same way as for a numerical function:†

$$x''(t) = [x'(t)]',\ x'''(t) = [x''(t)]', \ldots,\ x^{(n)}(t) = [x^{(n-1)}(t)]', \ldots$$

b. Let $x(t)$ be a function defined on an interval $[a,b]$, with values in a Banach space $\mathbf{X}$, and suppose $x(t)$ has continuous derivatives on $[a,b]$ up to order $n+1$ inclusive. Then we have *Taylor's formula*

$$x(b)-x(a) = x'(a)(b-a) + \frac{1}{2}x''(a)(b-a)^2 + \cdots + \frac{1}{n!}x^{(n)}(a)(b-a)^n + Q_n, \tag{20}$$

† We can also define the *differential of order n* of $x(t)$ as the function $d^n x = x^{(n)}(t)\,(dt)^n$, just as in Vol. 1, Chapter 8, Problem 18. The reader should verify that the invariance of the first differential under the change of variables $t=t(\tau)$ discussed in Sec. 1.61e breaks down for higher differentials, unless t is a linear function $A\tau+B$. He should also write Taylor's formula (20) in terms of higher differentials.

where the remainder Q_n can be written in the form

$$Q_n = \frac{1}{n!} \int_a^b x^{(n+1)}(t)(b-t)^n \, dt.$$

The proof of (20) is carried out in the same way as for numerical functions (Vol. 1, Sec. 9.52a), with the help of formula (19) for integration by parts. The remainder Q_n satisfies the estimate

$$\|Q_n\| \leqslant \max_{a \leqslant t \leqslant b} \|x^{(n+1)}(t)\| \frac{1}{n!} \int_a^b (b-t)^n \, dt = \max_{a \leqslant t \leqslant b} \|x^{(n+1)}(t)\| \frac{(b-a)^{n+1}}{n!}.$$

1.65. Sequences and series of functions with values in X

a. Let $x_n(t)$ be a sequence of functions of the argument $t \in [a,b]$, taking values in a Banach space $\mathbf{X}$. Then $x_n(t)$ is said to *converge on* $[a,b]$ to the *limit (function)* $x(t)$ if, given any $t \in [a,b]$,

$$\lim_{n \to \infty} \|x(t) - x_n(t)\| = 0.$$

Moreover, $x_n(t)$ is said to *converge uniformly on* $[a,b]$ to the limit $x(t)$ if

$$\lim_{n \to \infty} \sup_{a \leqslant t \leqslant b} \|x(t) - x_n(t)\| = 0,$$

i.e., if, given any $\varepsilon > 0$, there exists an integer $N > 0$ such that $\|x(t) - x_n(t)\| < \varepsilon$ for all $n > N$ and all $t \in [a,b]$. According to Vol. 1, Sec. 5.95b, the limit of a uniformly convergent sequence of continuous functions is itself a continuous function.

b. THEOREM. *Let $x_n(t)$ be a sequence of functions integrable on $[a,b]$ which converges uniformly on $[a,b]$ to a limit function $x(t)$. Then $x(t)$ is also integrable on $[a,b]$. Moreover, the formula*

$$\lim_{n \to \infty} \int_a^\tau x_n(t) \, dt = \int_a^\tau x(t) \, dt$$

holds uniformly for all $\tau \in [a,b]$, and in particular,

$$\lim_{n \to \infty} \int_a^b x_n(t) \, dt = \int_a^b x(t) \, dt.$$

Proof. The exact analogue of the proof for the case of numerical functions (Vol. 1, Sec. 9.102). ∎

c. THEOREM. *Let $x_n(t)$ be a sequence of piecewise smooth functions on $[a,b]$, which converges for at least one point $t_0 \in [a,b]$, and suppose the sequence of derivatives $x_n'(t)$ converges uniformly on $[a,b]$ to a piecewise continuous limit function $g(t)$. Then the sequence $x_n(t)$ converges uniformly on $[a,b]$ to a piecewise smooth function $x(t)$ with* *derivative*

$$x'(t) = \lim_{n\to\infty} x_n'(t) = g(t)$$

at every continuity point of $g(t)$.

Proof. Again the exact analogue of the proof for the case of numerical functions (Vol. 1, Sec. 9.106). ∎

d. A series of functions

$$x_1(t) + x_2(t) + \cdots + x_n(t) + \cdots, \tag{21}$$

each defined on an interval $[a,b]$ and taking values in a space $\mathbf{X}$, is said to *converge on* $[a,b]$ if the sequence of partial sums

$$s_1(t) = x_1(t), \ldots, s_n(t) = x_1(t) + \cdots + x_n(t), \ldots \tag{22}$$

of (21) converges on $[a,b]$. The limit

$$s(t) = \lim_{n\to\infty} s_n(t)$$

is then called the *sum* of the series (21). By the same token, the series (21) is said to *converge uniformly on* $[a,b]$ if the sequence (22) converges uniformly on $[a,b]$. The preceding two theorems lead at once to sufficient conditions for term-by-term integration and differentiation of series of functions with values in $\mathbf{X}$ (state these conditions).

1.66. Analytic functions with values in X

a. Let $x(\zeta)$ be a function with values in a complex normed linear space $\mathbf{X}$, defined on an open set G in the plane of the complex variable $\zeta = \xi + i\eta$. Then $x(t)$ is said to be *differentiable at a point* $\zeta_0 \in G$ if the limit

$$x'(\zeta_0) = \lim_{\zeta\to\zeta_0} \frac{x(\zeta) - x(\zeta_0)}{\zeta - \zeta_0},$$

called the *derivative of* $x(\zeta)$ *at the point* ζ_0 (with respect to the complex variable ζ), exists. A function $x(\zeta)$ is said to be *analytic on an open set* G (in particular, on a domain G†) if it is differentiable at every point $\zeta_0 \in G$.

b. Much of the ordinary theory of analytic functions (Vol. 1, Chapter 10) carries over to the case of analytic functions taking values in the space $\mathbf{X}$. For example, the key concept of the integral along a curve in the complex plane is introduced in the usual way, as follows: Let L be a piecewise smooth path in G, with equation

$$\zeta = \zeta(t) \qquad (a \leqslant t \leqslant b),$$

† I.e., on an open *connected* set G.

let

$$\Pi=\{a=t_0<t_1<t_2<\cdots<t_n=b\}$$

be a partition of the interval $[a,b]$, and let

$$\zeta_k=\zeta(t_k) \qquad (k=0,1,\ldots, n)$$

be the corresponding points of L. Then the *integral of* $x(\zeta)$ *along* L is defined as the limit

$$\int_L x(\zeta)\, d\zeta = \lim_{d(\Pi)\to 0} \sum_{k=0}^{n-1} x(\zeta_k)\Delta\zeta_k \qquad (\Delta\zeta_k=\zeta_{k+1}-\zeta_k).$$

The existence of this integral for every piecewise continuous function $x(\zeta)$ with values in a complete complex normed linear space $\mathbf{X}$ is proved in just the same way as for a (complex-valued) numerical function (cf. Vol. 1, Sec. 10.21). We then have the following generalization of *Cauchy's theorem* (Vol. 1, Sec. 10.32), proved in virtually the same way: *If* $x(\zeta)$ *is analytic on a simply connected domain* G, *then*

$$\int_L x(\zeta)\, d\zeta=0$$

for every piecewise smooth closed path $L\subset G$.

c. Cauchy's theorem in turn leads first to *Cauchy's formula*

$$x(\zeta_0)=\frac{1}{2\pi i}\oint_L \frac{x(\zeta)}{\zeta-\zeta_0}\, d\zeta,$$

where $x(\zeta)$ is analytic on a simply connected domain G containing a point ζ_0 and a piecewise smooth closed path L surrounding ζ_0,† and then to the various implications of Cauchy's formula (Vol. 1, Sec. 10.34ff). In particular, a function $x(\zeta)$ analytic on a domain G has derivatives of all orders, and has a Taylor series expansion

$$x(\zeta)=\sum_{n=0}^{\infty} a_n(\zeta-\zeta_0)^n, \tag{23}$$

where

$$a_n=\frac{x^{(n)}(\zeta_0)}{n!} \qquad (n=0,1,2,\ldots),$$

valid for every disk $|\zeta-\zeta_0|<\rho$ contained in G. The radius of convergence R

† A closed path L is said to "surround" a point ζ_0 (or a set E) if L has no multiple points and if ζ_0 (or E) belongs to the interior of L.

of this series is just the distance from the point ζ_0 to the nearest singular point of $x(\zeta)$, i.e., to the nearest point at which $x(\zeta)$ ceases to be differentiable, and is given by the Cauchy-Hadamard formula

$$R = \frac{1}{\overline{\lim\limits_{n\to\infty}} \sqrt[n]{\|a_n\|}}$$

(see Sec. 1.39i). To get the higher derivatives of $x(\zeta)$, we differentiate the series (23) term by term, obtaining

$$x'(\zeta) = \sum_{n=1}^{\infty} n a_n (\zeta - \zeta_0)^{n-1},$$

$$\cdots$$

$$x^{(k)}(\zeta) = \sum_{n=k}^{\infty} n(n-1)\cdots(n-k+1) a_n (\zeta-\zeta_0)^{n-k},$$

$$\cdots$$

1.7. Continuous Linear Operators

1.71. According to the definition given in Sec. 1.15a, a mapping $\mathbf{A}$ of a linear space $\mathbf{X}$ into a linear space $\mathbf{Y}$ (both over the same field K) is said to be a *linear operator* if it satisfies the condition

$$\mathbf{A}(\alpha_1 x_1 + \alpha_2 x_2) = \alpha_1 \mathbf{A} x_1 + \alpha_2 \mathbf{A} x_2$$

for every $x_1, x_2 \in \mathbf{X}$ and $\alpha_1, \alpha_2 \in K$. If the space $\mathbf{Y}$ is one-dimensional (so that $\mathbf{Y} = K$), the operator $\mathbf{A}$ is said to be a *linear functional* (Sec. 1.15d). Here we will be concerned with linear operators mapping a normed linear space $\mathbf{X}$ into a normed linear space $\mathbf{Y}$, beginning with the case where both spaces are over the field of real numbers R.

a. In keeping with the general definition of a continuous function (Vol. 1, Sec. 5.11a), a linear operator $\mathbf{A}$ mapping a normed linear space $\mathbf{X}$ into a normed linear space $\mathbf{Y}$ is said to be *continuous* at $x = x_0 \in \mathbf{X}$ if, given any $\varepsilon > 0$, there exists a $\delta > 0$ such that $|x - x_0| < \delta$ implies $|\mathbf{A}x - \mathbf{A}x_0| < \varepsilon$.† It should be noted that the operator $\mathbf{A}$ is continuous at $x = x_0$ if and only if $\mathbf{A}x_n \to \mathbf{A}x_0$ (in $\mathbf{Y}$) for every sequence $x_n \to x_0$ (in $\mathbf{X}$).

b. A linear operator $\mathbf{A}$ mapping a space $\mathbf{X}$ into a space $\mathbf{Y}$ is said to be *bounded* if it is bounded on the unit ball of the space $\mathbf{X}$, i.e., if $|x| \leqslant 1$ implies $|\mathbf{A}x| \leqslant M$

† Or, equivalently, such that $|x - x_0| \leqslant \delta$ implies $|\mathbf{A}x - \mathbf{A}x_0| \leqslant \varepsilon$.

for some fixed positive constant M. In this case, the quantity

$$\|\mathbf{A}\| = \sup_{|x| \leqslant 1} |\mathbf{A}x| \tag{1}$$

is called the *norm* of the operator $\mathbf{A}$. Given any vector $x \in \mathbf{X}$, we have

$$\left|\frac{x}{|x|}\right| = 1,$$

which implies

$$\left|\mathbf{A}\frac{x}{|x|}\right| \leqslant \|\mathbf{A}\|,$$

and hence

$$|\mathbf{A}x| \leqslant \|\mathbf{A}\|\,|x|. \tag{2}$$

c. THEOREM. *If a linear operator* $\mathbf{A}$ *is bounded, then* $\mathbf{A}$ *is continuous at every point* x_0 *of the space* $\mathbf{X}$.

Proof. Let $\mathbf{A}$ be bounded, with norm $\|\mathbf{A}\|$. Then, given any $\varepsilon > 0$,

$$|x - x_0| < \frac{\varepsilon}{\|\mathbf{A}\|}$$

implies

$$|\mathbf{A}x - \mathbf{A}x_0| = |\mathbf{A}(x - x_0)| \leqslant \|\mathbf{A}\|\,|x - x_0| < \varepsilon. \quad \blacksquare$$

d. THEOREM. *If a linear operator* $\mathbf{A}$ *is continuous at any point* x_0 *of the space* $\mathbf{X}$, *then* $\mathbf{A}$ *is bounded.*

Proof. Let $\delta > 0$ be such that $|x - x_0| \leqslant \delta$ implies $|\mathbf{A}x - \mathbf{A}x_0| \leqslant 1$, and, given any z in the unit ball (so that $|z| \leqslant 1$), let $x = x_0 + \delta z$. Then clearly

$$|x - x_0| = \delta|z| \leqslant \delta,$$

and hence

$$|\mathbf{A}x - \mathbf{A}x_0| = |\mathbf{A}(x - x_0)| = \delta|\mathbf{A}z| \leqslant 1,$$

which implies

$$|\mathbf{A}z| \leqslant \frac{1}{\delta}. \quad \blacksquare$$

e. COROLLARY. *Let* $\mathbf{A}$ *be a linear operator which is continuous at some point of the space* $\mathbf{X}$. *Then* $\mathbf{A}$ *is continuous on the whole space* $\mathbf{X}$.†

† I.e., at every point of $\mathbf{X}$.

Proof. An immediate consequence of the preceding two theorems. ∎

f. THEOREM. *Let* **A** *be a continuous linear operator mapping a Banach space* **X** *into a Banach space* **Y**. *Then* **A** *has the following properties*:

(a) *If the series*

$$\sum_{n=1}^{\infty} x_n = x$$

converges in **X**, *then*

$$\sum_{n=1}^{\infty} \mathbf{A}x_n = \mathbf{A}x,$$

i.e., **A** *can be applied term by term to any convergent series*;
(b) *If* $x(t)$ *is a piecewise continuous function on the interval* $a \leqslant t \leqslant b$, *with values in* **X**, *then*

$$\mathbf{A}\int_a^b x(t)\, dt = \int_a^b [\mathbf{A}x(t)]\, dt;$$

(c) *If* $x(t)$ *is a function with values in* **X** *which is differentiable at* $t = t_0$, *then*

$$\mathbf{A}x'(t_0) = [\mathbf{A}x(t)]'_{t=t_0}.$$

Proof. The proof of all three properties is essentially the same. The sum of the series, the integral

$$\int_a^b x(t)\, dt,$$

and the derivative $x'(t_0)$ are all results of certain linear operations and limiting processes. But a continuous linear operator **A** "commutes" with both linear operations and limiting processes, in the sense that it does not matter whether the operator is applied before or after the linear operations and limiting processes are carried out. For example, to prove Property c, we note that

$$\mathbf{A}x'(t_0) = \mathbf{A}\lim_{t\to t_0} \frac{x(t)-x(t_0)}{t-t_0} = \lim_{t\to t_0} \mathbf{A}\left[\frac{x(t)-x(t_0)}{t-t_0}\right]$$
$$= \lim_{t\to t_0} \frac{\mathbf{A}x(t)-\mathbf{A}x(t_0)}{t-t_0} = [\mathbf{A}x(t)]'_{t=t_0},$$

and similarly for Properties a and b (give the details). ∎

g. If the operators **A**, $\mathbf{A}_1$, $\mathbf{A}_2$ mapping a normed linear space **X** into a normed linear space **Y** are bounded, then so are the sum $\mathbf{A}_1 + \mathbf{A}_2$ and the product $\alpha\mathbf{A}$ (where α is any real number), since, according to (2),

$$|(\mathbf{A}_1+\mathbf{A}_2)x|=|\mathbf{A}_1x+\mathbf{A}_2x|\leqslant|\mathbf{A}_1x|+|\mathbf{A}_2x|\leqslant\|\mathbf{A}_1\|+\|\mathbf{A}_2\|,$$
$$|\alpha\mathbf{A}x|=|\alpha||\mathbf{A}x|\leqslant|\alpha|\,\|\mathbf{A}\|$$

if $|x|\leqslant 1$. It follows at once from these formulas and the definition of the norm of a bounded operator that

$$\|\mathbf{A}_1+\mathbf{A}_2\|=\sup_{|x|\leqslant 1}|(\mathbf{A}_1+\mathbf{A}_2)x|\leqslant\|\mathbf{A}_1\|+\|\mathbf{A}_2\|,$$
$$\|\alpha\mathbf{A}\|=\sup_{|x|\leqslant 1}|\alpha\mathbf{A}x|=|\alpha|\sup_{|x|\leqslant 1}|\mathbf{A}x|=|\alpha|\,\|\mathbf{A}\|.$$

Thus the space $\mathbf{L}(\mathbf{X},\mathbf{Y})$ of all bounded linear operators mapping $\mathbf{X}$ into $\mathbf{Y}$ can be regarded as a normed linear space, equipped with the norm (1).

h. Let $\mathbf{B}$ be a bounded linear operator mapping a normed linear space $\mathbf{X}$ into a normed linear space $\mathbf{Y}$, and let $\mathbf{A}$ be a bounded linear operator mapping $\mathbf{Y}$ into a normed linear space $\mathbf{Z}$. Then, as in Sec. 1.15g, we can define the operator $\mathbf{P}=\mathbf{AB}$ mapping $\mathbf{X}$ into $\mathbf{Z}$. The operator $\mathbf{P}$ is also bounded. In fact, given any $x\in\mathbf{X}$, we have

$$|\mathbf{AB}x|\leqslant\|\mathbf{A}\|\,|\mathbf{B}x|\leqslant\|\mathbf{A}\|\,\|\mathbf{B}\|\,|x|.$$

Therefore $\mathbf{P}=\mathbf{AB}$ is a bounded operator, and

$$\|\mathbf{AB}\|\leqslant\|\mathbf{A}\|\,\|\mathbf{B}\|.$$

In particular, if $\mathbf{A}$ maps $\mathbf{X}$ into itself, then

$$\|\mathbf{A}^2\|=\|\mathbf{AA}\|\leqslant\|\mathbf{A}\|^2,$$
$$\|\mathbf{A}^3\|=\|\mathbf{A}^2\mathbf{A}\|\leqslant\|\mathbf{A}^2\|\,\|\mathbf{A}\|\leqslant\|\mathbf{A}\|^3,$$
$$\ldots$$
$$\|\mathbf{A}^n\|=\|\mathbf{A}^{n-1}\mathbf{A}\|\leqslant\|\mathbf{A}^{n-1}\|\,\|\mathbf{A}\|\leqslant\|\mathbf{A}^{n-2}\|\,\|\mathbf{A}\|^2\leqslant\cdots\leqslant\|\mathbf{A}\|^n.$$

i. As an example, we now calculate the norm of a special linear operator, defined on the space $R^s(a,b)$ of continuous real functions on the interval $a\leqslant t\leqslant b$. Let $D(t,\lambda)$ be a continuous real function of $t\in[a,b]$ for every value of a parameter λ in some set Λ, and suppose the quantity

$$D=\sup_{\lambda\in\Lambda}\int_a^b|D(t,\lambda)|\,dt$$

is finite. Given any $x(t)\in R^s(a,b)$, let

$$y(\lambda)=\mathbf{A}x(t)=\int_a^b D(t,\lambda)x(t)\,dt, \tag{3}$$

so that the operator **A** carries every function $x(t)$ into a function $y(\lambda)$ defined on the set Λ. The function $y(\lambda)$ is bounded, since

$$|y(\lambda)| = |\mathbf{A}x(t)| = \left| \int_a^b D(t,\lambda)x(t)\,dt \right| \leqslant \max_{a \leqslant t \leqslant b} |x(t)| \int_a^b |D(t,\lambda)|\,dt \leqslant D\|x\|. \qquad (4)$$

Thus (3) defines an operator mapping the space $R^s(a,b)$ into the space $R(\Lambda)$ of bounded real functions $y(\lambda)$. The natural norm in the space $R(\Lambda)$ is just

$$\|y\| = \sup_{\lambda \in \Lambda} |y(\lambda)|.$$

The operator **A** is obviously linear. Moreover, it follows from (4) that **A** is bounded, with norm no greater than D, i.e., $\|\mathbf{A}\| \leqslant D$.

THEOREM. *The norm of* **A** *is precisely equal to* D:

$$\|\mathbf{A}\| = D.$$

Proof. For every fixed $\lambda \in \Lambda$, consider the function

$$x_n(t,\lambda) = u_n(D(t,\lambda)),$$

where

$$u_n(\tau) = \begin{cases} -1 & \text{if} \quad \tau \leqslant -1/n, \\ n\tau & \text{if} \quad -1/n \leqslant \tau \leqslant 1/n, \\ +1 & \text{if} \quad \tau \geqslant 1/n \end{cases}$$

is the continuous function shown in Figure 10. Clearly, $x_n(t,\lambda)$ is itself continuous in t, and hence is an element of the space $R^s(a,b)$ for every fixed $\lambda \in \Lambda$. Moreover, the product $D(t,\lambda)x_n(t,\lambda)$ is a nonnegative function, equal to $|D(t,\lambda)|$ if $|D(t,\lambda)| \geqslant 1/n$ and no greater than $|D(t,\lambda)|$ otherwise. Since

$$\begin{aligned} \mathbf{A}x_n(t,\lambda) &= \int_a^b D(t,\lambda)x_n(t,\lambda)\,dt \geqslant \int_{|D(t,\lambda)| \geqslant 1/n} |D(t,\lambda)|\,dt \\ &\geqslant \int_a^b |D(t,\lambda)|\,dt - \frac{1}{n}(b-a), \end{aligned}$$

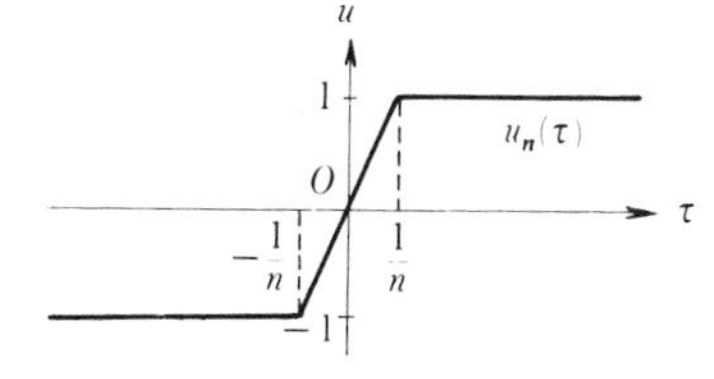

Figure 10

we have

$$\sup_{n,\lambda} |\mathbf{A}x_n(t,\lambda)| \geqslant \sup_{\lambda} \int_a^b |D(t,\lambda)| \, dt.$$

But $\|x_n(t,\lambda)\| \leqslant 1$, and hence

$$\|\mathbf{A}\| = \sup_{\|x\| \leqslant 1} \|\mathbf{A}x(t)\| \geqslant \sup_{n,\lambda} |\mathbf{A}x_n(t,\lambda)| \geqslant \sup_{\lambda} \int_a^b |D(t,\lambda)| \, dt = D.$$

Together with $\|\mathbf{A}\| \leqslant D$, this implies $\|\mathbf{A}\| = D$, i.e.,

$$\|\mathbf{A}\| = \sup_{\lambda} \int_a^b |D(t,\lambda)| \, dt. \quad \blacksquare \tag{5}$$

j. Let $D(t)$ be a continuous linear function of $t \in [a,b]$. Then the formula

$$y = \mathbf{A}x(t) = \int_a^b D(t)x(t) \, dt$$

defines a linear functional $\mathbf{A}$ on the space $R^s(a,b)$. This functional is a special case of the operator (3), corresponding to the case where the set Λ of possible parameters λ consists of a single point. Applying the preceding theorem, we find that the norm of the functional (3′) is just

$$\|\mathbf{A}\| = \int_a^b |D(t)| \, dt. \tag{5′}$$

1.72. The open mapping theorem

a. Let $y=f(x)$ be a function defined on a set X and taking values in a set Y. Then the set of all points $y=f(x) \in Y$, where x varies over some subset $E \subset X$, is called the *image* of the subset E (under the mapping f) and is denoted by $f(E)$. Similarly, the set of all points $x \in X$ for which $y=f(x)$ takes values in some subset $F \subset Y$ is called the *preimage* (or *inverse image*) of the subset F and is denoted by $f^{-1}(F)$.

If X and Y are metric spaces and if $y=f(x)$ is a continuous function, then the *preimage* $f^{-1}(G)$ of any open subset $G \subset Y$ is an open set in X (Vol. 1, Sec. 5.14a). However, the *image* $f(G)$ of an open subset $G \subset X$ need not be an open set in Y. For example, if X is the line $-\infty < x < \infty$ and Y the line $-\infty < y < \infty$ and if the function $y=f(x)$ is constant, say c, then the image of any open set $G \subset X$ (or, for that matter, of *any* set $G \subset X$) is the single point $y=c$, which is not an open set in Y. The continuous image of an open set may still fail to be open, even if we impose the extra requirement that the function $f(x)$ map the space X *onto the whole space* Y. For example, consider the continuous function

$$f(x)=\begin{cases}(x+1)^3 & \text{if } x\leqslant -1,\\ 0 & \text{if } -1\leqslant x\leqslant 1,\\ (x-1)^3 & \text{if } x\geqslant 1.\end{cases}$$

This function maps the whole x-axis onto the whole y-axis, but again carries the open interval $-1<x<1$ into a single point, namely $y=0$.

Next we examine the effect of assuming that $y=f(x)$ is a continuous *one-to-one* mapping of the space X onto the space Y. Even in this case, the continuous image of an open set may fail to be open. For example, let X be the space $D_1(a,b)$ of all continuously differentiable functions $x(t)$ on the interval $[a,b]$ (see Sec. 1.25a), with its natural metric, and let Y be the subset of the space $R^s(a,b)$ of all continuous functions on $[a,b]$ (with its natural metric) which are also continuously differentiable; note that we are justified in regarding the latter set as a metric space in its own right. Consider the mapping $y=f(x)$ which carries every function $x=x(t)\in D_1(a,b)$ into the same function $y=y(t)\equiv x(t)\in R^s(a,b)$. This mapping is continuous, since convergence of $x_n(t)$ to $x(t)$ in $D_1(a,b)$ implies convergence of $y_n(t)\equiv x_n(t)$ to $y(t)\equiv x(t)$ in $R^s(a,b)$. Moreover, the mapping $y=f(x)$ is obviously one-to-one. Nevertheless, the image of an open set in X, e.g., of the open unit ball V in $D_1(a,b)$, is not an open set in Y, since every neighborhood of a point $y_0(t)\in f(V)$, being defined only by the inequality $\max|y(t)-y_0(t)|<\varepsilon$, contains functions with an arbitrarily large derivative $y'(t)$.

b. It is now clear that the conditions of the following theorem due to Banach are essential:

THEOREM (**Open mapping theorem**). *Let* $\mathbf{A}$ *be a continuous linear operator establishing a one-to-one mapping of a complete normed linear space* $\mathbf{X}$ *onto a complete normed linear space* $\mathbf{Y}$. *Then* $\mathbf{A}$ *carries every open set* $G\subset\mathbf{X}$ *into an open set* $\mathbf{A}G\subset\mathbf{Y}$.†

Proof. Let V_r denote the ball $\{x:|x|<r\}$. We first show that the closure in $\mathbf{Y}$ of the set $\mathbf{A}V_1$ contains some ball $\{y:|y|<\rho\}$ in the space $\mathbf{Y}$. By hypothesis,

$$\mathbf{Y}=\mathbf{AX}=\mathbf{A}\left(\bigcup_{n=1}^{\infty}V_n\right)=\bigcup_{n=1}^{\infty}\mathbf{A}V_n,$$

and hence, a fortiori,

$$\mathbf{Y}=\bigcup_{n=1}^{\infty}\overline{\mathbf{A}V_n}.$$

Therefore, by Baire's theorem (Vol. 1, Sec. 3.75a), there is an integer $N>0$

† $\mathbf{A}G$ is the image of the set G under the mapping $\mathbf{A}$, this notation being more concise than $\mathbf{A}(G)$.

such that the set $\overline{\mathbf{A}V_N}$ contains some ball

$$\{y:|y-y_0|<\varepsilon\}. \tag{6}$$

Since the set $\overline{\mathbf{A}V_N}$ is centrally symmetric (why?), it also contains the ball

$$\{y:|y+y_0|<\varepsilon\}. \tag{6'}$$

Moreover, since the set $\overline{\mathbf{A}V_N}$ is convex,† it also contains the ball $W_\varepsilon = \{y:|y|<\varepsilon\}$ lying in the convex hull of the balls (6) and (6′). Given any $\delta>0$, it follows from an obvious similarity argument that $W_\delta \subset \overline{\mathbf{A}V_{N\delta/\varepsilon}}$, and in particular that $W_{\varepsilon/N} \subset \overline{\mathbf{A}V_1}$. Thus $\overline{\mathbf{A}V_1}$ contains some open ball in $\mathbf{Y}$, as asserted.

Next we show that the set $\mathbf{A}V_1$ itself (without the closure) contains the ball $W_{\varepsilon/2N}$. Suppose $y \in W_{\varepsilon/2N}$. Then, since $W_{\varepsilon/2N} \subset \overline{\mathbf{A}V_{1/2}}$, as just shown, we can find a point $y_1 \in \mathbf{A}V_{1/2}$ arbitrarily close to the point y. In particular, we can choose y_1 such that $|y-y_1|<\varepsilon/4N$. But then $y-y_1 \in W_{\varepsilon/4N} \subset \overline{\mathbf{A}V_{1/4}}$, and hence, in the same way, we can find a point $y_2 \in \mathbf{A}V_{1/4}$ such that $|y-y_1-y_2| <\varepsilon/8N$. Continuing this process, we construct a point y_n for every $n=1,2,\ldots$ such that $|y-y_1-y_2-\cdots-y_n|<\varepsilon/2^{n+1}N$. Thus

$$y=\sum_{n=1}^{\infty} y_n,$$

by construction, while on the other hand, $y_n=\mathbf{A}x_n$ where $x_n \in V_{1/2^n}$, so that $|x_n|<1/2^n$. Since the space $\mathbf{X}$ is complete, it follows from Theorem 1.37c that the series $x_1+x_2+\cdots$ is convergent, with sum

$$x=\sum_{n=1}^{\infty} x_n,$$

say. But the operator $\mathbf{A}$ is continuous, and hence

$$\mathbf{A}x=\mathbf{A}\sum_{n=1}^{\infty} x_n=\sum_{n=1}^{\infty} \mathbf{A}x_n=\sum_{n=1}^{\infty} y_n=y$$

(see Theorem 1.71f, Property a), where

$$|x| \leqslant \sum_{n=1}^{\infty} |x_n| \leqslant \sum_{n=1}^{\infty} \frac{1}{2^n}=1.$$

Thus $\mathbf{A}V_1$ contains the ball $W_{\varepsilon/2N}$, as asserted.

Finally, given any $\delta>0$, a similarity argument again shows that $W_\delta \subset \mathbf{A}V_{2N\delta/\varepsilon}$. Therefore $|x-x_0|<\delta$ implies $|\mathbf{A}x-\mathbf{A}x_0|=|\mathbf{A}(x-x_0)|<2N\delta/\varepsilon$, so that the image $\mathbf{A}U$ of the ball $U=\{x: |x-x_0|<\delta\}$ contains the ball $\{y: |y-\mathbf{A}x_0|<2N\delta/\varepsilon\}$. But then the image $\mathbf{A}G$ of an arbitrary open set $G \subset \mathbf{X}$ is an open set in $\mathbf{Y}$ (why?). ∎

† A linear operator carries convex sets into convex sets (why?), and the closure of a convex set is itself convex (Sec. 1.62i).

c. COROLLARY. *Let* $\mathbf{A}$ *be the same as in the preceding theorem. Then the inverse mapping* $\mathbf{A}^{-1}$ *is also continuous.*

Proof. The inverse mapping $\mathbf{A}^{-1}$ is uniquely defined, and is obviously a linear operator, like $\mathbf{A}$ itself. By the open mapping theorem, the preimage of any open set $G \subset \mathbf{X}$ under the mapping $\mathbf{A}^{-1}$ is an open set $\mathbf{A}G \subset \mathbf{Y}$. In particular, the preimage of any ball $\{x:|x|<\varepsilon\}$ contains some ball $\{y:|y|<\delta\}$. But this implies the continuity of $\mathbf{A}^{-1}$. ∎

d. COROLLARY. *Given a linear space* $\mathbf{L}$ *equipped with two norms* $|x|_1$ *and* $|x|_2$ *such that* $\mathbf{L}$ *is complete with respect to both norms, suppose there exists a positive constant* c_1 *such that* $|x|_2 \geqslant c_1|x|_1$ *for every* $x \in \mathbf{L}$. *Then there exists a positive constant* c_2 *such that* $|x|_1 \geqslant c_2|x|_2$ *for every* $x \in \mathbf{L}$, *so that the norms* $|x|_1$ *and* $|x|_2$ *are equivalent (by Corollary 1.35c).*

Proof. Let $\mathbf{X}$ be the normed linear space obtained by equipping $\mathbf{L}$ with the norm $|x|_2$, let $\mathbf{Y}$ be the normed linear space obtained by equipping $\mathbf{L}$ with the norm $|x|_1$, and let $\mathbf{A}$ be the identity operator mapping every element $x \in \mathbf{X}$ into the same element $x \in \mathbf{Y}$. The inequality $|x|_2 \geqslant c_1|x|_1$ implies the continuity of $\mathbf{A}$. But then the inverse mapping is also continuous, by the preceding corollary, and hence there exists a constant $c_2>0$ such that $|x|_1 \geqslant c_2|x|_2$ for every $x \in \mathbf{L}$ (see Theorem 1.71d). ∎

e. Suppose a complete normed linear space $\mathbf{X}$ is the direct sum of two closed subspaces $\mathbf{X}_1$ and $\mathbf{X}_2$ (Sec. 1.14h), so that every vector $x \in \mathbf{X}$ has a unique representation

$$x=x_1+x_2, \qquad x_1 \in \mathbf{X}_1,\ x_2 \in \mathbf{X}_2$$

in terms of its "components" x_1 and x_2. Let $\mathbf{P}_1$ be the operator carrying every vector $x \in \mathbf{X}$ into x_1 and $\mathbf{P}_2$ the operator carrying x into x_2. Then $\mathbf{P}_1$ is called the *projection operator onto the subspace* $\mathbf{X}_1$ and $\mathbf{P}_2$ the *projection operator onto the subspace* $\mathbf{X}_2$. The operators $\mathbf{P}_1$ and $\mathbf{P}_2$ are obviously linear, but it is hardly obvious that they are also continuous. To prove the continuity of $\mathbf{P}_1$ and $\mathbf{P}_2$, we use Corollary 1.72d, together with the completeness of the space $\mathbf{X}$ and the fact that the subspaces $\mathbf{X}_1$ and $\mathbf{X}_2$ are closed:

THEOREM. *The projection operators* $\mathbf{P}_1$ *and* $\mathbf{P}_2$ *are continuous.*

Proof. Besides the original norm $|x| \equiv |x|_1$, we introduce a new norm

$$|x|_2=|x_1|_1+|x_2|_1$$

in the space $\mathbf{X}$. Clearly $|x|_2$ satisfies all the norm axioms, and moreover $|x|_1 \leqslant |x_1|_1+|x_2|_1=|x|_2$. The space $\mathbf{X}$ is complete with respect to the norm $|x|_2$. To see this, suppose $x^{(n)}$ is a fundamental sequence with respect to the norm $|x|_2$. Then it follows from the formula

$$|x^{(m)}-x^{(n)}|_2=|x_1^{(m)}-x_1^{(n)}|_1+|x_2^{(m)}-x_2^{(n)}|_1$$

that the sequences $x_1^{(n)}$ and $x_2^{(n)}$ are fundamental with respect to the norm $|x|_1$. But $\mathbf{X}$ is complete with respect to the norm $|x|_1$, and hence the limits

$$\lim_{n\to\infty} x_1^{(n)}=x_1, \lim_{n\to\infty} x_2^{(n)}=x_2$$

exist in $\mathbf{X}$. Furthermore,

$$x_1 \in \mathbf{X}_1, x_2 \in \mathbf{X}_2,$$

since $\mathbf{X}_1$ and $\mathbf{X}_2$ are closed. Writing $x=x_1+x_2$, we have

$$|x-x^{(n)}|_2=|x_1-x_1^{(n)}|_1+|x_2-x_2^{(n)}|_1\to 0$$

as $n\to\infty$. Hence x is the limit of the sequence $x^{(n)}$ with respect to the norm $|x|_2$, and the completeness of $\mathbf{X}$ is proved.

Applying Corollary 1.72d, we now find that the norms $|x|_1$ and $|x|_2$ are equivalent. In particular, there exists a constant $c>0$ such that

$$|x|_2=|x_1|_1+|x_2|_1\leqslant c|x|_1=c|x|$$

for every $x \in \mathbf{X}$. But then

$$|\mathbf{P}_1 x|=|x_1|_1\leqslant c|x|, |\mathbf{P}_2 x|=|x_2|_1\leqslant c|x|,$$

which proves the continuity of the operators $\mathbf{P}_1$ and $\mathbf{P}_2$. ∎

f. Again let $\mathbf{X}$ be a complete normed linear space which is the direct sum of two closed subspaces $\mathbf{X}_1, \mathbf{X}_2$, and let $\mathbf{P}_1, \mathbf{P}_2$ be the corresponding projection operators. Given a continuous linear operator $\mathbf{A}_1$ defined on $\mathbf{X}_1$ and a continuous linear operator $\mathbf{A}_2$ defined on $\mathbf{X}_2$, let $\mathbf{A}$ be the operator on the whole space $\mathbf{X}$ defined by the formula

$$\mathbf{A}x=\mathbf{A}(x_1+x_2)=\mathbf{A}_1x_1+\mathbf{A}_2x_2.$$

The operator $\mathbf{A}$ is obviously linear.

THEOREM. *The operator* $\mathbf{A}$ *is continuous on* $\mathbf{X}$.

Proof. Clearly

$$\mathbf{A}x=\mathbf{A}_1x_1+\mathbf{A}_2x_2=\mathbf{A}_1\mathbf{P}_1x+\mathbf{A}_2\mathbf{P}_2x.$$

But the operators $\mathbf{P}_1$ and $\mathbf{P}_2$ are bounded on $\mathbf{X}$, by the preceding theorem and Theorem 1.71d, and hence

$$|\mathbf{A}x|\leqslant \|\mathbf{A}_1\||\mathbf{P}_1x|+\|\mathbf{A}_2\||\mathbf{P}_2x|\leqslant \|\mathbf{A}_1\|\,\|\mathbf{P}_1\||x|+\|\mathbf{A}_2\|\,\|\mathbf{P}_2\||x|=c|x|.$$ ∎

1.73. Convergence of sequences of linear operators

a. In Sec. 1.71g we equipped the space $\mathbf{L}(\mathbf{X},\mathbf{Y})$ of all bounded linear operators mapping a normed linear space $\mathbf{X}$ into a normed linear space $\mathbf{Y}$ with the norm

$$\|\mathbf{A}\| = \sup_{\|x\| \leqslant 1} |\mathbf{A}x|.$$

Accordingly, a sequence of operators $\mathbf{A}_1$, $\mathbf{A}_2, \ldots$ is said to *converge* in this norm to an operator $\mathbf{A}$ if

$$\lim_{n \to \infty} \|\mathbf{A} - \mathbf{A}_n\| = 0,$$

i.e., if, given any $\varepsilon > 0$, there is an integer $N > 0$ such that

$$\sup_{|x| \leqslant 1} |\mathbf{A}x - \mathbf{A}_n x| < \varepsilon$$

for all $n > N$.

b. THEOREM. *If the space* $\mathbf{Y}$ *is complete, then so is the space* $\mathbf{L}(\mathbf{X},\mathbf{Y})$.

Proof. Let $\mathbf{A}_1, \mathbf{A}_2, \ldots$ be a fundamental sequence of bounded linear operators mapping $\mathbf{X}$ into $\mathbf{Y}$. Then, given any $\varepsilon > 0$, there is an integer $N > 0$ such that

$$\|\mathbf{A}_m - \mathbf{A}_n\| < \varepsilon$$

for all $m,n > N$. Hence, given any $x \in \mathbf{X}$, it follows from the inequality (2), p. 98, that

$$|\mathbf{A}_m x - \mathbf{A}_n x| \leqslant \|\mathbf{A}_m - \mathbf{A}_n\| \, |x| < \varepsilon |x| \tag{7}$$

for all $m,n > N$, so that the vectors $\mathbf{A}_n x$ form a fundamental sequence in the space $\mathbf{Y}$. But $\mathbf{Y}$ is complete, and hence there is a vector $y \in \mathbf{Y}$ such that

$$y = \lim_{n \to \infty} \mathbf{A}_n x. \tag{8}$$

Let $\mathbf{A}$ be the operator defined by the equation $y = \mathbf{A}x$, where y is given by (8). Then, as we now show, $\mathbf{A}$ is a bounded linear operator, equal to the limit of the sequence $\mathbf{A}_1$, $\mathbf{A}_2, \ldots$ in the norm of the space $\mathbf{L}(\mathbf{X},\mathbf{Y})$.

The fact that $\mathbf{A}$ is linear follows from the formula

$$\begin{aligned}\mathbf{A}(\alpha_1 x_1 + \alpha_2 x_2) &= \lim_{n \to \infty} \mathbf{A}_n(\alpha_1 x_1 + \alpha_2 x_2) = \lim_{n \to \infty} (\alpha_1 \mathbf{A}_n x_1 + \alpha_2 \mathbf{A}_n x_2) \\ &= \alpha_1 \lim_{n \to \infty} \mathbf{A}_n x_1 + \alpha_2 \lim_{n \to \infty} \mathbf{A}_n x_2 = \alpha_1 \mathbf{A}x_1 + \alpha_2 \mathbf{A}x_2.\end{aligned}$$

Moreover, if $|x| \leqslant 1$, then

$$|\mathbf{A}x - \mathbf{A}_n x| = |(\mathbf{A} - \mathbf{A}_n)x| = \lim_{m \to \infty} |(\mathbf{A}_m - \mathbf{A}_n)x| \leqslant \varepsilon \tag{9}$$

for all $n > N$, because of (7). Therefore $\mathbf{A}-\mathbf{A}_n$ is a bounded operator, and hence so is $\mathbf{A}$. Finally, (9) shows that

$$\|\mathbf{A}-\mathbf{A}_n\| \leqslant \varepsilon$$

for all $n > N$, so that $\mathbf{A}_n \to \mathbf{A}$ in the norm of the space $\mathbf{L}(\mathbf{X},\mathbf{Y})$. ∎

c. If $\mathbf{Y}$ is the real line R, then the space $\mathbf{L}(\mathbf{X},\mathbf{Y}) = \mathbf{L}(\mathbf{X},R)$ is always complete. This space (the space of all continuous linear functionals on the space $\mathbf{X}$) is called the *conjugate space* of $\mathbf{X}$ and is denoted by $\mathbf{X}^*$.

d. The space $\mathbf{L}(\mathbf{X},\mathbf{X})$ of all bounded linear operators mapping a Banach space $\mathbf{X}$ into itself is complete. This space will henceforth be denoted simply by $\mathbf{L}(\mathbf{X})$.

e. There are cases where the operators $\mathbf{A}$, $\mathbf{A}_1$, $\mathbf{A}_2$,... have the property that $\mathbf{A}_n x \to \mathbf{A}x$ for every $x \in \mathbf{X}$, but $\|\mathbf{A}-\mathbf{A}_n\|$ does not approach zero as $n \to \infty$ (for an example of this behavior, see Sec. 1.75h). Henceforth we will say that a sequence of operators $\mathbf{A}_1$, $\mathbf{A}_2$,... with the property that $\mathbf{A}_n x \to \mathbf{A}x$ for every $x \in \mathbf{X}$ *converges strongly* to the operator $\mathbf{A}$, and $\mathbf{A}$ itself will be called the *strong limit* of the sequence $\mathbf{A}_1$, $\mathbf{A}_2$,...

f. A sequence of (bounded) linear operators $\mathbf{A}_1$, $\mathbf{A}_2$,... is said to be *uniformly bounded* if there exists a constant $M > 0$ such that

$$\sup_n \|\mathbf{A}_n\| \leqslant M. \tag{10}$$

Later we will need the following

LEMMA. *Let $\mathbf{A}_1$, $\mathbf{A}_2$,... be a uniformly bounded sequence of linear operators, all defined on the same normed linear space $\mathbf{X}$, such that*

$$\mathbf{A}x = \lim_{n\to\infty} \mathbf{A}_n x \tag{11}$$

for all x in some set E which is dense in $\mathbf{X}$. Then (11) holds for all $x \in \mathbf{X}$.

Proof. Given any $x \in \mathbf{X}$, there is a sequence x_k of elements of E such that

$$x = \lim_{k\to\infty} x_k$$

(why?). Given any $\varepsilon > 0$, we first find a positive integer K such that

$$|x - x_K| < \frac{\varepsilon}{3M},$$

where M is the constant figuring in (10), and then a positive integer N such that

$$|\mathbf{A}x_K - \mathbf{A}_n x_K| < \frac{\varepsilon}{3}$$

for all $n > N$. It follows that

$$|\mathbf{A}x - \mathbf{A}_n x| \leqslant |\mathbf{A}x - \mathbf{A}x_K| + |\mathbf{A}x_K - \mathbf{A}_n x_K| + |\mathbf{A}_n x_K - \mathbf{A}_n x|$$
$$\leqslant \|\mathbf{A}\| |x - x_K| + \frac{\varepsilon}{3} + \|\mathbf{A}_n\| |x_K - \mathbf{A}_n x| \leqslant M\frac{\varepsilon}{3M} + \frac{\varepsilon}{3} + M\frac{\varepsilon}{3M} = \varepsilon$$

for all $n > N$. Therefore (11) holds for all $x \in \mathbf{X}$. ▮

1.74. The principle of uniform boundedness

a. THEOREM (**Banach-Steinhaus**). *Let* $\mathbf{A}_1, \mathbf{A}_2, \ldots$ *be a sequence of continuous linear operators, mapping a Banach space* $\mathbf{X}$ *into a normed linear space* $\mathbf{Y}$, *which is not uniformly bounded, so that*

$$\sup_n \|\mathbf{A}_n\| = \infty. \tag{12}$$

Then every ball $U_\rho(x_0) = \{x \in \mathbf{X} : |x - x_0| \leqslant \rho\}$ *contains a point* x *such that*

$$\sup_n |\mathbf{A}_n x| = \infty. \tag{13}$$

Proof. According to (12),

$$\sup_{\substack{n \geqslant 1 \\ |x| \leqslant 1}} |\mathbf{A}_n x| = \infty,$$

and hence, by an obvious similarity argument,

$$\sup_{\substack{n \geqslant 1 \\ |x| \leqslant \rho}} |\mathbf{A}_n x| = \infty, \tag{14}$$

since otherwise

$$\sup_n |\mathbf{A}_n x_0| < \infty, \qquad \sup_n |\mathbf{A}_n x| < \infty$$

for every $x \in U_\rho(x_0)$, and hence

$$\sup_n |\mathbf{A}_n(x - x_0)| = \sup_n |\mathbf{A}_n x - \mathbf{A}_n x_0| < \infty$$

for every $x \in U_\rho(x_0)$. But this is impossible, because of (14), since $x \in U_\rho(x_0)$ implies $|x - x_0| \leqslant \rho$.

Given any ball $U_\rho(x_0)$, we now choose an element $x_1 \in U_\rho(x_0)$ such that one of the operators, which we denote by $\mathbf{A}_1$, gives a result whose norm exceeds 1:

$$|\mathbf{A}_1 x_1| > 1.$$

Since $\mathbf{A}_1$ is a continuous operator, the original ball $U_\rho(x_0)$ contains another ball $U_{\rho_1}(x_1)$ such that the inequality

$$|\mathbf{A}_1 x| > 1$$

holds for all $x \in U_{\rho_1}(x_1)$. We now choose an element $x_2 \in U_{\rho_1}(x_1)$ and an operator $\mathbf{A}_2$, say, such that

$$|\mathbf{A}_2 x_2| > 2,$$

afterwards choosing a new ball $U_{\rho_2}(x_2) \subset U_{\rho_1}(x_1)$ of radius $\rho_2 < \frac{1}{2}\rho_1$ such that

$$|\mathbf{A}_2 x| > 2$$

for all $x \in U_{\rho_2}(x_2)$. Continuing this process, we get a sequence of nested balls (cf. Vol. 1, Sec. 3.74a) whose radii $\rho_1, \rho_2, \ldots$ approach zero.† Since $\mathbf{X}$ is complete, there is a point x contained in all the balls (Vol. 1, Sec. 3.74c). For this point x we have

$$|\mathbf{A}_1 x| > 1, \ |\mathbf{A}_2 x| > 2, \ldots, \ |\mathbf{A}_n x| > n, \ldots$$

in keeping with (13). ∎

b. COROLLARY (**Principle of uniform boundedness**). *Let $\mathbf{A}_1, \mathbf{A}_2, \ldots$ be a sequence of continuous linear operators, mapping a Banach space $\mathbf{X}$ into a normed linear space $\mathbf{Y}$, such that the sequence $\mathbf{A}_1 x, \mathbf{A}_2 x, \ldots$ is bounded for every $x \in \mathbf{X}$. Then the sequence of operators $\mathbf{A}_1, \mathbf{A}_2, \ldots$ is uniformly bounded.*

Proof. An immediate consequence of the preceding theorem. ∎

c. COROLLARY. *Let $\mathbf{A}_1, \mathbf{A}_2, \ldots$ be a sequence of continuous linear operators, mapping a Banach space $\mathbf{X}$ into a normed linear space $\mathbf{Y}$, such that the sequence $\mathbf{A}_1 x, \mathbf{A}_2 x, \ldots$ converges to a limit $y \in \mathbf{Y}$ for every $x \in \mathbf{X}$. Let $\mathbf{A}$ be the operator defined by the equation $y = \mathbf{A}x$, where*

$$y = \lim_{n \to \infty} \mathbf{A}_n x.$$

Then $\mathbf{A}$ is a continuous linear operator mapping $\mathbf{X}$ into $\mathbf{Y}$.

Proof. The operator $\mathbf{A}$ is linear for the same reason as in the proof of Theorem 1.73b. Since the sequence $\mathbf{A}_1 x, \mathbf{A}_2 x, \ldots$ converges for every $x \in \mathbf{X}$, and hence is certainly bounded for every $x \in \mathbf{X}$, it follows from the preceding corollary that the sequence of operators $\mathbf{A}_1, \mathbf{A}_2, \ldots$ is uniformly bounded:

$$\sup_n \|\mathbf{A}_n\| \leqslant M.$$

† Note that $\rho_n < \frac{1}{2}\rho_{n-1} < \cdots < \frac{1}{2^{n-1}}\rho_1$.

But then

$$\sup_{|x|\leqslant 1} |\mathbf{A}_n x| = \|\mathbf{A}_n\| \leqslant M \qquad (n=1,2,\ldots),$$

and hence

$$\|\mathbf{A}\| = \sup_{|x|\leqslant 1} |\mathbf{A}x| = \sup_{|x|\leqslant 1} \lim_{n\to\infty} |\mathbf{A}_n x| \leqslant M,$$

so that the operator $\mathbf{A}$ is bounded and hence continuous. We note, in passing, that $\mathbf{A}$ is the strong limit of the sequence $\mathbf{A}_1, \mathbf{A}_2, \ldots$ (Sec. 1.73e), and that

$$\|\mathbf{A}\| \leqslant \sup_n \|\mathbf{A}_n\| \tag{15}$$

(why?). ∎

d. The estimate (15) can be sharpened. Let $c = \varliminf \|\mathbf{A}_n\|$. Then, given any $\varepsilon > 0$, let $\mathbf{A}_{n_k}$ $(k=1,2,\ldots)$ be a subsequence such that

$$\|\mathbf{A}_{n_k}\| < c + \varepsilon.$$

Since obviously

$$\mathbf{A}x = \lim_{k\to\infty} \mathbf{A}_{n_k} x$$

for every $x \in \mathbf{X}$, we have

$$\|\mathbf{A}\| \leqslant \sup_k \|\mathbf{A}_{n_k}\| < c + \varepsilon,$$

as before, and hence

$$\|\mathbf{A}\| \leqslant \varliminf_{n\to\infty} \|\mathbf{A}_n\|, \tag{16}$$

since ε is arbitrary. There are actually cases where the sign $\leqslant$ in (16) becomes $<$ (an example will be given in Sec. 1.75h).

e. LEMMA. *Let $\mathbf{A}_1$, $\mathbf{A}_2, \ldots$ be a sequence of continuous linear operators, mapping a Banach space $\mathbf{X}$ into a normed linear space $\mathbf{Y}$, which converges strongly to an operator $\mathbf{A}$, and let $x_1, x_2, \ldots$ be a sequence of vectors which converges (in the norm of $\mathbf{X}$) to a vector x. Then*

$$\mathbf{A}x = \lim_{n\to\infty} \mathbf{A}_n x_n.$$

Proof. By the principle of uniform boundedness, we have

$$\sup_n \|\mathbf{A}_n\| \leqslant M < \infty.$$

Therefore

$$|\mathbf{A}x-\mathbf{A}_n x| \leqslant |\mathbf{A}x-\mathbf{A}_n x| + |\mathbf{A}_n x-\mathbf{A}_n x_n| \leqslant |\mathbf{A}x-\mathbf{A}_n x| + M|x-x_n|,$$

where both terms on the right approach zero as $n\to\infty$. ▮

f. In all the above considerations we can replace the sequence of operators $\mathbf{A}_n$ by an operator-valued function $\mathbf{A}(t)$ defined on some set T, at the same time replacing convergence as $n\to\infty$ by convergence in some direction S defined on the set T (Vol. 1, Sec. 4.12).

Some important applications of the principle of uniform boundedness will be given below.

1.75. The space of bounded sequences and its subspaces

a. Now let $\mathbf{X}$ denote the linear space of all bounded real sequences $x=(\xi_1,\ldots,\xi_n,\ldots)$, with the usual (component-by-component) operations and a norm given by the formula

$$\|x\| = \sup_n |\xi_n|. \tag{17}$$

In this case, the axioms of a normed linear space are satisfied in an obvious way. Moreover, *the space* $\mathbf{X}$ *is complete.* This is easy to prove directly, but instead we need only refer to Theorem 1.23e, which implies the completeness of the space $R^s(M)$ of all bounded continuous real functions on the metric space M consisting of the positive integers 1,2,... with the usual metric of the real line.

b. Let $f_1,\ldots,f_n,\ldots$ be a sequence of real numbers such that

$$\sum_{n=1}^{\infty} |f_n| < \infty,$$

and let $x=(\xi_1,\ldots,\xi_n,\ldots)$ be any element of $\mathbf{X}$. Then the formula

$$f(x) = \sum_{n=1}^{\infty} f_n \xi_n \tag{18}$$

obviously defines a linear functional on the space $\mathbf{X}$, satisfying the inequality

$$|f(x)| \leqslant \sup_n |\xi_n| \sum_{n=1}^{\infty} |f_n|. \tag{19}$$

It follows from (19) that the functional f is bounded on the unit ball of the space $\mathbf{X}$ and hence is continuous. Moreover, for the norm of f we have the estimate

$$\|f\| \leqslant \sum_{n=1}^{\infty} |f_n|. \tag{20}$$

Now let

$$x_0 = (\operatorname{sgn} f_1, \ldots, \operatorname{sgn} f_n, \ldots),$$

where

$$\operatorname{sgn} x = \begin{cases} 1 & \text{if } x > 0, \\ 0 & \text{if } x = 0, \\ -1 & \text{if } x < 0. \end{cases}$$

Clearly x_0 belongs to the unit ball in $\mathbf{X}$. Moreover,

$$f(x_0) = \sum_{n=1}^{\infty} f_n \operatorname{sgn} f_n = \sum_{n=1}^{\infty} |f_n|,$$

and hence

$$\|f\| = \sup_{|x| \leqslant 1} |f(x)| \geqslant |f(x_0)| = \sum_{n=1}^{\infty} |f_n|. \tag{21}$$

Comparing (20) and (21), we get

$$\|f\| = \sum_{n=1}^{\infty} |f_n|. \tag{22}$$

c. Not every continuous linear functional on the space $\mathbf{X}$ is of the form (18). But (18) does give the general form of a continuous linear functional on some subspaces of $\mathbf{X}$. Anticipating Theorem 1.75f, we now describe such a subspace.

Let $\mathbf{X}_0$ denote the set of all elements $x = (\xi_1, \ldots, \xi_n, \ldots) \in \mathbf{X}$ such that

$$\lim_{n \to \infty} \xi_n = 0. \tag{23}$$

Then $\mathbf{X}_0$ is obviously a subspace of the space $\mathbf{X}$. Moreover, $\mathbf{X}_0$ *is closed*. In fact, let

$$x_m = (\xi_1^{(m)}, \ldots, \xi_n^{(m)}, \ldots) \qquad (m = 1, 2, \ldots)$$

be a sequence of vectors in $\mathbf{X}_0$ converging to a vector

$$x = (\xi_1, \ldots, \xi_n, \ldots)$$

as $m \to \infty$. Then, given any $\varepsilon > 0$, we first find an integer $p > 0$ such that

$$\|x_p - x\| = \sup_n |\xi_n^{(p)} - \xi_n| < \frac{\varepsilon}{2},$$

and then an integer $q > 0$ such that

$$|\xi_n^{(p)}| < \frac{\varepsilon}{2}$$

for all $n > q$. It follows that

$$|\xi_n| \leqslant |\xi_n - \xi_n^{(p)}| + |\xi_n^{(p)}| < \varepsilon$$

for all $n > q$, so that (23) holds. But then $x \in \mathbf{X}_0$, so that $\mathbf{X}_0$ is indeed closed.

The fact that $\mathbf{X}_0$ is closed in a complete space $\mathbf{X}$ implies that $\mathbf{X}_0$ is itself complete, regarded as a normed linear space in its own right (Vol. 1, Sec. 3.73b).

d. Next let

$$e_n = (0, \ldots, 0, 1, 0, \ldots),$$

where the 1 appears in the nth place. Then

$$\begin{aligned} \left\| x - \sum_{k=1}^{n} \xi_k e_k \right\| &= \|(\xi_1, \ldots, \xi_n, \xi_{n+1}, \ldots) - (\xi_1, \ldots, \xi_n, 0, \ldots)\| \\ &= \|(0, \ldots, 0, \xi_{n+1}, \xi_{n+2}, \ldots)\| \end{aligned} \tag{24}$$

for every $x = (\xi_1, \xi_2, \ldots) \in \mathbf{X}_0$. Given any $\varepsilon > 0$, let the integer $N > 0$ be such that $|\xi_n| < \varepsilon$ for all $n > N$. It then follows from (17) and (24) that

$$\left\| x - \sum_{k=1}^{n} \xi_k e_k \right\| < \varepsilon,$$

which in turn implies the convergence of the series

$$x = \sum_{k=1}^{\infty} \xi_k e_k$$

in the norm of the space $\mathbf{X}_0$. In particular, we see that the set of vectors $x = (\xi_1, \xi_2, \ldots)$ with all components ξ_n equal to zero after a certain value of the index n is dense in the space $\mathbf{X}_0$.

e. LEMMA. *Let the sequence $f_1, \ldots, f_n, \ldots$ be such that the right-hand side of (18) converges for every $x = (\xi_1, \ldots, \xi_n, \ldots) \in \mathbf{X}_0$. Then the series*

$$\sum_{n=1}^{\infty} |f_n| \tag{25}$$

also converges.

Proof. Consider the linear functionals

$$\varphi_n(x) = \sum_{k=1}^{n} f_k \xi_k \quad (n = 1, 2, \ldots).$$

By hypothesis, these functionals approach a limit as $n \to \infty$ for every $x \in \mathbf{X}_0$. Therefore, by Corollary 1.74b, the functionals φ_n are uniformly bounded:

$$\|\varphi_n\| \leqslant M < \infty \qquad (n=1,2,\ldots).$$

But

$$\|\varphi_n\| = \sum_{k=1}^{n} |f_k|,$$

because of (22), and hence

$$\sum_{k=1}^{n} |f_k| \leqslant M$$

for all $n=1,2,\ldots$, a fact which obviously implies the convergence of the series (25). ∎

f. THEOREM. *Every continuous linear functional on the space* $\mathbf{X}_0$ *is of the form* (18).

Proof. Let f be any continuous linear functional on the space $\mathbf{X}_0$. Since

$$x = \sum_{n=1}^{\infty} \xi_n e_n$$

holds for every $x=(\xi_1,\ldots,\xi_n,\ldots) \in \mathbf{X}_0$ (Sec. 1.75d), and since the functional f is continuous, we have

$$f(x) = f\left(\sum_{n=1}^{\infty} \xi_n e_n\right) = \sum_{n=1}^{\infty} f(\xi_n e_n) = \sum_{n=1}^{\infty} \xi_n f(e_n)$$

(see Theorem 1.71f, Property a), or

$$f(x) = \sum_{n=1}^{\infty} f_n \xi_n$$

after setting $f(e_n) = f_n$. Note that the series (25) converges, because of the lemma. ∎

g. The norm $\|f\|_0$ of a functional f on the space $\mathbf{X}_0$ coincides with its norm

$$\|f\| = \sum_{k=1}^{\infty} |f_k|$$

on the whole space $\mathbf{X}$. To see this, we first note the obvious inequality

$$\|f\|_0 \leqslant \|f\|. \tag{26}$$

On the other hand, applying the functional f to the vector

$$x_n = (\operatorname{sgn} f_1,\ldots, \operatorname{sgn} f_n, 0, 0, \ldots) \in \mathbf{X}_0,$$

we get

$$f(x_n) = \sum_{k=1}^{n} f_k \operatorname{sgn} f_k = \sum_{k=1}^{n} |f_k|,$$

and hence

$$\|f\|_0 \geqslant \sum_{k=1}^{n} |f_k|$$

for all $n=1,2,\ldots$, so that

$$\|f\|_0 \geqslant \sum_{k=1}^{\infty} |f_k| = \|f\|. \tag{27}$$

Comparing (26) and (27), we finally get

$$\|f\|_0 = \|f\| = \sum_{k=1}^{\infty} |f_k|.$$

h. Consider the functional

$$g_n(x) = \xi_n, \tag{28}$$

equal to the nth component of the vector $x = (\xi_1, \ldots, \xi_n, \ldots) \in \mathbf{X}_0$. This functional is of the form (18) if we set $f_n = 1, f_k = 0$ for $k \neq n$. Hence, in particular,

$$\|g_n\| = 1 \qquad (n=1,2,\ldots).$$

Clearly

$$\lim_{n\to\infty} g_n(x) = \lim_{n\to\infty} \xi_n = 0$$

for every $x \in \mathbf{X}_0$. Thus the sequence of functionals g_n converges strongly to 0 as $n \to \infty$, while the sequence of norms $\|g_n\|$ approaches 1 and not 0 as $n \to \infty$.

i. Next let $\mathbf{X}_1$ be the set of all vectors $x = (\xi_1, \ldots, \xi_n, \ldots) \in \mathbf{X}$ such that the sequence ξ_n approaches a finite limit as $n \to \infty$. Then $\mathbf{X}_1$ is obviously a subspace of $\mathbf{X}$, which contains both the subspace $\mathbf{X}_0$ and the one-dimensional subspace of $\mathbf{X}$ consisting of all vectors of the form λe (λ real), where e is the vector

$$e = (1, 1, \ldots, 1, \ldots)$$

with every component equal to 1; in fact, $\mathbf{X}_1$ is obviously the direct sum of these two subspaces. The subspace $\mathbf{X}_1$ is closed in the space $\mathbf{X}$. This is easily proved directly, but instead we need only refer to Theorem 1.23e, noting that the space $\mathbf{X}_1$ can be regarded as the space of all bounded continuous real functions on the metric space M consisting of the positive integers $1,2,\ldots$ and the symbol ∞, with a metric such that the points $1,2,\ldots$ are isolated and

$$\infty = \lim_{n\to\infty} n$$

(cf. Vol. 1, Secs. 3.35e, 3.73a, and 3.73c).

j. We can now easily give an example of a continuous linear functional on the space $\mathbf{X}_1$ which *cannot* be represented in the form (18), i.e., the functional

$$L(x) = \lim_{n\to\infty} \xi_n. \tag{29}$$

Suppose (29) can be represented in the form (18), with certain numbers $f_1,\dots,f_n,\dots$, so that

$$L(x) = \sum_{n=1}^{\infty} f_n\xi_n = \lim_{n\to\infty} \xi_n.$$

Then, setting $x=e_n$, we have

$$L(e_n)=f_n=0 \qquad (n=1,2,\dots).$$

Hence, choosing

$$x=e=(1,1,\dots,1,\dots),$$

we get

$$L(e) = \sum_{n=1}^{\infty} f_n\cdot 1=0,$$

contrary to the fact that

$$L(e) = \lim_{n\to\infty} 1=1.$$

It follows that (29) is not a functional of the form (18). However, (29) is the strong limit of a sequence of functionals of the form (18), since obviously

$$L(x) = \lim_{n\to\infty} g_n(x) \tag{30}$$

for all $x \in \mathbf{X}_1$, where g_n is the functional (28).

1.76. Generalized limits and Toeplitz's theorem

a. Again let $\mathbf{X}$ be the space of all bounded real sequences $x=(\xi_1,\dots, \xi_n,\dots)$, and let $\mathbf{X}_1$ be the subspace of $\mathbf{X}$ consisting of all convergent real sequences. We now pose the problem of enlarging the concept of a convergent sequence to include sequences which are divergent in the usual sense. In other words, our objective is to assign a number to every sequence ξ_n in some *closed* linear subspace $\mathbf{X}'\subset\mathbf{X}$ containing $\mathbf{X}_1$, where this number, denoted by LIM ξ_n and called the *generalized limit* of ξ_n, has the following natural properties:

(a) LIM $(\alpha\xi_n+\beta\eta_n)=\alpha$ LIM $\xi_n+\beta$ LIM η_n for arbitrary sequences ξ_n, η_n in $\mathbf{X}'$ and arbitrary real α, β;
(b) LIM $\xi_n=\lim \xi_n$ for every convergent sequence ξ_n;
(c) LIM ξ_n is a continuous linear functional on $\mathbf{X}'$.

b. By the *Cesàro limit* of a sequence ξ_n, denoted by C-lim ξ_n, we mean the limit in the ordinary sense (provided it exists) of the sequence of arithmetic means of ξ_n, i.e.,

$$C\text{-lim } \xi_n = \lim_{n\to\infty} \frac{\xi_1 + \cdots + \xi_n}{n}.$$

It can be shown that the Cesàro limit has Properties a–c on its domain of definition, and hence is a generalized limit. We need not stop to prove this here, since it is an immediate consequence of a more general theorem to be proved in a moment. Note that the Cesàro limit can exist in cases where the ordinary limit fails to exist. Thus, for example,

$$C\text{-lim } (0,1,0,1,\ldots) = \tfrac{1}{2}.$$

c. Functionals of the form (18) are unsuitable for constructing generalized limits, since, as already noted, even the ordinary limit (29) on the subspace $\mathbf{X}_1$ cannot be represented in this form. However, according to (30), the ordinary limit is the strong limit of special functionals of the form (18). This suggests the possibility of representing a generalized limit as the strong limit of a sequence of functionals

$$T_k(x) = \sum_{n=1}^{\infty} t_{kn}\xi_n \qquad (k=1,2,\ldots) \tag{31}$$

of the form (18). Specifying these functionals is equivalent to specifying an infinite "Toeplitz matrix"

$$T = \|t_{kn}\| \qquad (k,n=1,2,\ldots),$$

whose rows "generate" the functionals. Suppose the ordinary limit

$$T(x) = \lim_{k\to\infty} T_k(x)$$

exists, where $x = (\xi_1, \ldots, \xi_n, \ldots)$ as usual. Then we call $T(x)$ the *T-limit* of the sequence ξ_n, denoted by T-lim ξ_n.

It is now natural to look for conditions on the matrix T such that the functional $T(x)$ has Properties a–c of generalized limits and is defined for all convergent sequences. A definitive answer to this problem is given by the following

THEOREM (**Toeplitz**). *The functional $T(x)$ is a generalized limit if and only if the following conditions are satisfied:*

(1) $\sum_{n=1}^{\infty} |t_{kn}| \leqslant M$, *where M is independent of k;*

(2) $\lim_{k\to\infty} \sum_{n=1}^{\infty} t_{kn} = 1$;

(3) $\lim_{k\to\infty} t_{kn} = 0 \qquad (n=1,2,\ldots)$.

Proof. First we verify the necessity of Conditions 1–3. Suppose $T(x)$ is a generalized limit. Then, in particular, $T(x)$ is defined for every sequence $x=(\xi_1,\ldots,\xi_n,\ldots)$ converging to zero, i.e., for every $x\in\mathbf{X}_0$, and hence so are all the functionals $T_k(x)$. Therefore every series

$$\sum_{n=1}^{\infty} |t_{kn}|$$

converges, because of Lemma 1.75e. Moreover, by Corollary 1.74b, the convergence of the sequence $T_k(x)$ for every $x\in\mathbf{X}_0$ implies the uniform boundedness of the functionals $T_k(x)$:

$$\|T_k\| \leqslant M < \infty. \tag{32}$$

But

$$\|T_k\| = \sum_{n=1}^{\infty} |t_{kn}|,$$

because of (22), and this, together with (32), implies Condition 1. To get Conditions 2 and 3, we need only take the T-limit of the vector $e=(1,1,\ldots)$ and then the T-limit of the vector e_n, in each case invoking Property b of generalized limits.

Next we prove that Conditions 1–3 are sufficient for the functional $T(x)$ to be a generalized limit. Let

$$x=(\xi_1,\ldots,\xi_n,\ldots), \quad y=(\eta_1,\ldots,\eta_n,\ldots)$$

be sequences for which the values

$$T(x) = \lim_{k\to\infty} T_k(x), \; T(y) = \lim_{k\to\infty} T_k(y)$$

are defined. Then the value

$$T(\alpha x+\beta y) = \lim_{k\to\infty} T_k(\alpha x+\beta y) = \alpha \lim_{k\to\infty} T_k(x) + \beta \lim_{k\to\infty} T_k(y) = \alpha T(x) + \beta T(y)$$

is also defined for arbitrary real α and β. Hence the domain of definition $\mathbf{X}_T$ of the functional $T(x)$ is a linear manifold and $T(x)$ is linear on $\mathbf{X}_T$, so that Property a of generalized limits is satisfied. Moreover, for $x=e=(1,1,\ldots)$ we have

$$T(e) = \lim_{k\to\infty} T_k(e) = \lim_{k\to\infty} \sum_{n=1}^{\infty} t_{kn} = 1, \tag{33}$$

because of Condition 2. But the space $\mathbf{X}_1$ of all convergent sequences is the direct sum of the subspace $\mathbf{X}_0$ of all sequences converging to zero and the subspace consisting of all multiples of e (Sec. 1.75i). Therefore, to verify Property b of generalized limits, we can now confine ourselves to vectors $x \in \mathbf{X}_0$. According to Condition 1, the functionals $T_k(x)$ are uniformly bounded on $\mathbf{X}_0$, and moreover, because of Condition 3,

$$\lim_{k\to\infty} T_k(x) = \lim_{k\to\infty} \sum_{n=1}^{\infty} t_{kn}\xi_n = \lim_{k\to\infty} \sum_{n=1}^{m} t_{kn}\xi_n = 0$$

for every vector of the form $x = (\xi_1, \ldots, \xi_m, 0, 0, \ldots)$. But the set of all vectors of this type is dense in $\mathbf{X}_0$ (see Sec. 1.75d), and hence, applying Lemma 1.73f, we find that

$$\lim_{k\to\infty} T_k(x) = \lim_{k\to\infty} \sum_{n=1}^{\infty} t_{kn}\xi_n = 0$$

for every $x \in \mathbf{X}_0$. This, together with (33), shows that $\mathbf{X}_T$ contains $\mathbf{X}_1$ and establishes Property b of generalized limits. As for the continuity of the functional $T(x)$ on the space $\mathbf{X}_T$ (Property c), it follows at once from Corollary 1.74c. In the present case, the estimate (16) takes the form

$$\|T\| \leqslant \varliminf_{k\to\infty} \|T_k\| = \varliminf_{k\to\infty} \sum_{n=1}^{\infty} |t_{kn}|.$$

Since $T(e) = 1$, as just shown, we can also estimate $\|T\|$ from below, obtaining $\|T\| \geqslant 1$.

To complete the proof, we need only show that the subspace $\mathbf{X}_T$ is closed in $\mathbf{X}$, as required by the definition of a generalized limit. Let $\overline{\mathbf{X}}_T$ be the closure of $\mathbf{X}_T$. Then $\mathbf{X}_T$ is dense in $\overline{\mathbf{X}}_T$, while the functionals $T_k(x)$ are uniformly bounded and convergent on $\mathbf{X}_T$. Therefore, by Lemma 1.73f again, the functionals $T_k(x)$ converge on $\overline{\mathbf{X}}_T$. But then $\mathbf{X}_T \supset \overline{\mathbf{X}}_T$, and hence $\overline{\mathbf{X}}_T = \mathbf{X}_T$, since obviously $\mathbf{X}_T \subset \overline{\mathbf{X}}_T$. ∎

1.77. Examples

a. The matrix

$$\begin{Vmatrix} 1 & 0 & 0 & \cdots \\ 0 & 1 & 0 & \cdots \\ 0 & 0 & 1 & \cdots \\ \cdot & \cdot & \cdot & \cdots \end{Vmatrix}$$

leads to the ordinary limit

$$\lim_{n\to\infty} \xi_n,$$

defined only on $\mathbf{X}_1$.

b. The Cesàro limit (Sec. 1.76b) is "generated" by the matrix

$$\left\| \begin{matrix} 1 & 0 & 0 & \cdots \\ \frac{1}{2} & \frac{1}{2} & 0 & \cdots \\ \frac{1}{3} & \frac{1}{3} & \frac{1}{3} & \cdots \\ \cdot & \cdot & \cdot & \cdots \end{matrix} \right\|,$$

which clearly satisfies all the conditions of Toeplitz's theorem, thereby proving that the Cesàro limit has all the properties of a generalized limit.

c. As a third example, we now describe a whole class of matrices including the preceding two as special cases. Given any sequence

$$p_0 > 0,\ p_1 \geqslant 0,\ p_2 \geqslant 0, \ldots,$$

let

$$P_n = \sum_{k=0}^{n} p_k \qquad (n = 0,1,2,\ldots).$$

Then the "Voronoi matrix"

$$\left\| \begin{matrix} 1 & 0 & 0 & \cdots & 0 & \cdots \\ \frac{p_1}{P_1} & \frac{p_0}{P_1} & 0 & \cdots & 0 & \cdots \\ \frac{p_2}{P_2} & \frac{p_1}{P_2} & \frac{p_0}{P_2} & \cdots & 0 & \cdots \\ \cdot & \cdot & \cdot & \cdots & \cdot & \cdots \\ \frac{p_n}{P_n} & \frac{p_{n-1}}{P_n} & \frac{p_{n-2}}{P_n} & \cdots & \frac{p_0}{P_n} & \cdots \\ \cdot & \cdot & \cdot & \cdots & \cdot & \cdots \end{matrix} \right\|$$

leads to a method of taking generalized limits, known as *Voronoi's method.* Here Conditions 1 and 2 are satisfied in an obvious way. Condition 3 with $n = 1$ is equivalent to the condition $p_n/P_n \to 0$. But

$$\frac{p_{n-m}}{P_n} \leqslant \frac{p_{n-m}}{P_{n-m}} \qquad (m = 0,1,2,\ldots),$$

and therefore $p_n/P_n \to 0$ implies Condition 3 for arbitrary $n = 1,2,\ldots$ Hence *the condition $p_n/P_n \to 0$ is necessary and sufficient for the Voronoi matrix to be a Toeplitz matrix.* If $p_0 = 1$, $p_1 = p_2 = \cdots = 0$, we get the ordinary limit, while if $p_0 = p_1 = p_2 = \cdots = p_n$, we get the Cesàro limit.

1.78. Further properties of T-limits

a. For some T-limits the space $\mathbf{X}_T$ may coincide with the space $\mathbf{X}_1$, as in Example 1.77a. Hence the question arises of how to distinguish these "uninteresting" cases of the generalized limit. In this regard, there is a theorem of Brudno,† which states that $\mathbf{X}_T = \mathbf{X}_1$ if and only if there exists a constant δ_0 such that

$$\overline{\lim_{n\to\infty}} |T(x)| \geqslant \delta_0 \overline{\lim_{n\to\infty}} |\xi_n| \tag{34}$$

for every $x = (\xi_1, \ldots, \xi_n, \ldots) \in \mathbf{X}$. But the condition (34) is hard to verify. There is another condition (but only a sufficient one) for equality of the spaces $\mathbf{X}_T$ and $\mathbf{X}_1$, due to Agnew,‡ involving the numbers t_{kn} themselves, namely $\mathbf{X}_T = \mathbf{X}_1$ if

$$\underline{\lim} \sum_n \{t_{nn} - \sum_{k\neq n} |t_{kn}|\} > 0.$$

b. On the other hand, it turns out (see Problem 8) that it is impossible to construct a matrix T such that $\mathbf{X}_T = \mathbf{X}$. Nevertheless, there is a general argument due to Banach,§ proving the existence of a generalized limit LIM ξ_n defined on the whole space $\mathbf{X}$ with the additional property that the value of the limit is invariant under shifts of the index, so that LIM $\xi_n =$ LIM ξ_{n+1}. However, this generalized limit is not given by an explicit formula.

c. For some matrices T the quantity $T(x)$ may fall outside the interval

$$\Delta(x) = [\underline{\lim}\ \xi_n, \overline{\lim}\ \xi_n]$$

containing all the limit points of the sequence ξ_n. For example, this happens for the matrix

$$\begin{Vmatrix} 2 & -1 & 0 & 0 & 0 & 0 & \cdots \\ 0 & 0 & 2 & -1 & 0 & 0 & \cdots \\ 0 & 0 & 0 & 0 & 2 & -1 & \cdots \\ \cdot & \cdot & \cdot & \cdot & \cdot & \cdot & \cdots \end{Vmatrix}$$

and the sequence $x = (0,1,0,1,0,1,\ldots)$. It is natural to look for conditions on the matrix T guaranteeing that all limit points of the sequence $T_1(x)$, $T_2(x)$, ... lie in $\Delta(x)$ for arbitrary $x \in \mathbf{X}$. It can be shown (see Problem 9) that the limit points of the sequence $T_1(x)$, $T_2(x)$,... all lie in $\Delta(x)$ for arbitrary $x \in \mathbf{X}$ if and only if

$$\lim_{n\to\infty} \| T_n \| = 1.$$

† A. L. Brudno, *Summation of bounded sequences by matrices* (in Russian), Mat. Sb., **16**, 191–245 (1945).

‡ R. P. Agnew, *Equivalence of methods for evaluation of sequences*, Proc. Amer. Math. Soc., **3**, 550–565 (1952).

§ S. Banach, *Théorie des Opérations Linéaires*, Chelsea Publ. Co., New York (1955), p. 34.

1.8. Normed Algebras

1.81.a. A normed linear space $\mathbf{U}$ which is simultaneously an algebra (Sec. 1.18a) is said to be a *normed algebra* if $x_n \to x$ (in the norm of $\mathbf{U}$) implies $x_n y \to xy$ for every $y \in \mathbf{U}$.

b. The set $\mathbf{L}(\mathbf{X})$ of all bounded linear operators acting in a Banach space $\mathbf{X}$ is both a complete normed linear space (see Theorem 1.73b) and an algebra (Secs. 1.19c, 1.71h). The fact that $\mathbf{L}(\mathbf{X})$ is a normed algebra follows from the inequality

$$\|\mathbf{AB}\| \leqslant \|\mathbf{A}\| \, \|\mathbf{B}\| \tag{1}$$

(Sec. 1.71h), since if $\mathbf{A}_n, \mathbf{B} \in \mathbf{X}$ and $\mathbf{A}_n \to \mathbf{B}$, then

$$\|\mathbf{A}_n\mathbf{B} - \mathbf{AB}\| = \|(\mathbf{A}_n - \mathbf{A})\mathbf{B}\| \leqslant \|\mathbf{A}_n - \mathbf{A}\| \, \|\mathbf{B}\| \to 0,$$

which implies $\mathbf{A}_n\mathbf{B} \to \mathbf{AB}$.

c. It turns out that an inequality of the type (1) can be achieved in every complete normed algebra after making a certain "renormalization" (this will be shown in Sec. 1.88). Hence instead of the condition for continuity of multiplication in the form "$x_n \to x$ implies $x_n y \to xy$ for every y," we can start from the stronger condition

$$|xy| \leqslant |x| \, |y| \tag{2}$$

for every $x, y \in \mathbf{U}$.

d. In what follows, we will assume in addition to the axiom (2) that every normed algebra under discussion has a unit e (Sec. 1.18c), where $|e| = 1$. The last assumption is automatically satisfied for the algebra of linear operators acting in a normed linear space, with the unit operator $\mathbf{E}$ (Sec. 1.15c) serving as the unit.†

1.82.a. Just as in every algebra, the unit of a normed algebra is an invertible element, since $ee = e$.

THEOREM. *In a complete normed algebra the whole ball* $\{x : |e - x| < 1\}$ *consists of invertible elements.*

Proof. According to (2),

$$|(e - x)^n| \leqslant |e - x|^n,$$

and hence the series

$$y = e + (e - x) + (e - x)^2 + \cdots \qquad (|e - x| < 1) \tag{3}$$

† Note that $\|\mathbf{E}\| = 1$ (why?).

converges, by Weierstrass' test (Theorem 1.37d). Multiplying (3) by $x=e-(e-x)$, we get

$$[e-(e-x)]y=y[e-(e-x)]$$
$$=[e+(e-x)+(e-x)^2+\cdots]-[(e-x)+(e-x)^2+\cdots]=e.$$

Hence the sum of the series (3) is just the inverse of the element x. ∎

b. It follows from the estimate

$$|e-y|=|(e-x)+(e-x)^2+\cdots|\leqslant\frac{|e-x|}{1-|e-x|}$$

that $x\to e$ implies $y\to e$. In this sense, we can say that the (nonlinear) operator carrying x into $y=x^{-1}$ is *continuous at* $x=e$.

1.83.a. THEOREM. *The set O of all invertible elements of a complete normed algebra* $\mathbf{U}$ *is an open set in* $\mathbf{U}$, *and the operator x^{-1} is continuous on the whole set O.*

Proof. Since $xx^{-1}=e$, it follows from (2) that

$$|(x+h)x^{-1}-e|=|hx^{-1}|\leqslant|h||x^{-1}|<1$$

for all h with $|h|<1/|x^{-1}|$. Hence the element $(x+h)x^{-1}$ is invertible, by the preceding theorem, i.e., there exists an element $z(h)$ such that $(x+h)x^{-1}x(h)=e$. But then $x+h$ is also invertible, with inverse $x^{-1}z(h)$. Moreover, if $h\to 0$, then $(x+h)x^{-1}\to xx^{-1}=e$, so that $z(h)\to e$, by Sec. 1.82b. It follows that $(x+h)^{-1}=x^{-1}z(h)\to x^{-1}$, which proves the continuity of the operator x^{-1} on O. ∎

b. As just shown, if x belongs to the set O, then so does some open ball of radius $r\geqslant 1/|x^{-1}|$ centered at x. But then $|x^{-1}|\geqslant 1/r$. Therefore, as x approaches the boundary (Sec. 1.21c) of the set O, in which case naturally $r\to 0$, the norm of the element x^{-1} increases without limit.

1.84. A noninvertible element z lying on the boundary of the set O is a "generalized divisor of zero," which means that there exists a sequence of elements $y_1, y_1, \ldots$ such that $|y_n|\geqslant c>0$ while $zy_n\to 0$. In fact, let $y_n=x_n^{-1}/|x_n^{-1}|$ where $x_n\in O$ and $x_n\to z$. Then

$$|zy_n|\leqslant|(z-x_n)y_n|+|x_ny_n|\leqslant|z-x_n|+1/|x_n^{-1}|\to 0$$

(Sec. 1.83b), as required.

If z is a generalized divisor of zero, then z is noninvertible, since otherwise $zy_n\to 0$ would imply $z^{-1}zy_n=y_n\to 0$. But noninvertible elements which are not limits of invertible elements need not be generalized divisors of zero (see Problem 10).

1.85. Banach algebras

a. By a *Banach algebra* we mean a *complete* complex normed algebra $\mathscr{B}$. If x is any element of a Banach algebra, then the element $e-\mu x$ is invertible for all sufficiently small complex μ (e.g., for $|\mu|<1/|x|$ if $x\neq 0$), with inverse

$$(e-\mu x)^{-1}=e+\mu x+\mu^2x^2+\cdots, \tag{4}$$

as in the proof of Theorem 1.82a. The radius of convergence R of the series (4) is actually equal to

$$R=\frac{1}{\overline{\lim\limits_{n\to\infty}}\sqrt[n]{|x^n|}}$$

(Sec. 1.66c). Note that (4) converges in the whole λ-plane if $R=\infty$.

b. The element $x-\lambda e$ is invertible for all sufficiently large $|\lambda|$, e.g., for $|\lambda|>|x|$, as follows at once from the formula

$$x-\lambda e=-\lambda(e-\lambda^{-1}x).$$

Let S_x be the set of all numbers λ for which the element $x-\lambda e$ is noninvertible. Then S_x is called the *spectrum* of the element x. Thus the function $(x-\lambda e)^{-1}$ is defined on the complement of S_x. According to Theorem 1.83a, this complement is an open set G in the λ-plane (why?), so that the spectrum S_x itself is a closed set. Moreover, it also follows from Theorem 1.83a that $(x-\lambda e)^{-1}$ is a continuous function of λ on G (with values in $\mathbf{U}$).

c. We can go even further and assert that $(x-\lambda e)^{-1}$ is analytic on G, i.e., differentiable at every $\lambda\in G$ (Sec. 1.66a). In fact†

$$\left[\frac{(x-(\lambda+h)e)^{-1}-(x-\lambda e)^{-1}}{h}\right](x-(\lambda+h)e)\,(x-\lambda e)$$
$$=\frac{(x-\lambda e)-(x-(\lambda+h)e)}{h}=e, \tag{5}$$

which shows that the element in square brackets is invertible, with inverse

$$(x-(\lambda+h)e)(x-\lambda e). \tag{6}$$

As $h\to 0$, (6) approaches $(x-\lambda e)^2$, and hence

$$\lim_{h\to 0}\frac{(x-(\lambda+h)e)^{-1}-(x-\lambda e)^{-1}}{h}=[(x-\lambda e)^2]^{-1} \tag{7}$$

† Note that elements of the form $x-\lambda e$ always commute, i.e., $(x-\lambda e)(x-\mu e)=(x-\mu e)(x-\lambda e)$.

exists. But the existence of (7) for all $\lambda \in G$ means that $(x-\lambda e)^{-1}$ is analytic on G, as asserted.

d. THEOREM. *Every element x in a Banach algebra $\mathscr{B}$ has a nonempty spectrum.*

Proof. Let Γ be a circle in the λ-plane with center at the point 0 and radius $r>|x|$. Consider the integral

$$I=\frac{1}{2\pi i}\oint_{\Gamma}(\lambda e-x)^{-1}\,d\lambda, \tag{8}$$

whose existence follows from the continuity of the function $(\lambda e-x)^{-1}=-(x-\lambda e)^{-1}$ on the curve Γ. To evaluate (8), we make the change of variable $\lambda^{-1}=\mu$ and use the expansion (4), obtaining

$$\begin{aligned}I&=\frac{1}{2\pi i}\oint_{|\lambda|=r}\lambda^{-1}(e-\lambda^{-1}x)^{-1}\,d\lambda=-\frac{1}{2\pi i}\oint_{|\mu|=1/r}(e-\mu x)^{-1}\frac{d\mu}{\mu}\\&=\frac{1}{2\pi i}\oint_{|\mu|=1/r}\sum_{n=0}^{\infty}\mu^n x^n\frac{d\mu}{\mu}=\frac{1}{2\pi i}\sum_{n=0}^{\infty}x^n\oint_{|\mu|=1/r}\mu^{n-1}\,d\mu=e\end{aligned}$$

$(x^0=e)$.† Suppose the spectrum of x is empty. Then the function $(x-\lambda e)^{-1}$ is analytic on the whole λ-plane, so that the integral (7) vanishes, by Cauchy's theorem (Sec. 1.66b), contrary to the fact that $I=e$. This contradiction shows that the spectrum of x must be nonempty. ∎

e. COROLLARY (**Gelfand-Mazur theorem**). *If the Banach algebra $\mathscr{B}$ is a field, i.e., if every nonzero element $x\in\mathscr{B}$ has an inverse, then $\mathscr{B}$ is the field of complex numbers.*

Proof. Given any element $x\in\mathscr{B}$, let λ be a number in the spectrum of x. Then $x-\lambda e$ has no inverse. Since 0 is the only element with no inverse, it follows that $x-\lambda e=0$ and hence $x=\lambda e$. ∎

1.86.a. Let $\mathscr{B}$ be the algebra of all linear operators acting in a finite-dimensional complex normed linear space $\mathbf{C}_n$, equipped with some norm (according to Sec. 1.39g, all norms in $\mathbf{C}_n$ are equivalent). In this case, every linear operator $\mathbf{A}$ is continuous, since the components of the vector $\mathbf{A}x$ are linear (and hence continuous) functions of the components of the vector x. Thus the algebra $\mathscr{B}$ coincides with the algebra $\mathbf{L}(\mathbf{C}_n)$ of all bounded linear operators acting in the space $\mathbf{C}_n$.

The spectrum of the element $\mathbf{A}\in\mathbf{L}(\mathbf{C}_n)$, in the sense of the definition of

† The last step follows at once from

$$\oint_{|\mu|=1/r}\mu^{n-1}\,d\mu=\frac{i}{r^n}\int_0^{2\pi}e^{in\theta}\,d\theta=\begin{cases}2\pi i & \text{if } n=0,\\ 0 & \text{otherwise}\end{cases}$$

(cf. Vol. 1, Sec. 10.33a).

Sec. 1.85b, is just the set of all (distinct) eigenvalues of the operator **A** (Sec. 1.15k). In fact, the operator $\mathbf{A}-\lambda\mathbf{E}$ is noninvertible if and only if

$$\det(A-\lambda E)=0,$$

where A is the matrix of the operator **A** and E is the $n\times n$ unit matrix. But this is precisely the characteristic equation determining the eigenvalues of A (Sec. 1.17e). Thus we see that the definition of the spectrum $S_{\mathbf{A}}$ of the operator **A** given in Sec. 1.85b coincides with that given in Sec. 1.19e (without the multiplicities r_k).

We have already used $S_{\mathbf{A}}$ (with the appropriate r_k) to construct an isomorphism between the polynomial algebra $\mathscr{P}(\mathbf{A})$ of Sec. 1.19e and the algebra $\mathscr{J}(S_{\mathbf{A}})$ of jets on $S_{\mathbf{A}}$. As in Sec. 1.19f, we can also use $S_{\mathbf{A}}$ to construct an epimorphism of $\mathscr{F}(S_{\mathbf{A}})$ into $\mathscr{P}(\mathbf{A})$, where $\mathscr{F}(S_{\mathbf{A}})$ is the algebra of all functions $f(\lambda)$ analytic on $S_{\mathbf{A}}$, i.e., analytic on some open set containing $S_{\mathbf{A}}$.† It turns out that analogous morphisms exist for the general case of an arbitrary Banach algebra:

b. THEOREM. *Given any Banach algebra $\mathscr{B}$ and any element $x\in\mathscr{B}$, with spectrum $S=S_x$, let $\mathscr{F}(S)$ be the algebra of all functions $f(\lambda)$ analytic on S. Then there exists a morphism of $\mathscr{F}(S)$ into $\mathscr{B}$ such that*

(a) *The function $f(\lambda)\equiv 1$ goes into the unit e;*
(b) *The function $f(\lambda)\equiv\lambda$ goes into the element x;*
(c) *Every sequence of functions $f_n(\lambda)\in\mathscr{F}(S)$ converging uniformly to a function $f(\lambda)\in\mathscr{F}(S)$ on an open set containing S goes into a sequence of elements $f_n\in\mathscr{B}$ converging in norm to the element $f\in\mathscr{B}$ corresponding to the function $f(\lambda)$.*

Proof. Let G be an open set whose boundary Γ consists of piecewise smooth closed Jordan paths, and suppose G contains S while $G\cup\Gamma$ is contained in the domain of analyticity of $f(\lambda)$. Then the required morphism is given by the formula‡

$$f=\frac{1}{2\pi i}\oint_{\Gamma}(\lambda e-x)^{-1}f(\lambda)\,d\lambda, \tag{9}$$

where, by Cauchy's theorem (Sec. 1.66b), the integral does not depend on the particular choice of the sets G and Γ. The fact that (9) maps the function $f(\lambda)\equiv 1$ into the element e has already been proved in Theorem 1.85d, and the fact that $f(\lambda)\equiv\lambda$ goes into the element x is proved in just the same way. Moreover, (9) obviously defines a linear mapping of $\mathscr{F}(S)$ into $\mathscr{B}$, and we

† The open set in question is called the *domain of analyticity* of $f(\lambda)$.
‡ Note the profound analogy between (9) and Cauchy's formula (Sec. 1.66c). In general, the integral in (9) is to be interpreted as a sum of integrals over the separate paths making up Γ.

must now verify that the product of two functions $f(\lambda)$ and $g(\lambda)$ goes into the product of the corresponding elements f and g.

To prove this, we start from the formula

$$(\lambda e-x)^{-1}(\mu e-x)^{-1}=\frac{(\lambda e-x)^{-1}-(\mu e-x)^{-1}}{\mu-\lambda} \tag{10}$$

obtained from (5) by replacing $\lambda+h$ by μ. Given two functions $f(\lambda)$ and $g(\lambda)$ analytic on S, let G_1 and G_2 be two open sets containing S, with boundaries Γ_1 and Γ_2 respectively (each made up of piecewise smooth closed Jordan paths), such that $G_1\cup\Gamma_1$ is contained in G_2 while $G_2\cup\Gamma_2$ is contained in the common domain of analyticity of $f(x)$ and $g(x)$. Integrating formula (10) multiplied by $f(\lambda)g(\lambda)/(2\pi i)^2$ first along Γ_1 and then along Γ_2, and using the fact that the order of integration can be reversed (Vol. 1, Sec. 10.24), we get

$$\begin{aligned}\frac{1}{2\pi i}\oint_{\Gamma_1}(\lambda e-x)^{-1}f(\lambda)\,d\lambda\,\frac{1}{2\pi i}\oint_{\Gamma_2}(\mu e-x)^{-1}g(\mu)\,d\mu\\ =\frac{1}{2\pi i}\oint_{\Gamma_1}(\lambda e-x)^{-1}f(\lambda)\left\{\frac{1}{2\pi i}\oint_{\Gamma_2}\frac{g(\mu)}{\mu-\lambda}\,d\mu\right\}d\lambda\\ +\frac{1}{2\pi i}\oint_{\Gamma_2}(\mu e-x)^{-1}g(\mu)\left\{\frac{1}{2\pi i}\oint_{\Gamma_1}\frac{f(\lambda)}{\lambda-\mu}\,d\lambda\right\}d\mu.\end{aligned} \tag{11}$$

The first integral in curly brackets on the right equals $g(\lambda)$, by Cauchy's formula, since λ lies inside Γ_2, while the second integral in curly brackets vanishes, since μ lies outside Γ_1. Thus (11) reduces to

$$\begin{aligned}\frac{1}{2\pi i}\oint_{\Gamma_1}(\lambda e-x)^{-1}f(\lambda)\,d\lambda\,\frac{1}{2\pi i}\oint_{\Gamma_2}(\mu e-x)^{-1}g(\mu)\,d\mu\\ =\frac{1}{2\pi i}\oint_{\Gamma_1}(\lambda e-x)^{-1}f(\lambda)\,g(\lambda)\,d\lambda,\end{aligned}$$

so that (9) does indeed map the product of the functions $f(\lambda)$ and $g(\lambda)$ into the product of the corresponding elements f and g.

We must still prove the last assertion of the theorem. Suppose the sequence of functions $f_n(\lambda)$ converges uniformly to a function $f(\lambda)$ on an open set G containing S, with a boundary Γ of the indicated type, so that

$$\lim_{n\to\infty}\sup_{\lambda\in G}|f(\lambda)-f_n(\lambda)|=0.$$

Then we have the estimate

$$\begin{aligned}\|f-f_n\|&=\left\|\frac{1}{2\pi i}\oint_{\Gamma}(\lambda e-x)^{-1}[f(\lambda)-f_n(\lambda)]\,d\lambda\right\|\\ &\leqslant\sup_{\lambda\in\Gamma}|f(\lambda)-f_n(\lambda)|\frac{1}{2\pi i}\oint_{\Gamma}\|(\lambda e-x)^{-1}\|\,|d\lambda|,\end{aligned}$$

and hence

$$\lim_{n\to\infty} \|f-f_n\|=0. \quad \blacksquare$$

It should be noted that the mapping (9) is in general not a monomorphism, and may well carry a function $f(\lambda)\not\equiv 0$ into the zero element of the algebra $\mathscr{B}$.

c. Let $f(x)$ be the element of $\mathscr{B}$ into which (9) maps the function $f(\lambda)\in\mathscr{F}(S)$. Then, in particular, the functions e^{tx}, $\cos tx$, and $\sin tx$ are defined for arbitrary $x\in\mathscr{B}$ (cf. Sec. 1.19f), and moreover

$$e^{(t_1+t_2)x}=e^{t_1x}e^{t_2x} \qquad (t_1,t_2\in C),$$

since (9) is a morphism. The last assertion of the theorem shows that these functions can be written in the form of power series:

$$e^{tx}=\sum_{n=0}^{\infty}\frac{t^n}{n!}x^n,$$

$$\cos tx=1-\frac{t^2}{2!}x^2+\frac{t^4}{4!}x^4-\cdots,$$

$$\sin tx=x-\frac{t^3}{3!}x^3+\frac{t^5}{5!}x^5-\cdots.$$

1.87. THEOREM. *Suppose (9) maps the function $f(\lambda)\in\mathscr{F}(S)$ into the element $f(x)\in\mathscr{B}$. Then the spectrum of $f(x)$ is just the set $S_{f(x)}=\{f(\lambda):\lambda\in S_x\}$, where S_x is the spectrum of x.*

Proof. Given any $\lambda_0\in S_x$, let $\mu_0=f(\lambda_0)$. Then the analytic function $f(\lambda)-\mu_0$ vanishes at $\lambda=\lambda_0$, and hence can be written in the form

$$f(\lambda)-\mu_0=(\lambda-\lambda_0)g(\lambda),$$

where $g(\lambda)$ is another function analytic on the domain of analyticity of $f(\lambda)$. Therefore

$$f(x)-\mu_0e=(x-\lambda_0e)g(x),$$

by the properties of the morphism (9). But if $f(x)-\mu_0e$ is invertible, then so is $x-\lambda_0e$ (with inverse $g(x)(f(x)-\mu_0e)^{-1}$), contrary to the assumption that $\lambda_0\in S_x$. It follows that $\mu_0\in S_{f(x)}$.

Conversely, suppose $\mu_0\in S_{f(x)}$. Then there exists a $\lambda_0\in S_x$ such that $f(\lambda_0)=\mu_0$. In fact, if $f(\lambda)-\mu_0$ were nonvanishing on S_x, then the function

$$g(\lambda)=\frac{1}{f(\lambda)-\mu_0}$$

would be analytic on S, and the corresponding element $g(x)\in\mathscr{B}$ would then be the inverse of $f(x)-\mu_0 e$, contrary to the assumption that $\mu_0 \in S_{f(x)}$. ∎

1.88. Let $\mathbf{U}$ be an algebra with continuous multiplication, i.e., an algebra such that $x_n \to x$ implies $x_n y \to xy$ for every $y \in \mathbf{U}$. Then, as we now show, $\mathbf{U}$ can be "renormalized" in such a way that the new norm satisfies the condition (2).

THEOREM. *Let* $\mathbf{U}$ *be a complete normed algebra with a unit e and a norm* $|x|_1$. *Then* $\mathbf{U}$ *can be equipped with an equivalent norm* $|x|_2$ *such that* $|e|_2 = 1$ *and* $|xy|_2 \leqslant |x|_2 |y|_2$ *for every* $x, y \in \mathbf{U}$.

Proof. Every element x of the algebra $\mathbf{U}$ generates an operator $\mathbf{A}_x$ of multiplication by x, in accordance with the formula

$$\mathbf{A}_x y = xy.$$

It follows from the properties of $\mathbf{U}$, in particular from the continuity of multiplication, that $\mathbf{A}_x$ is a continuous linear operator. Let $\mathbf{L}(\mathbf{U})$ be the algebra of all continuous linear operators acting in $\mathbf{U}$. Then the operators of the form $\mathbf{A}_x$ form a subalgebra $\mathbf{V} \subset \mathbf{L}(\mathbf{U})$, with the identity operator $\mathbf{E} = \mathbf{A}_e$ serving as the unit.

By the associativity of multiplication, we have

$$\mathbf{A}_x(yz) = x(yz) = (xy)z = (\mathbf{A}_x y)z. \tag{12}$$

It is not hard to see that this property characterizes the operators of the subalgebra $\mathbf{V}$. In fact, if the operator $\mathbf{A}$ is such that $\mathbf{A}(yz) = (\mathbf{A}y)z$ for all $y, z \in \mathbf{U}$, then writing $\mathbf{A}e = x$, we have $\mathbf{A}y = \mathbf{A}(ey) = (\mathbf{A}e)y = xy$, i.e., $\mathbf{A}$ is the operator of multiplication by x.

Next we use (12) to show that *the subalgebra* $\mathbf{V}$ *is closed in the algebra* $\mathbf{L}(\mathbf{U})$. Suppose the operators $\mathbf{A}_1, \mathbf{A}_2, \ldots$ in $\mathbf{V}$ converge (in the norm of $\mathbf{L}(\mathbf{U})$) to the operator $\mathbf{A}$. Then $\mathbf{A}_n x$ converges to $\mathbf{A}x$ for every $x \in \mathbf{U}$. By the continuity of multiplication, we have

$$\mathbf{A}(xy) = \lim \mathbf{A}_n(xy) = \lim (\mathbf{A}_n x)y = (\lim \mathbf{A}_n x)y = (\mathbf{A}x)y,$$

and hence $\mathbf{A} \subset \mathbf{V}$, by the criterion (12).

Since the algebra $\mathbf{L}(\mathbf{U})$ is complete (Theorem 1.73b), the subalgebra $\mathbf{V} \subset \mathbf{L}(\mathbf{U})$, regarded as a normed linear space in its own right, is also complete, being closed in $\mathbf{L}(\mathbf{U})$. There are now two norms in the algebra $\mathbf{U}$, the original norm $|x|_1$ and the norm

$$|x|_2 = \|\mathbf{A}_x\| = \sup_{|y|_1 \leqslant 1} |\mathbf{A}_x y|_1 = \sup_{|y|_1 \leqslant 1} |xy|_1,$$

and the algebra $\mathbf{U}$ is complete with respect to both norms. Moreover

$$|e|_2 = \|\mathbf{A}_e\| = \|\mathbf{E}\| = 1,$$

$$|x|_2 = \sup_{|y|_1 \leqslant 1} |xy|_1 \geqslant \left| x \frac{e}{|e|_1} \right|_1 = \frac{|x|_1}{|e|_1},$$

and hence

$$|x|_2 \geqslant c_1 |x|_1 \qquad (c_1 = 1/|e|_1).$$

It follows from Corollary 1.72d that the norms $|x|_1$ and $|x|_2$ are equivalent. Furthermore

$$|xy|_2 = \|\mathbf{A}_{xy}\| \leqslant \|\mathbf{A}_x\| \, \|\mathbf{A}_y\| = |x|_2 |y|_2,$$

by a familiar property of continuous linear operators (Sec. 1.71h). ∎

1.9. Spectral Properties of Linear Operators

1.91. Every bounded linear operator **A** acting in a Banach space **X** belongs to the algebra **L**(**X**) of all bounded linear operators acting in **X**. As an element of this algebra, **A** has a spectrum $S_{\mathbf{A}}$ (Sec. 1.85b), where $S_{\mathbf{A}}$ is the set of all complex numbers λ for which $\mathbf{A} - \lambda\mathbf{E}$ fails to have a bounded inverse. In the case of a finite-dimensional space $\mathbf{X} = \mathbf{C}_n$, the spectrum of the operator **A** consists of a finite number, say m, of distinct points, namely the eigenvalues of **A** (Sec. 1.86a), and we can then represent $\mathbf{C}_n$ as a direct sum of m invariant subspaces, in each of which **A** has a spectrum consisting of a single point. In fact, in this case we can actually give a complete description of **A** (see Sec. 1.17f). On the other hand, if **X** is infinite-dimensional, the spectrum of **A** is a nonempty compact set in the complex λ-plane, contained in the disk $|\lambda| \leqslant \|\mathbf{A}\|$, or, more exactly, in the disk

$$|\lambda| \leqslant \overline{\lim} \sqrt[n]{\|\mathbf{A}^n\|},$$

and in general nothing more can be said.†

1.92.a. If **X** is infinite-dimensional, a number λ belonging to the spectrum of an operator **A** need not be an eigenvalue of **A**. The more natural concept in the infinite-dimensional case is not that of an eigenvalue, but rather that of a *generalized eigenvalue*, defined as any number λ for which there exists a sequence of vectors $x_1, x_2, \ldots$ such that $|x_n| \geqslant c > 0$ and $\mathbf{A}x_n - \lambda x_n \to 0$ as $n \to \infty$. Every ordinary eigenvalue of an operator is obviously a generalized eigenvalue.

b. THEOREM. *Every generalized eigenvalue of a bounded linear operator* **A** *belongs to the spectrum of* **A**.

† See Problem 11, where we construct an operator with any given compact set in the plane as its spectrum.

Proof. If $\mathbf{A}x_n - \lambda x_n \to 0$ for some sequence x_n and if the operator $\mathbf{A} - \lambda\mathbf{E}$ has a bounded inverse, then

$$x_n = (\mathbf{A} - \lambda\mathbf{E})^{-1}(\mathbf{A} - \lambda\mathbf{E})x_n \to 0. \quad \blacksquare$$

c. THEOREM. *Every boundary point of the spectrum of a bounded linear operator* $\mathbf{A}$ *is a generalized eigenvalue.*

Proof. Let λ be a boundary point of the spectrum of $\mathbf{A}$. Since $\mathbf{A} - \lambda E$ is a limit of invertible operators $\mathbf{A} - \mu E$, where $\mu \notin S_{\mathbf{A}}$, it follows from Sec. 1.84 that $\mathbf{A} - \lambda\mathbf{E}$ is a generalized divisor of zero, i.e., there exists a sequence of operators $\mathbf{B}_n$ such that $\|\mathbf{B}_n\| \geqslant c > 0$, while $(\mathbf{A} - \lambda\mathbf{E})\mathbf{B}_n \to 0$ in the algebra $\mathbf{L}(\mathbf{X})$. For every operator $\mathbf{B}_n$ we find a vector y_n such that $|y_n| \leqslant 1$, $|\mathbf{B}_n y_n| \geqslant c/2$. Writing $x_n = \mathbf{B}_n y_n$, we then have $|x_n| \geqslant c/2$, while

$$|(\mathbf{A} - \lambda\mathbf{E})x_n| = |(\mathbf{A} - \lambda\mathbf{E})\mathbf{B}_n y_n| \leqslant \|(\mathbf{A} - \lambda\mathbf{E})\mathbf{B}_n\| |y_n| \to 0. \quad \blacksquare$$

It should be noted, however, that interior points of the spectrum of the operator $\mathbf{A}$ may fail to be generalized eigenvalues (see Problem 10).

1.93. The study of linear operators can often be simplified with the help of the following

THEOREM. *Suppose the spectrum* $S = S_{\mathbf{A}}$ *of a bounded linear operator* $\mathbf{A}$ *acting in a Banach space* $\mathbf{X}$ *is the union of two nonintersecting closed sets* S_1 *and* S_2. *Then the space* $\mathbf{X}$ *can be represented as the direct sum of two closed subspaces* $\mathbf{X}_1$ *and* $\mathbf{X}_2$, *both invariant under the operator* $\mathbf{A}$, *such that the spectrum of* $\mathbf{A}$ *coincides with* S_1 *in the subspace* $\mathbf{X}_1$ *and with* S_2 *in the subspace* $\mathbf{X}_2$.

Proof. Let G be a disconnected open set whose boundary Γ consists of two nonintersecting piecewise smooth closed paths Γ_1 and Γ_2, where Γ_1 surrounds S_1 and Γ_2 surrounds S_2, while $G \cup \Gamma$ is contained in the domain of analyticity of $f(\lambda)$. Then, by an obvious modification of formula (9), p. 127, the formula

$$f(\mathbf{A}) = \frac{1}{2\pi i} \oint_{\Gamma} (\lambda\mathbf{E} - \mathbf{A})^{-1} f(\lambda)\, d\lambda,$$

where by the integral over $\Gamma = \Gamma_1 \cup \Gamma_2$ is meant the sum of the corresponding integrals over Γ_1 and Γ_2, establishes a morphism of the algebra $\mathscr{F}(S)$ of all functions analytic on S into the algebra $\mathbf{L}(\mathbf{X})$ of all bounded linear operators acting in $\mathbf{X}$. Let H_1 and H_2 be nonintersecting open sets, the first containing Γ_1 (and hence G_1 and S_1), the second containing Γ_2 (and hence G_2 and S_2). Then the functions

$$e_1(\lambda) = \begin{cases} 1 & \text{if } x \in H_1, \\ 0 & \text{if } x \in H_2, \end{cases} \qquad e_2(\lambda) = \begin{cases} 0 & \text{if } x \in H_1, \\ 1 & \text{if } x \in H_2 \end{cases}$$

belong to the algebra $\mathcal{F}(S)$, and moreover have the obvious properties

$$e_1(\lambda)+e_2(\lambda)\equiv 1 \text{ (on } H=H_1\cup H_2),$$
$$e_1^2(\lambda)=e_1(\lambda),\ e_2^2(\lambda)=e_2(\lambda),$$
$$e_1(\lambda)e_2(\lambda)=e_2(\lambda)e_1(\lambda)=0.$$

Let $\mathbf{E}_1$ and $\mathbf{E}_2$ be the linear operators corresponding to the functions $e_1(\lambda)$ and $e_2(\lambda)$. Then, by the properties of the morphism, we have

$$\mathbf{E}_1+\mathbf{E}_2=\mathbf{E}, \qquad \mathbf{E}_1^2=\mathbf{E}_1, \qquad \mathbf{E}_2^2=\mathbf{E}_2, \qquad \mathbf{E}_1\mathbf{E}_2=\mathbf{E}_2\mathbf{E}_1=\mathbf{0}.$$

Now let $\mathbf{X}_1$ be the set of solutions (in the space $\mathbf{X}$) of the equation $\mathbf{E}_1x=x$, and let $\mathbf{X}_2$ be the set of solutions of the equation $\mathbf{E}_2x=x$. In particular, every vector of the form $x=\mathbf{E}_1y$ with arbitrary $y\in\mathbf{X}$ is a solution of the equation $\mathbf{E}_1x=x$, since $\mathbf{E}_1(\mathbf{E}_1y)=\mathbf{E}_1^2y=\mathbf{E}_1y$. Obviously $\mathbf{X}_1$ and $\mathbf{X}_2$ are subspaces of the space $\mathbf{X}$, which are closed because of the continuity of the operators $\mathbf{E}_1$ and $\mathbf{E}_2$. If $z\in\mathbf{X}_1\cap\mathbf{X}_2$, then $z=\mathbf{E}_1z=\mathbf{E}_2z$. But then $\mathbf{E}_1z=\mathbf{E}_1(\mathbf{E}_1z)=\mathbf{E}_1(\mathbf{E}_2z)=0$, and hence $z=\mathbf{E}_1z=0$. Thus the intersection of the spaces $\mathbf{X}_1$ and $\mathbf{X}_2$ contains only the zero vector. Applying the expansion $\mathbf{E}=\mathbf{E}_1+\mathbf{E}_2$ to any vector $y\in\mathbf{X}$, we get $y=\mathbf{E}_1y+\mathbf{E}_2y$, where the first term belongs to $\mathbf{X}_1$ and the second to $\mathbf{X}_2$. It follows that the space $\mathbf{X}$ is the direct sum of two subspaces $\mathbf{X}_1$ and $\mathbf{X}_2$. Moreover, if $x\in\mathbf{X}_1$, then $\mathbf{A}x=\mathbf{A}(\mathbf{E}_1x)=\mathbf{E}_1(\mathbf{A}x)$, and hence $\mathbf{A}x$ also belongs to the subspace $\mathbf{A}_1$, i.e., $\mathbf{A}_1$ is invariant under the operator $\mathbf{A}$. Similarly, the subspace $\mathbf{X}_2$ is also invariant under the operator $\mathbf{A}$.

We must still prove the last assertion of the theorem. Let $\mathbf{A}_1=\mathbf{A}\mathbf{E}_1$, so that $\mathbf{A}_1$ coincides with $\mathbf{A}$ on the subspace $\mathbf{X}_1$ and equals $\mathbf{0}$ on $\mathbf{X}_2$. The operator $\mathbf{A}$ can be written in the form

$$\mathbf{A}_1=\frac{1}{2\pi i}\oint_\Gamma \lambda e_1(\lambda)(\lambda\mathbf{E}-\mathbf{A})^{-1}\,d\lambda,$$

where Γ can be replaced by Γ_1 since the function $e_1(\lambda)$ vanishes on Γ_2. This, together with the fact that $e_1(\lambda)\equiv 1$ on Γ_1, gives

$$\mathbf{A}_1=\frac{1}{2\pi i}\oint_{\Gamma_1} \lambda(\lambda\mathbf{E}-\mathbf{A})^{-1}\,d\lambda.$$

Furthermore

$$\mathbf{A}_1-\mu\mathbf{E}_1=\frac{1}{2\pi i}\oint_{\Gamma_1} (\lambda-\mu)(\lambda\mathbf{E}-\mathbf{A})^{-1}\,d\lambda$$

for arbitrary complex μ. Assuming that μ lies outside the contour Γ_1, we now construct the operator

$$\mathbf{B}_\mu=\frac{1}{2\pi i}\oint_{\Gamma_1} \frac{(\lambda\mathbf{E}-\mathbf{A})^{-1}}{\lambda-\mu}\,d\lambda.$$

Since (9′) is a morphism, the last two formulas imply

$$(\mathbf{A}_1-\mu\mathbf{E}_1)\mathbf{B}_\mu=\frac{1}{2\pi i}\oint_{\Gamma_1}(\lambda-\mu)\frac{(\lambda\mathbf{E}-\mathbf{A})^{-1}}{\lambda-\mu}\,d\lambda=\frac{1}{2\pi i}\oint_{\Gamma_1}(\lambda\mathbf{E}-\mathbf{A})^{-1}\,d\lambda=\mathbf{E}_1.$$

It follows that the operator $\mathbf{A}-\mu\mathbf{E}$ is invertible in the space $\mathbf{X}_1$, and hence that the spectrum of $\mathbf{A}$ in the subspace $\mathbf{X}_1$ can consist only of points of the set S_1. Similarly, the spectrum of the operator $\mathbf{A}$ in the subspace $\mathbf{X}_2$ can consist only of points of the set S_2.

Finally we show that the spectrum of the operator $\mathbf{A}$ in the subspace $\mathbf{X}_1$ (or $\mathbf{X}_2$) contains *all* points of the set S_1 (or S_2). Suppose $\lambda_0 \in S_1$. Then the operator $\mathbf{A}-\lambda_0\mathbf{E}$ is invertible in the space $\mathbf{X}_2$, as just shown, and hence there is a bounded linear operator $\mathbf{B}_2$ such that $(\mathbf{A}-\lambda_0\mathbf{E})\mathbf{B}_2x=x$ for all $x\in\mathbf{X}_2$. If the operator $\mathbf{A}-\lambda_0\mathbf{E}$ were also invertible in the subspace $\mathbf{X}_1$, we could find another bounded linear operator $\mathbf{B}_1$ such that $(\mathbf{A}-\lambda_0\mathbf{E})\mathbf{B}_1x=x$ for all $x\in\mathbf{X}_1$. Then $\mathbf{A}-\lambda_0\mathbf{E}$ would have as its inverse the linear operator $\mathbf{B}$ defined on the whole space $\mathbf{X}$ by the formula

$$\mathbf{B}x=\mathbf{B}(x_1+x_2)=\mathbf{B}_1x_1+\mathbf{B}_2x_2 \qquad (x_1\in\mathbf{X}_1,\ x_2\in\mathbf{X}_2),$$

where $\mathbf{B}$ is itself bounded, by Theorem 1.72f. But this is impossible, since $\lambda_0\in S_{\mathbf{A}}$. Therefore $\mathbf{A}-\lambda_0\mathbf{E}$ is not invertible in $\mathbf{X}_1$, i.e., λ_0 belongs to the spectrum of $\mathbf{A}$ in the subspace $\mathbf{X}_1$. ∎

1.94. Completely continuous operators

a. Let $\mathbf{L}(\mathbf{X})$ be the space of all bounded linear operators acting in a normed linear space $\mathbf{X}$. Then an operator $\mathbf{A}\in\mathbf{L}(\mathbf{X})$ is said to be *completely continuous* if it maps every bounded set $Q\subset\mathbf{X}$ into a precompact set (Sec. 1.24a).

b. Completely continuous operators have properties closely resembling those of operators acting in a finite-dimensional space. In fact, *every* linear operator acting in a finite-dimensional space is completely continuous (why?). Examples of completely continuous operators acting in the infinite-dimensional space $C^s(a,b)$ will be given in Secs. 1.98 and 1.99.

c. The identity operator $\mathbf{E}$ in an infinite-dimensional space is not completely continuous, since it maps the unit ball into itself, and hence into a set which is not precompact (see Theorem 1.36b).

1.95. Operations on completely continuous operators

a. THEOREM. *The sum $\mathbf{A}_1+\mathbf{A}_2$ of two completely continuous operators $\mathbf{A}_1$ and $\mathbf{A}_2$ is a completely continuous operator.*

Proof. Given any bounded set $Q\subset\mathbf{X}$, let x_n be any sequence of points in Q.

Since $\mathbf{A}_1$ is completely continuous, x_n contains a subsequence x_n' such that the sequence $\mathbf{A}_1 x_n'$ is fundamental. Moreover, since $\mathbf{A}_2$ is completely continuous, x_n' in turn contains a "finer" subsequence x_n'' such that the sequence $\mathbf{A}_2 x_n''$ is fundamental. But then the sequence $(\mathbf{A}_1 + \mathbf{A}_2)x_n''$ is obviously also fundamental. ∎

b. THEOREM. *The product (in either order) of a completely continuous operator* **A** *and a bounded operator* **B** *is a completely continuous operator.*

Proof. Given any bounded set $Q \subset \mathbf{X}$, the set $\mathbf{B}Q$ is bounded, and hence $\mathbf{AB}Q$ is precompact. Therefore the operator **AB** is completely continuous. On the other hand, **B** maps any fundamental sequence into a fundamental sequence (why?), and hence maps the precompact set $\mathbf{A}Q$ into a precompact set. Therefore the operator **BA** is also completely continuous. ∎

c. COROLLARY. *If a completely continuous operator* **A** *acting in a space* **X** *is invertible, then* **X** *is finite-dimensional.*

Proof. The operator $\mathbf{E} = \mathbf{A}\mathbf{A}^{-1}$ is completely continuous, by the preceding theorem, and hence **X** must be finite-dimensional, by Sec. 1.94c. ∎

d. THEOREM. *Let* $\mathbf{A}_n$ *be a sequence of completely continuous operators which converges in norm to an operator* **A** *(so that* $\|\mathbf{A} - \mathbf{A}_n\| \to 0$ *as* $n \to \infty$*). Then* **A** *is also a completely continuous operator.*

Proof. Given any $\varepsilon > 0$, let n be such that $\|\mathbf{A} - \mathbf{A}_n\| < \varepsilon$. Let $Q \subset \mathbf{X}$ be any bounded set, so that Q is contained in some ball $|x| \leqslant c$. Then the set $\mathbf{A}_n Q$ is an εc-net for the set $\mathbf{A}Q$ (Vol. 1, Sec. 3.93c). But $\mathbf{A}_n Q$ is precompact, and hence so is $\mathbf{A}Q$ (Vol. 1, Sec. 3.95). Therefore **A** is completely continuous. ∎

1.96. The spectrum of a completely continuous operator. In each of the following three lemmas, **A** denotes a completely continuous operator acting in a Banach space **X**:

a. LEMMA. *Every nonzero generalized eigenvalue of* **A** *is also an ordinary eigenvalue.*

Proof. Let λ be a generalized eigenvalue of **A**, so that there exists a (bounded) sequence of vectors $x_1, x_2, \ldots$ such that $|x_n| \geqslant c > 0$ and $(\mathbf{A} - \lambda\mathbf{E})x_n = q_n \to 0$ as $n \to \infty$. By the precompactness of the set of points $\mathbf{A}x_n$ $(n = 1, 2, \ldots)$, there exists a subsequence $x_{n_1}, x_{n_2}, \ldots$ such that the sequence $\mathbf{A}x_{n_1}, \mathbf{A}x_{n_2}, \ldots$ has a limit

$$z = \lim_{k \to \infty} \mathbf{A}x_{n_k}$$

in the space **X**. But then $\lambda x_{n_k} = \mathbf{A}x_{n_k} - q_{n_k}$ also has the limit z, and in particular

$$|z| = \lim_{k \to \infty} |\lambda x_{n_k}| \geqslant \lambda c > 0.$$

Moreover, since $\lambda \neq 0$,

$$\lim_{k\to\infty} x_{n_k} = \frac{z}{\lambda}, \qquad z = \lim_{k\to\infty} \mathbf{A}x_{n_k} = \frac{\mathbf{A}z}{\lambda},$$

and hence $\mathbf{A}z = \lambda z$. ∎

b. LEMMA. *The operator* **A** *has only a finite number of distinct eigenvalues outside any given disk* $|\lambda| \leqslant c$ $(c > 0)$.

Proof. Suppose **A** has a countable number of distinct eigenvalues $\lambda_1, \ldots, \lambda_n, \ldots$ such that

$$|\lambda_n| \geqslant c \qquad (n = 1, 2, \ldots),$$

and let $e_1, e_2, \ldots$ be the corresponding eigenvectors. Since eigenvectors corresponding to distinct eigenvalues are linearly independent (Theorem 1.15j), the linear manifold L_{n-1} spanned by the vectors $e_1, \ldots, e_{n-1}$ is a proper subspace of the linear manifold L_n spanned by the vectors $e_1, \ldots, e_n$. Hence, by Lemma 1.36a, there exists a vector $h_n \in L_n$ such that $|h_n| = 1$ and $|h_n - x| > \frac{1}{2}$ for all $x \in L_{n-1}$. Writing $h_n = x_0 + \alpha e_n$ where $x_0 \in L_{n-1}$, we then have

$$\mathbf{A}h_n = \mathbf{A}(x_0 + \alpha e_n) = \mathbf{A}x_0 + \alpha\lambda_n e_n = \mathbf{A}x_0 + \lambda_n(h_n - x_0) = (\mathbf{A}x_0 - \lambda_n x_0) + \lambda_n h_n,$$

and hence, if $m \leqslant n - 1$,

$$\begin{aligned} |\mathbf{A}h_n - \mathbf{A}h_m| &= |(\mathbf{A}x_0 - \lambda_n x_0 - \mathbf{A}h_m) + \lambda_n h_n| \\ &= |\lambda_n| \left| h_n - \frac{1}{|\lambda_n|}(\mathbf{A}h_m + \lambda_n x_0 - \mathbf{A}x_0) \right| > \frac{1}{2}|\lambda_n| \geqslant \frac{1}{2}c, \end{aligned}$$

since $\mathbf{A}h_m + \lambda_n x_0 - \mathbf{A}x_0 = \lambda_m h_m \in L_m \subset L_{n-1}$. But then the sequence $\mathbf{A}h_n$ cannot contain a convergent subsequence, contrary to the assumption that **A** is completely continuous. ∎

c. LEMMA. *The spectrum of* **A** *has only a finite number of points outside any given disk* $|\lambda| \leqslant c$ $(c > 0)$, *and these points are all eigenvalues of* **A**.

Proof. Every boundary point of the spectrum of the operator **A** is a generalized eigenvalue of **A** (Theorem 1.92c), and hence, by Lemma 1.96a, is an ordinary eigenvalue of **A**. Therefore it follows from Lemma 1.96b that the spectrum of **A** can have only a finite number of boundary points $\lambda_1, \ldots, \lambda_n$ (say) outside the disk $|\lambda| \leqslant c$, all eigenvalues of **A**. These points make up the entire spectrum of **A** outside the disk $|\lambda| \leqslant c$. In fact, suppose there were another point λ_0 in the spectrum with $|\lambda_0| \leqslant c$. Then we could draw a straight line from λ_0 to infinity without passing through either the disk $|\lambda| \leqslant c$ or any of the points $\lambda_1, \ldots, \lambda_n$. The last point on this line would then be a boundary point of the spectrum distinct from $\lambda_1, \ldots, \lambda_n$, which is impossible. ∎

d. THEOREM. *Let* **A** *be a completely continuous operator acting in a Banach space* **X**. *Then the spectrum of* **A** *consists of no more than a countable number of isolated eigenvalues with the single limit point* 0.

Proof. According to Lemma 1.96c, the spectrum of **A** has only a finite number of points outside any disk $|\lambda| \leqslant c$, and hence we can count all the points of the spectrum "in the order of their approach to 0." ∎

e. If **X** is infinite-dimensional, then, by Corollary 1.95c, the point 0 is always a point of the spectrum of **A** (but not necessarily an eigenvalue); the rest of the spectrum may then consist of a countable set of points, a finite set of points, or even no (nonzero) points at all. In the latter case, it can be shown that

$$\overline{\lim_{n\to\infty}} \sqrt[n]{\|\mathbf{A}^n\|} = 0$$

(cf. Sec. 1.85a), and the operator **A** might be called a "generalized nilpotent operator." It will be recalled that in the finite-dimensional case, a nilpotent operator is defined as an operator such that $\mathbf{A}^m = 0$ for some positive integer m (Lin. Alg., Sec. 6.11). The structure of a nilpotent operator in a finite-dimensional space can be described in complete detail; in fact, it is given by a Jordan matrix all of whose diagonal elements vanish (Lin Alg., Sec. 6.14). In the infinite-dimensional case, the structure of a generalized nilpotent operator is still not completely known.†

1.97.a. Let $\lambda \neq 0$ be a point of the spectrum $S_{\mathbf{A}}$ of a completely continuous operator **A** acting in a Banach space **X**. According to Lemma 1.96c, λ is isolated (Sec. 1.21a), and hence we can apply Theorem 1.93. As a result, we get the following "spectral decomposition" of the operator **A**: The space **X** can be represented as the direct sum of two closed subspaces $\mathbf{P}_\lambda$ and $\mathbf{Q}_\lambda$, both invariant under the operator **A**, such that the spectrum of **A** in $\mathbf{P}_\lambda$ consists of the point λ alone, while the spectrum of **A** in $\mathbf{Q}_\lambda$ is just the set $S_{\mathbf{A}}$ minus the single point λ. Naturally, the operator **A** remains completely continuous in each of the subspaces $\mathbf{P}_\lambda$ and $\mathbf{Q}_\lambda$. Moreover **A** is invertible in $\mathbf{P}_\lambda$, since the spectrum of **A** in $\mathbf{P}_\lambda$ does not contain the point 0. But then the subspace $\mathbf{P}_\lambda$ is finite-dimensional, by Corollary 1.95c. It follows that *every point* $\lambda \neq 0$ *of the spectrum of* **A** *corresponds to a finite-dimensional invariant subspace,* in which the structure of **A** has the familiar description given in Sec. 1.17f.

b. THEOREM (**Fredholm alternative**). *Let* **A** *be a completely continuous operator acting in a Banach space* **X**, *and let* μ *be any complex number. Then there are just*

† See M. S. Brodskiĭ, *Triangular and Jordan Representations of Linear Operators*, American Mathematical Society, Providence, R.I. (1971).

two possibilities: Either the equation

$$(\mathbf{E}-\mu\mathbf{A})x=y$$

with y any element of $\mathbf{X}$ *is uniquely solvable for x, or the homogeneous equation*

$$(\mathbf{E}-\mu\mathbf{A})x=0$$

has a nonzero solution.

Proof. If $\mu=0$, then obviously the first possibility holds (choose $x=y$). If $\mu\neq 0$, let $\lambda=1/\mu$. Then the equation $(\mathbf{E}-\mu\mathbf{A})x=y$ is equivalent to the equation $(\mathbf{A}-\lambda\mathbf{E})x=-\lambda y$. If λ does not belong to the spectrum of $\mathbf{A}$, then $\mathbf{A}-\lambda\mathbf{E}$ is invertible, and the first possibility holds (choose $x=-(\mathbf{A}-\lambda\mathbf{E})^{-1}\lambda y$). If λ belongs to the spectrum of $\mathbf{A}$, then, since $\lambda\neq 0$, λ is an eigenvalue of $\mathbf{A}$ (by Lemma 1.96c), and hence the second possibility holds. ∎

c. The only property of completely continuous operators used in the proof of the Fredholm alternative is that every nonzero number λ belonging to the spectrum of $\mathbf{A}$ is an eigenvalue of $\mathbf{A}$. However, there exists a much larger class of operators with the same property, for example, the class of all operators $\mathbf{A}$ such that $\mathbf{A}^m$ is completely continuous for some positive integer m (see Problem 13).

1.98. The Fredholm operator

a. As usual, let $C^s(a,b)$ be the Banach space of all continuous complex functions $x=x(t)$ defined on the interval $[a,b]$, with norm

$$\|x\|=\sup_t |x(t)|.$$

Let $K(t,s)$ be a continuous complex function of two real variables t and s, both varying in the interval $[a,b]$. Then the formula

$$y(t)=\mathbf{A}_F x(t)=\int_a^b K(t,s)x(s)\,ds \tag{1}$$

defines a linear operator $\mathbf{A}_F$, called the *Fredholm operator*, carrying every function $x(t)\in C^s(a,b)$ into another function $y(t)\in C^s(a,b)$.† It follows from the obvious inequality

$$|y(t)|\leqslant\sup_s |x(s)|\int_a^b |K(t,s)|\,ds$$

that $\mathbf{A}_F$ is a bounded operator, with norm no greater than

$$\sup_t\int_a^b |K(t,s)|\,ds.$$

† The continuity of $y(t)$ follows from the continuity of $x(t)$ and $K(t,s)$ (cf. Vol. 1, Sec. 9.111).

b. THEOREM. *The Fredholm operator* $\mathbf{A}_F$ *is completely continuous.*

Proof. Suppose $x(t)$ belongs to a bounded set $Q \subset C^s(a,b)$, so that $\|x\| \leqslant M$, say, for all $x(t) \in Q$. If $|t'-t''| \leqslant \delta$, then (1) implies

$$|y(t')-y(t'')| \leqslant \sup|x(t)| \int_a^b |K(t',s)-K(t'',s)|\, ds \leqslant \sup|x(t)|\omega_K(\delta)(b-a),$$

where

$$\omega_K(\delta) = \sup_{\substack{|t'-t''|\leqslant\delta \\ a\leqslant s,t\leqslant b}} |K(t',s)-K(t'',s)|.$$

Hence for the modulus of continuity $\omega_y(\delta)$ of the function $y(t)$ we have the estimate

$$\omega_y(\delta) = \sup_{|t'-t''|\leqslant\delta} |y(t')-y(t'')| \leqslant M\omega_K(\delta)(b-a),$$

independent of the choice of the function $x(t) \in Q$. But

$$\lim_{\delta\to 0} \omega_K(\delta) = 0,$$

by the (uniform) continuity of $K(t,s)$ (cf. Vol. 1, Sec. 5.17c), and hence also

$$\lim_{\delta\to 0} \omega_y(\delta) = 0,$$

so that the set $\mathbf{A}_F Q \subset C^s(a,b)$ is equicontinuous. Moreover, $\mathbf{A}_F Q$ is obviously uniformly bounded. Therefore, by Arzelà's theorem (Sec. 1.24b), the set $\mathbf{A}_F Q$ is precompact for every bounded set $Q \subset C^s(a,b)$, i.e., the operator $\mathbf{A}_F$ is completely continuous. ∎

c. Thus $\mathbf{A}_F$ has all the properties of completely continuous operators given above. In particular, $\mathbf{A}_F$ satisfies the Fredholm alternative, which now goes as follows: *Either the equation*

$$x(t) - \mu \int_a^b K(t,s)x(s)\, ds = y(t)$$

with $y(t)$ *any element of* $C^s(a,b)$ *is uniquely solvable for* $x(t)$, *or the homogeneous equation*

$$x(t) - \mu \int_a^b K(t,s)x(s)\, ds = 0$$

has a nonzero solution.

1.99. The Volterra operator

a. Again let $K(t,s)$ be a continuous complex function of two real variables s

and t, both varying in the interval $[a,b]$, and consider the formula

$$y(t)=\mathbf{A}_V x(t)=\int_a^t K(t,s)x(s)\,ds, \tag{2}$$

differing from (1) only by having a variable upper limit of integration t rather than a fixed upper limit b. Then (2) defines a linear operator $\mathbf{A}_V$, called the *Volterra operator*, carrying every function $x(t)\in C^s(a,b)$ into another function $y(t)\in C^s(a,b)$. A trivial modification of the proof of Theorem 1.98b shows that the Volterra operator $\mathbf{A}_V$ is completely continuous, just like the Fredholm operator $\mathbf{A}_F$. However, there is a basic difference between the spectrum of $\mathbf{A}_V$ and that of $\mathbf{A}_F$, as shown by the following

THEOREM. *The spectrum of the Volterra operator $\mathbf{A}_V$ consists of the single point* 0.

Proof. The spectrum of $\mathbf{A}_V$ contains the point 0, since the space $C^s(a,b)$ is infinite-dimensional (see Sec. 1.96e). Suppose the spectrum of $\mathbf{A}_V$ contains a point $\lambda\neq 0$. Then λ is an eigenvalue of $\mathbf{A}_V$, i.e., there is a function $x_0(t)\in C^s(a,b)$ where $x_0(t)\not\equiv 0$ such that

$$\mathbf{A}_V x_0(t)=\int_a^t K(t,s)x_0(s)\,ds=\lambda x_0(t). \tag{3}$$

Setting $t=a$, we get $\lambda x_0(a)=0$ and hence $x_0(a)=0$. Without loss of generality, we can assume that $x_0(t)$ does not vanish identically in any interval of the form $a\leqslant t\leqslant a+\delta$ $(\delta>0)$, since otherwise we need only shift the lower limit of integration to the nearest point for which the assumption is true. Hence the function

$$m(\delta)=\sup_{a\leqslant t\leqslant a+\delta}|x(t)| \qquad (\delta>0)$$

is nonzero and approaches 0 as $\delta\to 0$. Moreover, given any $\delta>0$, we can find a point $t_\delta\in[a,a+\delta]$ such that $|x_0(t_\delta)|=m(\delta)$. It then follows from (3) that

$$|\lambda x_0(t_\delta)|=|\lambda|m(\delta)\leqslant \max_{a\leqslant s\leqslant a+\delta}|x_0(s)|\int_a^{t_\delta}|K(t,s)|\,ds\leqslant M\delta m(\delta),$$

where

$$M=\sup_{t,s}|K(t,s)|.$$

Dividing by $m(\delta)$, we get

$$|\lambda|\leqslant M\delta$$

for every $\delta>0$, contrary to the assumption that $\lambda\neq 0$. ∎

b. COROLLARY. *Given any* $y(t) \in C^s(a,b)$, *the Volterra integral equation*

$$(\mathbf{E} - \mu \mathbf{A}_V)x(t) = x(t) - \mu \int_a^t K(t,s)x(s)\, ds = y(t)$$

has a unique solution for every $y(t) \in C^s(a,b)$, *which can be written in the form of the series*

$$x(t) = (\mathbf{E} - \mu \mathbf{A}_V)^{-1} y(t) = y(t) + \mu \mathbf{A}_V y(t) + \mu^2 \mathbf{A}_V^2 y(t) + \cdots + \mu^n \mathbf{A}_V^n y(t) + \cdots.$$

Proof. Recall the proof of Fredholm's alternative and Sec. 1.85a. ▮

c. It can be shown that the operator $\mathbf{A}_V^n$ $(n=1,2,\ldots)$ is also a Volterra operator, whose "kernel" $K_n(t,s)$ can be calculated recursively from the formula

$$K_n(t,s) = \int_s^t K_{n-1}(t,\sigma) K_1(\sigma,s)\, d\sigma,$$

where $K_1(t,s) = K(t,s)$.†

Problems

1. Consider the following three spaces of functions on the real line:

(a) The set of all bounded continuous functions;
(b) The set of all continuous functions $f(x)$ such that $f(x) \to 0$ as $|x| \to \infty$;
(c) The set of all continuous functions, each of which vanishes outside some closed interval.

Suppose each of these spaces is equipped with the metric

$$\rho(f,g) = \sup |f(x) - g(x)|.$$

Which of the spaces are complete?

2. In the space $R^s(0,\infty)$ of all bounded continuous functions on the half-line $0 \leqslant t < \infty$, with the norm

$$\|x(t)\| = \sup_t |x(t)|,$$

find a set of functions $\{x_\alpha(t): \alpha \in A\}$ with the power of the continuum such that $\|x_\alpha(t)\| = 1$ for every $\alpha \in A$, while $\|x_\alpha(t) - x_\beta(t)\| \geqslant 1$ if $\alpha \neq \beta$.

Comment. It follows that $R^s(0,\infty)$ is not separable.

† See e.g., W. V. Lovitt, *Linear Integral Equations*, Dover Publications, Inc., New York (1950), Sec. 9.

3. Verify that the functional

$$F(y)=\int_0^{1/2} y(x)\,dx-\int_{1/2}^{0} y(x)\,dx$$

is continuous on the space $R^s(0,1)$. Show that the least upper bound of the values of $F(y)$ on the closed unit ball in $R^s(0,1)$ equals 1, but that this least upper bound is not achieved for any element of the ball.

4. Suppose it is known that the parallelogram equality, p. 58, holds for every pair of elements x, y in a normed linear space $\mathbf{X}$. Prove that the norm in $\mathbf{X}$ is then generated by the scalar product

$$(x,y)=\tfrac{1}{4}(|x+y|^2-|x-y|^2).$$

5. Let $B(Q)$ be the algebra of all polynomials $p(z)$ with complex coefficients defined on the disk $Q=\{z:|z|\leqslant 1\}$, with norm $\|p(z)\|=\max|p(z)|$. Then $B(Q)$ contains 1 and separates any two points of the compactum Q. Prove that $B(Q)$ does not satisfy Stone's theorem for a complex algebra (Theorem 1.53b), i.e., $B(Q)$ is not dense in the space $C^s(Q)$ of all continuous complex functions on Q.

6. If Q is a metric space, the set $J(F)$ of all functions $f(x)\in R^s(Q)$ vanishing on some closed set $F\subset Q$ forms a closed ideal in the normed algebra $R^s(Q)$ of all continuous real functions on Q. Prove that if Q is a compactum, then every closed ideal $J\subset R^s(Q)$ coincides with the ideal $J(F)$ for suitable $F\subset Q$.

7. Given a metric space P and a compactum Q, let $E\subset P^s(Q)$ be an equicontinuous set of functions such that the values of all functions $x(t)\in E$ for every fixed $t\in Q$ belong to some precompactum $P_t\subset P$. Prove that there exists a precompactum $P_0\subset P$ such that P_0 contains all values of all functions $x(t)\in E$ at all points $t\in Q$.

8. Let $T=\|t_{kn}\|$ be a matrix satisfying the conditions of Toeplitz's theorem (Sec. 1.76c). Construct a sequence ξ_n, made up of the numbers 1 and -1, for which the T-limit does not exist (cf. Sec. 1.78b).

9 (*Robinson's theorem*). Prove that the interval $[\underline{\lim}\ T_n(x), \overline{\lim}\ T_n(x)]$ is contained in the interval $[\underline{\lim}\ \xi_n, \overline{\lim}\ \xi_n]$ for every bounded sequence $x=(\xi_1,\ldots,\xi_n,\ldots)$ if and only if

$$\lim_{n\to\infty}\|T_n\|=1$$

(cf. Sec. 1.78c).

10. Let $C=C^s(Q)$ be the algebra of all continuous complex functions $f(z)$ on the circle $|z|=1$ (with the usual norm), and let Z be the algebra of all functions $\varphi(z)$ analytic on the open disk $|z|<1$ and continuous on the closed disk $|z|\leqslant 1$, with the same norm $\|\varphi\|=\sup|\varphi(z)|$. Prove that

(a) The mapping carrying each function $\varphi(z)\in Z$ into the boundary function $\varphi(e^{it})\in C$ is a monomorphism of Z into C, so that the algebra Z can be regarded as a subalgebra of the algebra C;
(b) Z is a closed subalgebra of C;
(c) If $\mathbf{A}$ is the operator of multiplication by z in the space C, then the spectrum of $\mathbf{A}$ in the space C is the circle $|z|=1$, while the spectrum of $\mathbf{A}$ in the space Z is the disk $|z|\leqslant 1$, with the values $|z|=1$ (and only these values) being generalized eigenvalues of $\mathbf{A}$ in Z;
(d) The element z is invertible in the algebra C, but is neither invertible in Z nor a generalized divisor of zero in Z.

11. Let Q be a compact set in the plane, and let $C=C^s(Q)$ be the space of all continuous complex functions on Q. Prove that the spectrum of the operator of multiplication by z is precisely the set Q.

12. Given an operator $\mathbf{A}$ acting in a Banach space $\mathbf{X}$, suppose it is known that the operator $p(\mathbf{A})$ is completely continuous for some polynomial $p(\lambda)$. Prove that all the points of the spectrum of $\mathbf{A}$ (with the possible exception of the zeros of $p(\lambda)$) are eigenvalues of $\mathbf{A}$.

13. Prove that the Fredholm alternative (Theorem 1.97b) holds for any operator $\mathbf{A}$ such that $\mathbf{A}^m$ is completely continuous for some integer $m>0$.

14 (*Hölder's inequality for functions*). Let $x(t)$ and $y(t)$ be arbitrary (piecewise) continuous real or complex functions on the interval $[a,b]$. Prove that

$$\left|\int_a^b x(t)y(t)\,dt\right|\leqslant\sqrt[p]{\int_a^b|x(t)|^p\,dt}\sqrt[q]{\int_a^b|y(t)|^q\,dt}$$

for arbitrary real numbers p and q such that

$$\frac{1}{p}+\frac{1}{q}=1\qquad(p>1,\ q>1).\tag{1}$$

15 (*Minkowski's inequality for functions*).† Let $x(t)$ and $y(t)$ be the same as in the preceding problem. Prove that

$$\sqrt[p]{\int_a^b|x(t)+y(t)|^p\,dt}\leqslant\sqrt[p]{\int_a^b|x(t)|^p\,dt}+\sqrt[p]{\int_a^b|y(t)|^p\,dt}$$

for every $p\geqslant 1$.

† See Secs. 1.26a, 133b, and 1.39c.

16 (*Hölder's inequality for vectors*). Let $x=(\xi_1,\ldots,\xi_n)$ and $y=(\eta_1,\ldots,\eta_n)$ be arbitrary vectors in R_n or C_n. Prove that

$$\left|\sum_{k=1}^{n}\xi_k\eta_k\right| \leqslant \sqrt[p]{\sum_{k=1}^{n}|\xi_k|^p}\sqrt[q]{\sum_{k=1}^{n}|\eta_k|^q}$$

for arbitrary real numbers p and q satisfying (1).

17 (*Minkowski's inequality for vectors*). Let x and y be the same as in the preceding problem. Prove that

$$\sqrt[p]{\sum_{k=1}^{n}|\xi_k+\eta_k|^p} \leqslant \sqrt[p]{\sum_{k=1}^{n}|\xi_k|^p}+\sqrt[p]{\sum_{k=1}^{n}|\eta_k|^p}$$

for every $p\geqslant 1$.

18. Prove that the space l_p (Secs. 1.33d, 1.39e) is complete for every $p\geqslant 1$.

19. Prove that the norm $|x|_p$ defined in Sec. 1.33c does not satisfy the triangle inequality if $p<1$.

20. Let

$$x_1(t),\ x_2(t),\ldots \tag{2}$$

be a sequence of functions defined on the interval $[a,b]$, with derivatives of all orders on $[a,b]$. Suppose there exists a sequence of positive constants A_k $(k=0,1,2,\ldots)$ such that

$$|x_n^{(k)}(t)|\leqslant A_k \qquad (n=1,2,\ldots;\ k=0,1,2,\ldots).$$

Prove that (2) contains a subsequence

$$x_{n_1}(t),\ x_{n_2}(t),\ldots \tag{3}$$

such that (3) and all the sequences of derivatives

$$x_{n_1}^{(k)}(t),\ x_{n_2}^{(k)}(t),\ldots \qquad (k=0,1,2,\ldots)$$

are uniformly convergent on $[a,b]$.

21. Prove the following variant of Arzelà's theorem (Sec. 1.24b), that does not require continuity of the functions $x(t)$ and compactness (or even metrizability) of the set Q on which the functions $x(t)$ are defined: Let $P(Q)$ be a set of bounded functions $x(t)$ defined on an arbitrary set Q and taking values in a metric space P, equipped with the metric $\rho(x,y)=\sup\rho\{x(t),y(t)\}$. Then a set $E\subset P(Q)$ is precompact if and only if, given any $\varepsilon>0$, there exists a decomposition of Q into a finite number of subsets $Q_1,\ldots,Q_n$ such that no function $x(t)\in E$ changes by more than ε on any of the subsets.

2 Differential Equations

2.1. Definitions and Examples

2.11.a. An equation for an unknown function

$$u = u(t) \qquad (a \leqslant t \leqslant b),$$

in which $u(t)$ is differentiated one or more times, is called a *differential equation*. Depending on the conditions of the problem, the function $u(t)$ may be a numerical function, a vector function with values in an n-dimensional space, or even a vector function with values in a normed linear space.

Every function $u(t)$ satisfying a given differential equation is called a *solution* of the equation. The set of all solutions of a differential equation is called a *general* solution, and the individual solutions themselves are known as *particular* solutions.

b. Thus the simplest differential equation

$$u'(t) = 0 \qquad (a \leqslant t \leqslant b) \tag{1}$$

has the general solution $u(t) \equiv \text{const}$. This constant is a number, a fixed vector, or a fixed element of a normed linear space $\mathbf{X}$, depending on whether it is specified that $u(t)$ is a numerical function, a vector function, or a function taking values in the space $\mathbf{X}$ (see Theorem 1.61f). The general solution of the differential equation

$$u'(t) = g(t) \qquad (a \leqslant t \leqslant b), \tag{2}$$

where $g(t)$ is a given (numerical or vector) function, can be written as the integral

$$u(t) = \int_{t_0}^{t} g(\tau)\, d\tau + \text{const},$$

provided that $g(t)$ is piecewise continuous (see Theorem 1.63b).

As we see from these two examples, there is a certain arbitrariness in the solutions of differential equations, so that extra conditions are needed to single out a unique solution. Usually this extra condition consists in specifying the value of the unknown function $u(t)$ for some value $t = t_0 \in [a,b]$, i.e., in specifying an "initial condition" $u(t_0) = u_0$. Note that equations (1) and (2) both have a unique solution, once $u(t_0)$ is specified.

c. More generally, a differential equation may be of the form

$$u'(t) = \varphi(t, u(t)) \qquad (a \leqslant t \leqslant b), \tag{3}$$

where $\varphi(t,x)$ is a function of two variables, with values in the same normed linear space $\mathbf{X}$ as $u(t)$ itself (note that $x \in \mathbf{X}$ also). Here the problem is to prove the existence of a solution $u(t)$ and its uniqueness for given $u(t_0)$.

d. It is often useful to give the general equation (3) a certain "kinematic" interpretation in the space $\mathbf{X}$. Consider a moving point in the space $\mathbf{X}$ which occupies the "position" $u=u(t)$ at each instant of time t. As t varies, the point describes a curve

$$u=u(t) \qquad (a\leqslant t\leqslant b)$$

in the space $\mathbf{X}$, and the curve (or the moving point) is said to have the variable "position vector" $u(t)$ as its *law of motion*. We can also interpret $u'(t)$ as the velocity vector of the moving point (the limit of the ratio of the traversed path Δu to the elapsed time Δt). Given any fixed $t\in[a,b]$, the function $\varphi(t,x)$ associates a vector in $\mathbf{X}$ with each vector $x\in\mathbf{X}$. Hence the solution $u(t)$ of the differential equation (3) can be interpreted as the law of motion of a moving point in the space $\mathbf{X}$, whose velocity at the time t and at the corresponding position $u\in\mathbf{X}$ is just the value $\varphi(t,u)$. Thus, at every instant of time t, the function $\varphi(t,u)$ specifies a field of vectors in $\mathbf{X}$, each describing the velocity of a moving point as it goes through the corresponding point $u\in\mathbf{X}$. With this picture, the solutions of (3) are just the possible "trajectories" of the moving point, often called the "integral curves" of (3).

e. We can also give equation (3) a purely geometric interpretation in the product space $R_1\times\mathbf{X}$ (where R_1 is the real line). Every function $u=u(t)$ determines a curve in the space $R_1\times\mathbf{X}$ ($t\in R_1$, $u\in\mathbf{X}$). The differential $u'(t)dt$ is the principal linear part of the increment of the function $u(t)$ when the argument changes from t to $t+dt$. Therefore, in the real case ($\mathbf{X}=R_1$), the derivative $u'(t)$ can be interpreted as the slope of the tangent, i.e., as the "rise" of the curve $u=u(t)$ in going from t to $t+dt$ divided by dt, to within terms of order higher than 1 in dt. In the general case, where the space $\mathbf{X}$ is arbitrary, the derivative has a similar meaning: The line

$$u-u_0=u'(t_0)(t-t_0) \qquad (u_0=u(t_0))$$

is the tangent to the curve $u=u(t)$ at the point (t_0,u_0), and we can still call $u'(t_0)$ the "slope" of the curve. The function $\varphi(t,x)$ specifies a line

$$u-x_0=\varphi(t_0,x_0)(t-t_0)$$

at every point (t_0,x_0) of the space $R_1\times\mathbf{X}$, and equation (3) requires that for every $t_0\in[a,b]$ the curve $u=u(t)$ have a tangent coinciding with the line

$$u-u_0=\varphi(t_0,u_0)(t-t_0) \qquad (u_0=u(t_0)).$$

f. Example. Consider the differential equation

$$u'(t)=v(u)$$

in the space $\mathbf{X}=R_2$, where $v(u)$ denotes the vector obtained by rotating any given vector $u \in R_2$ through 90° in the positive direction. In the kinematic interpretation, the point in the plane R_2 must move in such a way that its velocity vector coincides with the vector $v(u)$ at every point u. It is obvious that the solution of the problem corresponds to motion along the circle centered at the origin, with a linear velocity of magnitude equal to the radius of the circle (see Figure 11).

In the geometric interpretation, we look for curves in the three-dimensional space $R_1 \times R_2$, whose tangents at every point (t_0,u_0) are given by the equation

$$u-u_0=v(u_0)(t-t_0).$$

The curves in question have the form of helices, winding around the t-axis (see Figure 12).

g. We will see later (Secs. 2.4–2.5) that both the system of equations

$$u_1'(t)=\varphi_1(t,u_1(t),\ldots,u_n(t)),$$

$$\ldots$$

$$u_n'(t)=\varphi_n(t,u_1(t),\ldots,u_n(t))$$

and the nth-order equation†

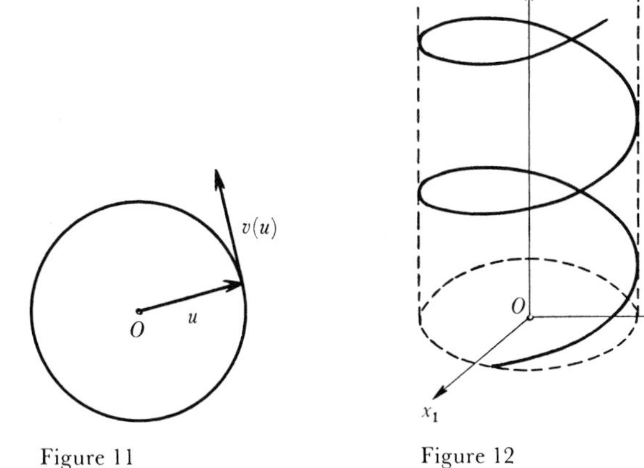

Figure 11

Figure 12

† By the *order* of a differential equation we mean the order of the highest derivative of the unknown function appearing in the equation.

$$u^{(n)} = \varphi(t,u(t),u'(t),\ldots, u^{(n-1)}(t))$$

can be reduced to an equation of the form (3).

h. The central theme of this chapter is the problem of the existence of solutions of differential equations and of their uniqueness under given supplementary conditions. In the next few pages, we will consider some particularly simple differential equations, whose solutions can be found in explicit form.

2.12. Consider the differential equation

$$u'(t) = A(t)u(t) \qquad (a \leqslant t \leqslant b), \tag{4}$$

called a *homogeneous linear equation*. We begin with the case where the unknown function $u(t)$ is a numerical function and the coefficient $A(t)$ is a given continuous numerical function. We will also assume that $u(t)$ satisfies the initial condition

$$u(t_0) = u_0. \tag{5}$$

The function $u(t) \equiv 0$ is an obvious solution of (4), but it does not satisfy the initial condition (5) if $u_0 \neq 0$. Thus we now look for other solutions of (4). If $u(t)$ is a solution which does not vanish identically, there is an interval on which $u(t) \neq 0$, say $u(t) > 0$. Then, dividing by $u(t)$, we get

$$\frac{u'(t)}{u(t)} = [\ln u(t)]' = A(t),$$

and hence

$$\ln u(t) = \int_{t_0}^{t} A(t)\, dt + C.$$

Setting $t = t_0$, we find that $C = \ln u_0$. Thus, finally,†

$$u(t) = \exp\left\{ \int_{t_0}^{t} A(\tau)\, d\tau \right\} u_0 \tag{6}$$

after getting rid of the logarithms. A direct check shows that (6) is actually the solution of (4), subject to the initial condition (5), even if $u(t)$ is no longer positive. Note that the solution (6) is defined for all $t \in [a,b]$, and is nonzero everywhere if $u_0 \neq 0$. If $A(t) \equiv A$ is a constant, then (6) takes the particularly simple form

$$u(t) = e^{(t-t_0)A} u_0. \tag{7}$$

† As usual, $\exp x \equiv e^x$.

2.13. Next consider the differential equation

$$u'(t)=\mathbf{A}(t)u(t) \qquad (a\leqslant t\leqslant b), \tag{8}$$

where this time $u(t)$ is a vector function taking values in some Banach space $\mathbf{X}$, and the coefficient $\mathbf{A}(t)$ is a bounded linear operator which depends continuously on t as a parameter† and maps the space $\mathbf{X}$ into itself for every $t\in[a,b]$. The considerations of Sec. 2.12 are no longer applicable, since division by $u(t)$ now makes no sense. However, it turns out that the final results (6) and (7), with A replaced by $\mathbf{A}$, continue to hold and can be given precise meaning.

Suppose first that the operator $\mathbf{A}(t)\equiv\mathbf{A}$ is independent of t (the general case will be considered in Sec. 2.18). Then we interpret $e^{(t-t_0)\mathbf{A}}$ as a function of the operator $\mathbf{A}$ in the sense of Sec. 1.86c:

$$e^{(t-t_0)\mathbf{A}}=\sum_{n=0}^{\infty}\frac{(t-t_0)^n}{n!}\mathbf{A}^n. \tag{9}$$

This function is defined for all real t and takes values in the space $\mathbf{L}(\mathbf{X})$ of all bounded linear operators mapping $\mathbf{X}$ into itself. Differentiating the series (9) term by term (Sec. 1.66c), we get

$$\frac{d}{dt}e^{(t-t_0)\mathbf{A}}=\sum_{n=0}^{\infty}\frac{n(t-t_0)^{n-1}}{n!}\mathbf{A}^n=\mathbf{A}e^{(t-t_0)\mathbf{A}}.$$

From this it is clear that

$$u(t)=e^{(t-t_0)\mathbf{A}}u_0 \tag{10}$$

is actually a solution of equation (8). Moreover, (10) obviously reduces to the vector $u(t_0)=u_0$ if $t=t_0$, and hence satisfies the initial condition (5).

To prove the uniqueness of the solution (10), we first prove the following

LEMMA. *If $\mathbf{A}(t)$ is a strongly differentiable operator function with derivative $\mathbf{A}'(t)$,‡ and if $x(t)$ is a differentiable vector function, then the vector function $y(t)=\mathbf{A}(t)x(t)$ is also differentiable, with derivative*

$$y'(t)=\mathbf{A}(t)x'(t)+\mathbf{A}'(t)x(t).$$

† For the precise definition of continuous dependence of $\mathbf{A}(t)$ on t, see Sec. 2.61.

‡ In keeping with Sec. 1.73e, we say that an operator function $\mathbf{A}(t)$ $(a\leqslant t\leqslant b)$ acting in the space $\mathbf{X}$ is *strongly differentiable*, with *derivative* $\mathbf{A}'(t)$ (itself an operator function acting in $\mathbf{X}$), if the limiting relation

$$\lim_{t\to t_0}\frac{\mathbf{A}(t)-\mathbf{A}(t_0)}{t-t_0}x=\mathbf{A}'(t_0)x$$

holds for every $x\in\mathbf{X}$ (and $t_0\in[a,b]$).

Proof. We have

$$\frac{\mathbf{A}(t)x(t)-\mathbf{A}(t_0)x(t_0)}{t-t_0}=\mathbf{A}(t)\frac{x(t)-x(t_0)}{t-t_0}+\frac{\mathbf{A}(t)-\mathbf{A}(t_0)}{t-t_0}x(t_0),$$

where the first term on the right approaches the limit $\mathbf{A}(t)x'(t)$ as $t\to t_0$, by Lemma 1.64e (cf. Sec. 1.64f), and the second term approaches $\mathbf{A}'(t_0)x(t_0)$, by hypothesis. ∎

THEOREM. *The unique solution of the differential equation* (8) *subject to the initial condition* $u(t_0)=u_0$ *is given by* (10).

Proof. Let $u(t)$ be any solution of (8) with the given value of $u(t_0)$, and define a new unknown function $v(t)$ by the formula

$$u(t)=e^{(t-t_0)\mathbf{A}}v(t), \tag{11}$$

or equivalently

$$v(t)=e^{-(t-t_0)\mathbf{A}}u(t).$$

Substituting $u(t)$ into (8) and using the lemma, we get

$$u'(t)=\mathbf{A}e^{(t-t_0)\mathbf{A}}v(t)+e^{(t-t_0)\mathbf{A}}v'(t)=\mathbf{A}e^{(t-t_0)\mathbf{A}}v(t),$$

which implies

$$e^{(t-t_0)\mathbf{A}}v'(t)=0.$$

Multiplying this by $e^{-(t-t_0)\mathbf{A}}$, we find that $v'(t)=0$, and hence

$$v(t)\equiv v(t_0)=u(t_0). \tag{12}$$

But substitution of (12) into (11) gives (10). ∎

2.14.a. Suppose $\mathbf{X}$ is the n-dimensional real space R_n, and let $t_0=0$ for simplicity. Choosing a basis $e_1,\dots, e_n$ in R_n, we expand the vector function $u(t)$ with respect to this basis, obtaining

$$u(t)=\sum_{k=1}^{n}u_k(t)e_k.$$

Then

$$u'(t)=\sum_{k=1}^{n}u_k'(t)e_k,$$

and the vector differential equation (8) is equivalent to the system of n scalar differential equations

$$\begin{aligned} u_1'(t)&=a_{11}u_1(t)+\cdots+a_{1n}u_n(t),\\ &\cdots\\ u_n'(t)&=a_{n1}u_1(t)+\cdots+a_{nn}u_n(t), \end{aligned} \tag{13}$$

with a constant real coefficient matrix $A = \|a_{jk}\|$ (the matrix of the operator **A**). The solution of this system is obtained by applying the operator $e^{t\mathbf{A}}$ to the initial vector $u_0 = u(0)$.

***b.** It turns out that the solution can be written in rather simple form if for the initial vector u_0 we successively choose the *vectors of the Jordan basis of the operator* **A** (Sec. 1.17g). The following notation will be used to denote these basis vectors:

(a) f_k denotes the basis vector corresponding to the one-element Jordan block $\|\lambda_k\|$.
(b) $f_k^1, \ldots, f_k^p$ denote the basis vectors corresponding to the $p \times p$ Jordan block

$$\begin{Vmatrix} \lambda_k & 1 & & & \\ & \lambda_k & 1 & & \\ & & \ddots & & \\ & & & \lambda_k & 1 \\ & & & & \lambda_k \end{Vmatrix} \tag{14}$$

(λ_k real).
(c) g_k, h_k denote the basis vectors corresponding to the 2×2 real Jordan block

$$\begin{Vmatrix} \sigma_k & \tau_k \\ -\tau_k & \sigma_k \end{Vmatrix}. \tag{15}$$

(d) $g_k^1, h_k^1, \ldots, g_k^p, h_k^p$ denote the basis vectors corresponding to the $2p \times 2p$ real Jordan block

$$\begin{Vmatrix} \sigma_k & \tau_k & 1 & 0 & & & & \\ -\tau_k & \sigma_k & 0 & 1 & & & & \\ & & \sigma_k & \tau_k & & & & \\ & & -\tau_k & \sigma_k & & & & \\ & & & & \ddots & & & \\ & & & & \sigma_k & \tau_k & 1 & 0 \\ & & & & -\tau_k & \sigma_k & 0 & 1 \\ & & & & & & \sigma_k & \tau_k \\ & & & & & & -\tau_k & \sigma_k \end{Vmatrix}. \tag{16}$$

It should be recalled that every λ_k is a root of the characteristic equation

$$\begin{vmatrix} a_{11}-\lambda & a_{12} & \cdots & a_{1n} \\ a_{21} & a_{22}-\lambda & \cdots & a_{2n} \\ \cdot & \cdot & \cdots & \cdot \\ a_{n1} & a_{n2} & \cdots & a_{nn}-\lambda \end{vmatrix} = 0 \tag{17}$$

(Sec. 1.17e). The numbers σ_k and τ_k are just the real and imaginary parts of the nonreal roots of (17).

Every block of the Jordan matrix of the operator **A** corresponds to an invariant subspace of **A** (of dimension 1, p, 2, or $2p$, as the case may be). Thus the operator $e^{t\mathbf{A}}$ carries every vector in any of these subspaces into another vector in the same subspace. Let

$$f_k(t),\ f_k^r(t),\ g_k(t),\ h_k(t),\ g_k^r(t),\ h_k^r(t)$$

be the solutions of the differential equation (8) corresponding to the initial vectors

$$f_k,\ f_k^r,\ g_k,\ h_k,\ g_k^r,\ h_k^r,$$

respectively. Then the nature of the solutions in the various invariant subspaces of **A** is easily described:

(a) In the one-dimensional invariant subspace generated by the vector f_k, the operator **A** reduces to multiplication by λ_k and the operator $e^{t\mathbf{A}}$ to multiplication by $e^{\lambda_k t}$, so that

$$f_k(t) = f_k e^{\lambda_k t}. \tag{18}$$

(b) In the p-dimensional invariant subspace corresponding to the Jordan block (14), we have

$$e^{tA} = \begin{Vmatrix} e^{\lambda_k t} & te^{\lambda_k t} & \dfrac{t^2}{2} e^{\lambda_k t} & \cdots & \dfrac{t^{p-1}}{(p-1)!} e^{\lambda_k t} \\ 0 & e^{\lambda_k t} & te^{\lambda_k t} & \cdots & \dfrac{t^{p-2}}{(p-2)!} e^{\lambda_k t} \\ \cdot & \cdot & \cdot & \cdots & \cdot \\ 0 & 0 & 0 & \cdots & e^{\lambda_k t} \end{Vmatrix}$$

(Sec. 1.19f), and hence

$$\begin{aligned} f_k^1(t) &= e^{t\mathbf{A}} f_k^1 = f_k^1 e^{\lambda_k t}, \\ f_k^2(t) &= e^{t\mathbf{A}} f_k^2 = [tf_k^1 + f_k^2] e^{\lambda_k t}, \\ &\cdots \\ f_k^p(t) &= e^{t\mathbf{A}} f_k^p = \left[\frac{t^{p-1}}{(p-1)!} f_k^1 + \frac{t^{p-2}}{(p-2)!} f_k^2 + \cdots + f_k^p\right] e^{\lambda_k t}. \end{aligned} \tag{19}$$

(c) In the two-dimensional invariant subspace corresponding to the Jordan block (15), we have†

$$e^{tA} = e^{\sigma t} \begin{Vmatrix} \cos \tau t & \sin \tau t \\ -\sin \tau t & \cos \tau t \end{Vmatrix}$$

† Here, and in similar situations below, we write σ and τ instead of σ_k and τ_k, to keep the notation simple.

(Sec. 1.19i), and hence

$$\begin{aligned} g_k(t) &= (g_k \cos \tau t - h_k \sin \tau t)e^{\sigma t}, \\ h_k(t) &= (g_k \sin \tau t + h_k \cos \tau t)e^{\sigma t}. \end{aligned} \tag{20}$$

(d) In the $2p$-dimensional invariant subspace corresponding to the Jordan block (16), we have

$$e^{tA} = e^{\sigma t} \left\| \begin{matrix} \cos \tau t & \sin \tau t & t \cos \tau t & t \sin \tau t & & & \\ -\sin \tau t & \cos \tau t & -t \sin \tau t & t \cos \tau t & & & \\ & & \cos \tau t & \sin \tau t & & & \\ & & -\sin \tau t & \cos \tau t & & & \\ & & & & \ddots & & \\ & & & & & \cos \tau t & \sin \tau t \\ & & & & & -\sin \tau t & \cos \tau t \end{matrix} \right\|$$

(Sec. 1.19i), and hence

$$\begin{aligned} g_k^1(t) &= [g_k^1 \cos \tau t - h_k^1 \sin \tau t]e^{\sigma t}, \\ h_k^1(t) &= [g_k^1 \sin \tau t + h_k^1 \cos \tau t]e^{\sigma t}, \\ &\cdots \\ g_k^p(t) &= \left[\frac{t^{p-1}}{(p-1)!}(g_k^1 \cos \tau t - h_k^1 \sin \tau t) + \cdots + (g_k^p \cos \tau t - h_k^p \sin \tau t) \right] e^{\sigma t}, \\ h_k^p(t) &= \left[\frac{t^{p-1}}{(p-1)!}(g_k^1 \sin \tau t + h_k^1 \cos \tau t) + \cdots + (g_k^p \sin \tau t + h_k^p \cos \tau t) \right] e^{\sigma t}. \end{aligned} \tag{21}$$

***2.15.** We have just constructed n different particular solutions of the differential equation

$$u'(t) = \mathbf{A}u(t) \qquad (a \leqslant t \leqslant b), \tag{8'}$$

in the space R_n, choosing the n basis vectors of the Jordan matrix of the operator $\mathbf{A}$ as the initial vectors. Suppose we expand each of these n solutions with respect to the original basis $e_1, \ldots, e_n$ in which (8′) takes the form of the system (13). Then we get

$$f_k(t) = \sum_{m=1}^{n} u_{km}(t)e_m, \qquad g_k(t) = \sum_{m=1}^{n} v_{km}(t)e_m, \qquad h_k(t) = \sum_{m=1}^{n} w_{km}(t)e_m,$$

$$f_k^r(t) = \sum_{m=1}^{n} u_{km}^r(t)e_m, \qquad g_k^r(t) = \sum_{m=1}^{n} v_{km}^r(t)e_m, \qquad h_k^r(t) = \sum_{m=1}^{n} w_{km}^r(t)e_m,$$

in terms of suitable scalar functions

$$u_{km}(t),\ v_{km}(t),\ w_{km}(t),\ u_{km}^r(t),\ v_{km}^r(t),\ w_{km}^r(t).$$

Similarly, the n initial vectors have expansions

$$f_k=\sum_{m=1}^{n} u_{km}e_m, \qquad g_k=\sum_{m=1}^{n} v_{km}e_m, \qquad h_k=\sum_{m=1}^{n} w_{km}e_m,$$
$$f_k^r=\sum_{m=1}^{n} u_{km}^r e_m, \qquad g_k^r=\sum_{m=1}^{n} v_{km}^r e_m, \qquad h_k^r=\sum_{m=1}^{n} w_{km}^r e_m,$$

in terms of suitable numbers

$$u_{km}, \quad v_{km}, \quad w_{km}, \quad u_{km}^r, \quad v_{km}^r, \quad w_{km}^r.$$

Then, since

$$f_k(t)=e^{\lambda_k t} f_k=e^{\lambda_k t}\sum_{m=1}^{n} u_{km}e_m$$

on the one hand, while

$$f_k(t)=\sum_{m=1}^{n} u_{km}(t)e_m$$

on the other, we have

$$u_{km}(t)=u_{km}e^{\lambda_k t},$$

and similarly

$$v_{km}(t)=[v_{km}\cos\tau t-w_{km}\sin\tau t]e^{\sigma t},$$
$$w_{km}(t)=[v_{km}\sin\tau t+w_{km}\cos\tau t]e^{\sigma t},$$
$$u_{km}^r(t)=\left[\frac{t^{r-1}}{(r-1)!}u_{km}^1+\frac{t^{r-2}}{(r-2)!}u_{km}^2+\cdots+u_{km}^r\right]e^{\lambda_k t},$$
$$v_{km}^r(t)=\left[\frac{t^{r-1}}{(r-1)!}(v_{km}^1\cos\tau t-w_{km}^1\sin\tau t)+\cdots+(v_{km}^r\cos\tau t-w_{km}^r\sin\tau t)\right]e^{\sigma t},$$
$$w_{km}^r(t)=\left[\frac{t^{r-1}}{(r-1)!}(v_{km}^1\sin\tau t+w_{km}^1\cos\tau t)+\cdots+(v_{km}^r\sin\tau t+w_{km}^r\cos\tau t)\right]e^{\sigma t}.$$

***2.16.** Using formulas (18)–(21) and the kinematic interpretation of Sec. 2.11d, we can get a clear picture of the integral curves $u=u(t)$ and of their asymptotic behavior as $t\to\infty$:

(a) Suppose the initial vector u_0 is a vector of the type f_k. Then there are three possibilities: (1) The solution $f_k(t)$ has the constant value f_k if $\lambda_k=0$; (2) As $t\to\infty$, $f_k(t)$ increases exponentially along the positive f_k-axis if $\lambda_k>0$; (3) As $t\to\infty$, $f_k(t)$ decreases exponentially to zero along the same axis if $\lambda_k<0$.
(b) Suppose the initial vector u_0 is a vector of the type f_k^r, where f_k^r is one of

the basis vectors of the p-dimensional invariant subspace corresponding to the Jordan block (19):

$$f_k^r(t) = \left[\frac{t^{r-1}}{(r-1)!} f_k^1 + \frac{t^{r-2}}{(r-2)!} f_k^2 + \cdots + t f_k^{r-1} + f_k^r \right] e^{\lambda_k t}.$$

Thus the position vector of the corresponding curve, which coincides with the vector f_k^r at the time $t=0$, in the course of time acquires components along the vectors $f_k^{r-1}, \ldots, f_k^2, f_k^1$, with the component along f_k^1 predominating for large t. If $\lambda_k \geqslant 0$, $r > 1$, then as $t \to \infty$, the curve moves away from the origin, and in the limit its tangent becomes parallel to the vector f_k^1 (this is easily verified by differentiation), the difference between the cases $\lambda_k = 0$ and $\lambda_k > 0$ being that the departure from the origin obeys a power law in the first case and an exponential law in the second. On the other hand, if $\lambda_k < 0$, then the curve approaches the origin as $t \to \infty$, and, because of the predominant role of the component along f_k^1, as $t \to \infty$ the curve enters an arbitrarily "narrow" cone with its vertex at the origin and its axis along the vector f_k^1; thus the curve approaches the origin as $t \to \infty$, becoming tangent to the vector f_k^1.

(c) Next suppose the initial vector u_0 is the basis vector g_k or h_k of the two-dimensional invariant subspace E_2 corresponding to the Jordan block (15). Then, examining (20), we see that the solution $u = u(t)$ in the plane E_2 can behave in three possible ways: (1) If $\sigma_k = 0$, u describes an ellipse with center at the origin; (2) If $\sigma_k > 0$, u describes an "expanding spiral" winding around the origin; (3) If $\sigma_k < 0$, u describes a "contracting spiral" winding around the origin, which it approaches as $t \to \infty$.

(d) Finally, suppose the initial vector u_0 is one of the vectors g_k^r or h_k^r in the $2p$-dimensional invariant subspace E_{2p} corresponding to the Jordan block (16). Then, examining (21), we see that the solution $u = u(t)$ in the space E_{2p} can behave in two possible ways: (1) If $\sigma_k > 0$, u describes an "expanding spiral" whose tangent in the limit as $t \to \infty$ approaches a position in the plane of the first two basis vectors g_k^1, h_k^1; (2) If $\sigma_k < 0$, u describes a "contracting spiral" which approaches the origin as $t \to \infty$, becoming tangent to the plane of the first two basis vectors.

In the general case, where the vector u_0 has components along several vectors of the Jordan basis, the corresponding motion is a geometric superposition of the types just described.

2.17. Higher-order equations

a. Consider the differential equation

$$y^{(n)}(t) = a_1(t) y(t) + \cdots + a_n(t) y^{(n-1)}(t), \tag{22}$$

called a *homogeneous linear equation of order n*. Making the substitutions

$$\begin{aligned} y(t) &= u_1(t), \\ y'(t) &= u_2(t), \\ &\cdots \\ y^{(n-1)}(t) &= u_n(t), \end{aligned} \tag{23}$$

we can reduce (22) to the following system of n first-order equations:

$$\begin{aligned} u_1'(t) &= u_2(t), \\ u_2'(t) &= u_3(t), \\ &\cdots \\ u_n'(t) &= a_1(t)u_1(t) + a_2(t)u_2(t) + \cdots + a_n(t)u_n(t). \end{aligned} \tag{24}$$

Conversely, starting from any solution $(u_1(t), \ldots, u_n(t))$ of the system (24), we can use the formulas (23) to determine a function $y(t) = u_1(t)$ and its derivatives, with the last equation of the system (24) guaranteeing that $y(t)$ satisfies equation (22).

***b.** Suppose the functions $a_1 = a_1(t), \ldots, a_n = a_n(t)$ appearing in (22) and (24) are all constants. In this case, the matrix of the operator **A** corresponding to the system (24) has the special form

$$A = \begin{Vmatrix} 0 & 1 & 0 & \cdots & 0 & 0 \\ 0 & 0 & 1 & \cdots & 0 & 0 \\ \cdot & \cdot & \cdot & \cdots & \cdot & \cdot \\ 0 & 0 & 0 & \cdots & 0 & 1 \\ a_1 & a_2 & a_3 & \cdots & a_{n-1} & a_n \end{Vmatrix}.$$

Let

$$f = \sum_{k=1}^{n} \xi_k e_k$$

be an eigenvector of **A**, corresponding to the eigenvalue λ. Then $\mathbf{A}f = \lambda f$, or

$$\begin{aligned} \xi_2 &= \lambda\xi_1, \\ \xi_3 &= \lambda\xi_2, \\ &\cdots \\ a_1\xi_1 + a_2\xi_2 + \cdots + a_n\xi_n &= \lambda\xi_n \end{aligned}$$

in terms of the components $\xi_1, \xi_2, \ldots, \xi_n$. Setting $\xi_1 = 1$, we get in turn

$$\xi_1 = 1,\ \xi_2 = \lambda,\ \xi_3 = \lambda^2,\ \ldots,\ \xi_n = \lambda^{n-1},$$

and then

$$a_1 + a_2\lambda + \cdots + a_n\lambda^{n-1} = \lambda^n. \tag{25}$$

Equation (25) is called the *characteristic equation* of equation (22), and is obtained from (22) by replacing $y^{(k)}$ by λ^k ($k=0,1,\ldots, n$). Note also that the roots of (25) are just the eigenvalues of the operator **A**. The eigenvector corresponding to the eigenvalue λ is collinear with the vector $(1,\lambda,\lambda^2,\ldots,\lambda^{n-1})$, and hence is uniquely determined (to within collinearity). Therefore every (real or complex) multiple root λ_k of equation (25) gives rise to a "full" $p \times p$ Jordan block

$$\begin{Vmatrix} \lambda_k & 1 & & & \\ & \lambda_k & & & \\ & & \ddots & & \\ & & & \lambda_k & 1 \\ & & & & \lambda_k \end{Vmatrix}.$$

In other words, in the present case the degrees of the elementary divisors equal the multiplicities of the roots, and the minimal annihilating polynomial of the matrix A coincides with its characteristic polynomial (see Lin. Alg., Sec. 6.52).

Using the results of Sec. 2.15, we can now write down n different particular solutions of the system (24), corresponding to the selection of the n Jordan basis vectors as the initial vectors. We will write down only the first components of these solutions, since, according to (23), only the first component $u_1(t)=y(t)$, giving the solution of the original equation (22), is of interest. Thus we have

(a) One solution

$$u_{k1}(t) = u_{k1}e^{\lambda_k t}$$

for every simple real root λ_k;

(b) p solutions

$$u_{k1}^r(t) = \left[\frac{t^{r-1}}{(r-1)!}u_{k1}^1 + \frac{t^{r-2}}{(r-2)!}u_{k1}^2 + \cdots + u_{k1}^r\right]e^{\lambda_k t}$$

($r=1,\ldots,p$) for every real root λ_k of multiplicity p;

(c) Two solutions

$$v_{k1}(t) = [v_{k1}\cos\tau_k t - w_{k1}\sin\tau_k t]e^{\sigma_k t},$$
$$w_{k1}(t) = [v_{k1}\sin\tau_k t + w_{k1}\cos\tau_k t]e^{\sigma_k t}$$

for every pair of simple imaginary roots $\lambda_k = \sigma_k \pm i\tau_k$;

(d) $2p$ solutions

$$v_{k1}^r(t)=\left[\frac{t^{r-1}}{(r-1)!}(v_{k1}^1\cos\tau_k t-w_{k1}^1\sin\tau_k t)+\cdots+(v_{k1}^r\cos\tau_k t-w_{k1}^r\sin\tau_k t)\right]e^{\sigma_k t},$$

$$w_{k1}^r(t)=\left[\frac{t^{r-1}}{(r-1)!}(v_{k1}^1\sin\tau_k t+w_{k1}^1\cos\tau_k t)+\cdots+(v_{k1}^r\sin\tau_k t+w_{k1}^r\cos\tau_k t)\right]e^{\sigma_k t}$$

$(r=1,\dots,p)$ for every pair of imaginary roots $\lambda_k=\sigma_k\pm i\tau_k$ of multiplicity p.

Replacing these solutions by certain of their linear combinations, we can write our n particular solutions in the following even simpler form:

(a′) One solution

$$e^{\lambda_k t}$$

for every simple real root λ_k;

(b′) p solutions

$$e^{\lambda_k t},\ te^{\lambda_k t},\dots,\ t^{p-1}e^{\lambda_k t}$$

for every real root λ_k of multiplicity p;

(c′) Two solutions

$$e^{\sigma_k t}\cos\tau_k t,\ e^{\sigma_k t}\sin\tau_k t$$

for every pair of simple imaginary roots $\lambda_k=\sigma_k\pm i\tau_k$;

(d′) $2p$ solutions

$$e^{\sigma_k t}\cos\tau_k t,\ e^{\sigma_k t}\sin\tau_k t,\ te^{\sigma_k t}\cos\tau_k t,\ te^{\sigma_k t}\sin\tau_k t,\dots,$$
$$t^{p-1}e^{\sigma_k t}\cos\tau_k t,\ t^{p-1}e^{\sigma_k t}\sin\tau_k t$$

for every pair of imaginary roots $\lambda_k=\sigma_k\pm i\tau_k$ of multiplicity p.

2.18.a. We now turn to the general case, where the operator $\mathbf{A}(t)$ in the equation

$$u'(t)=\mathbf{A}(t)u(t)\qquad(a\leqslant t\leqslant b)\tag{26}$$

actually depends (continuously) on the parameter t. It will be recalled that the solution of (26), in the case where $\mathbf{A}(t)$ reduces to a numerical function $A(t)$, is given by the formula

$$u(t)=\exp\left\{\int_{t_0}^t A(\tau)\,d\tau\right\}u(t_0)$$

(Sec. 2.12). We might, of course, try forming the operator

$$\mathbf{W}(t)=\int_{t_0}^t\mathbf{A}(\tau)\,d\tau$$

and then the expression

$$\exp\left\{\int_{t_0}^{t} \mathbf{A}(\tau)\, d\tau\right\} u(t_0) = e^{\mathbf{W}(t)} u(t_0). \tag{27}$$

However, (27) will not in general be a solution of equation (26). In fact, in trying to differentiate the expression $e^{\mathbf{W}(t)}$ with respect to t, we encounter the difficulty that in transforming the difference

$$e^{\mathbf{W}(t+\Delta t)} - e^{\mathbf{W}(t)} = e^{\mathbf{W}(t) + \Delta t\, \overline{\mathbf{A}}(t;\Delta t)} - e^{\mathbf{W}(t)},$$

where

$$\overline{\mathbf{A}}(t;\Delta t) = \frac{1}{\Delta t}\int_{t}^{t+\Delta t} \mathbf{A}(\tau)\, d\tau$$

denotes the mean value of $\mathbf{A}(t)$ on the interval $[t, t+\Delta t]$, we cannot write

$$e^{\mathbf{W}(t) + \Delta t\, \overline{\mathbf{A}}(t;\Delta t)} = e^{\mathbf{W}(t)} e^{\Delta t\, \mathbf{A}(t;\Delta t)},$$

since the formula $e^{\mathbf{A}+\mathbf{B}} = e^{\mathbf{A}} e^{\mathbf{B}}$ is in general false unless $\mathbf{A}$ and $\mathbf{B}$ are *commuting* operators (in which case $\mathbf{AB} = \mathbf{BA}$), and the operators $\mathbf{W}(t)$ and $\overline{\mathbf{A}}(t;\Delta t)$ do not in general commute. Thus the derivative of $e^{\mathbf{W}(t)}$ will not in general coincide with $e^{\mathbf{W}(t)}\mathbf{W}'(t) = e^{\mathbf{W}(t)}\mathbf{A}(t)$.

Nevertheless, there is still a sense in which the expression (27) is the solution of equation (26). This consists in interpreting (27) as a "product integral," as we now describe.

***b.** Let

$$\Pi = \{a = t_0 \leqslant \tau_0 \leqslant t_1 \leqslant \tau_1 \leqslant t_2 \leqslant \cdots \leqslant t_{n-1} \leqslant \tau_{n-1} \leqslant t_n = t\}$$

be a partition of the interval $a \leqslant \tau \leqslant t$, with marked points $\tau_0, \tau_1, \ldots, \tau_{n-1}$ and parameter

$$d(\Pi) = \max\{\Delta t_0, \Delta t_1, \ldots, \Delta t_{n-1}\} \qquad (\Delta t_k = t_{k+1} - t_k).$$

Consider the operator

$$e^{\mathbf{A}(\tau_{n-1})\Delta t_{n-1}} e^{\mathbf{A}(\tau_{n-2})\Delta t_{n-2}} \cdots e^{\mathbf{A}(\tau_0)\Delta t_0}. \tag{28}$$

If the operators $\mathbf{A}(\tau)$ corresponding to different values of τ commute, we can write (28) in the form

$$e^{\mathbf{A}(\tau_0)\Delta t_0 + \cdots + \mathbf{A}(\tau_{n-1})\Delta t_{n-1}},$$

an expression which approaches the limit

$$\exp\left\{\int_{t_0}^{t} \mathbf{A}(\tau)\, d\tau\right\} \tag{29}$$

as $d(\Pi)\to 0$. However, as already noted, the operator (29) does not lead to the solution of equation (26) in the case of noncommuting $\mathbf{A}(\tau)$. Nevertheless, it turns out (see Problem 16) that the solution of (26) in the case of noncommuting $\mathbf{A}(\tau)$ can be found by applying a certain operator to the initial vector $u(t_0)$, namely the operator obtained by taking the limit of the operator (28) as $d(\Pi)\to 0$. This limiting operator is again denoted by (29), and is called the *product integral.*

To within small quantities of higher order, we can write

$$e^{\mathbf{A}(\tau_k)\Delta t_k} \approx \mathbf{E}+\mathbf{A}(\tau_k)\Delta t_k.$$

It can be shown (see Problem 14) that we get the same limit (29) by starting from the product

$$[\mathbf{E}+\mathbf{A}(\tau_{n-1})\Delta t_{n-1}][\mathbf{E}+\mathbf{A}(\tau_{n-2})\Delta t_{n-2}]\cdots[\mathbf{E}+\mathbf{A}(\tau_0)\Delta t_0]$$

instead of (28). For this reason, the product integral is sometimes denoted by

$$\prod_{t_0}^{t} [\mathbf{E}+\mathbf{A}(t)]\ dt.$$

Although product integrals can be used to estimate solutions, their domain of applicability is small compared with that of general existence theory. Therefore proofs of the above assertions will not be given here.

2.2. The Fixed Point Theorem

Later in this chapter we will consider the basic theorems on the existence and uniqueness of solutions of ordinary differential equations.† All these theorems are based on an important general principle of analysis, called the *fixed point theorem.*

2.21. Definition. Let M be a set and $\mathbf{A}$ a mapping of M into itself, i.e., a rule associating a point $y=\mathbf{A}x \in M$ with each point $x \in M$.‡ Then by a *fixed point of the mapping* $\mathbf{A}$ we mean any point $x \in M$ which is mapped into itself, i.e., any point $x \in M$ such that $\mathbf{A}x=x$.

For example, if M is a disk in the plane and $\mathbf{A}$ the mapping of M into itself which consists of rotating M through $90°$ about its center O, then O is the only fixed point of $\mathbf{A}$. If the mapping of M into itself consists of a twofold contraction of M with O as the center of similitude, followed by a transla-

† In contradistinction to *partial* differential equations, which involve *partial* derivatives (Vol. 1, Sec. 10.16a).

‡ We continue to use operator notation, but it should be noted that the operator $\mathbf{A}$ is in general no longer linear. Note that $\mathbf{A}$ may only map M into a proper subset of itself.

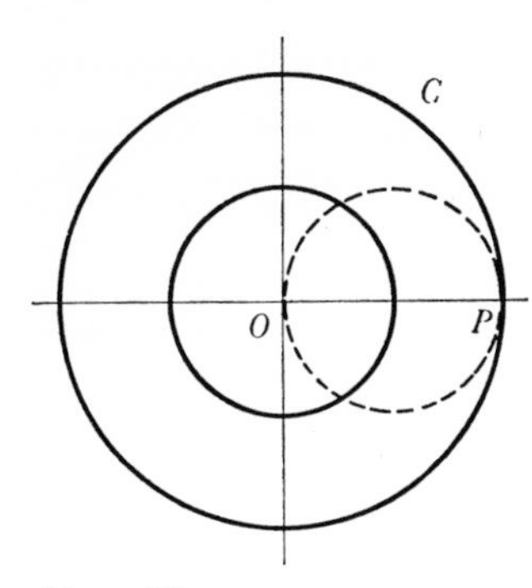

Figure 13

tion of the smaller disk so obtained until it becomes tangent to the boundary C of the original disk (see Figure 13), then the point of tangency P is the only fixed point of the resulting mapping, despite the fact that P is not a fixed point of either of the separate mappings making up the total mapping (it is the result that counts, and not the steps). If the mapping $\mathbf{A}$ carries the *circle* C into itself, by rotating C through $90°$ about its center, then $\mathbf{A}$ has no fixed points at all.

It is clearly of interest to find general (sufficient) conditions for the existence of fixed points. We now give one of the simplest theorems guaranteeing the existence of a fixed point, in fact of a *unique* fixed point, under certain assumptions about the set M and the mapping $\mathbf{A}$.

2.22. Definition. Let M be a metric space with distance ρ, and let $\mathbf{A}$ be a mapping of M into itself. Then $\mathbf{A}$ is said to be a *contraction mapping* if there exists a constant θ $(0 \leqslant \theta < 1)$ such that the inequality

$$\rho(\mathbf{A}x, \mathbf{A}y) \leqslant \theta \rho(x, y)$$

holds for every pair of points $x, y \in M$.

THEOREM (**Picard-Banach fixed point theorem**). *Let* $\mathbf{A}$ *be a contraction mapping of a complete metric space* M *into itself. Then* $\mathbf{A}$ *has a unique fixed point.*

Proof. Given any point $x_0 \in M$, we construct the sequence of points

$$x_1 = \mathbf{A}x_0,\ x_2 = \mathbf{A}x_1 = \mathbf{A}^2 x_0, \ldots,\ x_n = \mathbf{A}x_{n-1} = \cdots = \mathbf{A}^n x_0, \ldots \tag{1}$$

Then this sequence is fundamental in M. In fact, for any $n \geqslant 1$,

$$\begin{aligned} \rho(x_n, x_{n+p}) &\leqslant \rho(x_n, x_{n+1}) + \rho(x_{n+1}, x_{n+2}) + \cdots + \rho(x_{n+p-1}, x_{n+p}) \\ &\leqslant (\theta^n + \theta^{n+1} + \cdots + \theta^{n+p-1}) \rho(x_0, x_1) \\ &\leqslant \theta^n (1 + \theta + \theta^2 + \cdots) \rho(x_0, x_1) = \frac{\theta^n}{1-\theta} \rho(x_0, x_1). \end{aligned} \tag{2}$$

But the right-hand side becomes arbitrarily small for sufficiently large n, so that (1) is a fundamental sequence, as asserted. Since M is complete, the limit

$$x=\lim_{n\to\infty} x_n \in M$$

exists.

To see that x is a fixed point, we note that

$$\rho(\mathbf{A}x,x_n)=\rho(\mathbf{A}x,\mathbf{A}x_{n-1})\leqslant\theta\rho(x,x_{n-1})\to 0$$

as $n\to\infty$, and hence

$$\mathbf{A}x=\lim_{n\to\infty} x_n=x.$$

Suppose there were another fixed point y, so that both $\mathbf{A}x=x$ and $\mathbf{A}y=y$. Then

$$\rho(x,y)=\rho(\mathbf{A}x,\mathbf{A}y)\leqslant\theta\rho(x,y).$$

If $\rho(x,y)\neq 0$, we can divide by $\rho(x,y)$, obtaining the contradiction $1\leqslant\theta<1$. It follows that $\rho(x,y)=0$, $x=y$, i.e., there is no fixed point other than x. ∎

2.23. Fixed points of two contraction mappings. Two mappings $\mathbf{A}$ and $\mathbf{B}$ of a metric space M into itself are said to be *ε-close* if

$$\rho(\mathbf{A}x,\mathbf{B}x)\leqslant\varepsilon$$

for every $x\in M$.

LEMMA. *Let* $\mathbf{A}$ *and* $\mathbf{B}$ *be two contraction mappings defined on a complete metric space, so that*

$$\rho(\mathbf{A}x,\mathbf{A}y)\leqslant\theta_{\mathbf{A}}\rho(x,y),\qquad \rho(\mathbf{B}x,\mathbf{B}y)\leqslant\theta_{\mathbf{B}}\rho(x,y)$$

for all $x,y\in M$, *where* $0\leqslant\theta_{\mathbf{A}}<1$, $0\leqslant\theta_{\mathbf{B}}<1$, *and suppose* $\mathbf{A}$ *and* $\mathbf{B}$ *are ε-close. Then the distance between the fixed point of* $\mathbf{A}$ *and that of* $\mathbf{B}$ *cannot exceed* $\varepsilon=1/(1-\theta)$, *where* $\theta=\min\{\theta_{\mathbf{A}},\theta_{\mathbf{B}}\}$.

Proof. Let x_0 be the fixed point of $\mathbf{A}$ and y_0 the fixed point of $\mathbf{B}$. Then y_0 is the limit of the sequence $\mathbf{B}x_0, \mathbf{B}^2x_0,\ldots, \mathbf{B}^nx_0,\ldots,$ as in the proof of the preceding theorem. Therefore, by the inequality (2),

$$\rho(x_0,\mathbf{B}^nx_0)\leqslant\frac{1}{1-\theta_{\mathbf{B}}}\rho(x_0,\mathbf{B}x_0)=\frac{1}{1-\theta_{\mathbf{B}}}\rho(\mathbf{A}x_0,\mathbf{B}x_0)\leqslant\frac{\varepsilon}{1-\theta_{\mathbf{B}}},$$

since $\mathbf{A}$ and $\mathbf{B}$ are ε-close. Taking the limit as $n\to\infty$, we get

$$\rho(x_0,y_0)\leqslant\frac{\varepsilon}{1-\theta_{\mathbf{B}}}.$$

The same argument, this time applied to the sequence $\mathbf{A}y_0, \mathbf{A}^2y_0,\ldots, \mathbf{A}^ny_0,\ldots$

which converges to x_0, shows that

$$\rho(x_0, y_0) \leqslant \frac{\varepsilon}{1-\theta_{\mathbf{A}}}.$$

Combining the last two inequalities, we get

$$\rho(x_0, y_0) \leqslant \frac{\varepsilon}{1-\theta}. \quad \blacksquare$$

2.3. Existence and Uniqueness of Solutions

2.31. Let $\mathbf{B}$ be a Banach space, and let $\varphi(t,x)$ be a mapping of $\mathbf{B}$ into itself, depending on a real parameter t $(a \leqslant t \leqslant b)$. Let $u(t)$ be a differentiable vector function with values in $\mathbf{B}$, defined on the same interval $[a,b]$. Then the derivative $u'(t)$ is also a vector function with values in $\mathbf{B}$, defined on $[a,b]$. Substituting the vector function $u(t)$ for the argument $x \in \mathbf{B}$ in $\varphi(t,x)$, we get a new vector function $\varphi(t,u(t))$ with values in $\mathbf{B}$, again defined on the interval $[a,b]$.

Our aim is to solve the differential equation

$$u'(t) = \varphi(t,u(t)), \tag{1}$$

with the initial condition

$$u(t_0) = u_0 \qquad (a \leqslant t_0 \leqslant b,\ u_0 \in \mathbf{B}). \tag{2}$$

Under certain natural assumptions about continuity, which will be made precise later, the problem of solving (1) subject to the initial condition (2) is equivalent to that of solving the integral equation

$$u(t) = u_0 + \int_{t_0}^{t} \varphi(\tau,u(\tau))\, d\tau, \tag{3}$$

since (3) is obtained from (1) and (2) by integration from t_0 to t, (2) is obtained from (3) by making the substitution $t = t_0$, and (1) is obtained from (3) by differentiation with respect to t. Thus the problem reduces to that of finding fixed points of the transformation

$$\mathbf{A}x(t) = u_0 + \int_{t_0}^{t} \varphi(\tau,x(\tau))\, d\tau \tag{4}$$

acting in the space of vector functions $x(t)$.

2.32. Naturally, at this juncture we think in terms of applying the Picard-Banach fixed point theorem. But to do so, we first need a complete metric space M and a contraction mapping $\mathbf{A}$ corresponding to our problem. For M we choose the space of all continuous vector functions $x(t)$ with values in

B, defined on some interval $[t_0-h, t_0+h]$,† where the value of h will be indicated later. We then equip M with the metric

$$\rho\{x_1(t),x_2(t)\} = \max_{|t-t_0|\leqslant h} \|x_1(t)-x_2(t)\|.$$

The resulting metric space M is *complete*, by Theorem 2.23e.

2.33. Next we assume (until Sec. 2.39a) that the vector function $\varphi(t,x)$ is continuous in both arguments t and x (jointly). This means that given any $t \in [a,b]$, $x \in \mathbf{B}$ and any $\varepsilon>0$, there exists a $\delta=\delta(\tilde{t},\tilde{x},\varepsilon)>0$ such that $|t-\tilde{t}|<\delta$, $\|x-\tilde{x}\|<\delta$ implies $\|\varphi(t,x)-\varphi(\tilde{t},\tilde{x})\|<\varepsilon$ (cf. Vol. 1, Sec. 5.18).

LEMMA. *The function $\varphi(t,x(t))$ is continuous in t for every $x(t) \in M$.*

Proof. Given any $t \in [a,b]$, $x \in \mathbf{B}$ and any $\varepsilon>0$, use the continuity of $\varphi(t,x)$ in t and x to choose $\delta_1>0$ such that $|t-\tilde{t}|<\delta_1$, $\|x-\tilde{x}\|<\delta_1$ implies $\|\varphi(t,x)-\varphi(\tilde{t},\tilde{x})\|<\varepsilon$. Then use the continuity of $x(t)$ to choose $\delta_2>0$ such that $|t-\tilde{t}|<\delta_2$ implies $\|x(t)-x(\tilde{t})\|<\delta_1$. But then, choosing $\delta=\min\{\delta_1,\delta_2\}$, we find that $|t-\tilde{t}|<\delta$ implies $\|\varphi(t,x(t))-\varphi(\tilde{t},x(\tilde{t}))\|<\varepsilon$. ∎

In particular, the integral in (4) exists for every $x(t) \in M$, so that (4) does in fact define a mapping of M into itself.

2.34. Before the fixed point theorem can be applied, we must still impose a condition on $\varphi(t,x)$ guaranteeing that **A** is a contraction mapping. To this end, we assume (until Sec. 2.39b) that $\varphi(t,x)$ satisfies a *Lipschitz condition in the argument* x, i.e., that there exists a constant $C>0$ such that the inequality

$$\|\varphi(t,x)-\varphi(t,\tilde{x})\|<C\|x-\tilde{x}\| \tag{5}$$

holds for all $t \in [a,b]$ and $x,\tilde{x} \in \mathbf{B}$.

LEMMA. *With the above assumptions on $\varphi(t,x)$, the mapping* **A** *defined by* (4) *is a contraction mapping, at least for sufficiently small h.*

Proof. Given any two points of the space M, i.e., any two vector functions $x(t)$ and $y(t)$ defined and continuous on the interval $[t_0-h, t_0+h]$, we have

$$\begin{aligned}\rho\{\mathbf{A}x(t), \mathbf{A}y(t)\} &= \max_{|t-t_0|\leqslant h} \|\mathbf{A}x(t)-\mathbf{A}y(t)\| \\ &\leqslant \max_{|t-t_0|\leqslant h} \left\| \int_{t_0}^{t} [\varphi(\tau,x(\tau))-\varphi(\tau,y(\tau))]\, d\tau \right\| \\ &\leqslant h \max_{|t-t_0|\leqslant h} \|\varphi(t,x(t))-\varphi(t,y(t))\| \\ &\leqslant Ch \max_{|t-t_0|\leqslant h} \|x(t)-y(t)\| = Ch\rho\{x(t),y(t)\}.\end{aligned}$$

† We replace $[t_0-h, t_0+h]$ by $[a, a+h]$ if $t_0=a$ and by $[b-h, b]$ if $t_0=b$.

But then we need only choose $h<1/C$ to make $\mathbf{A}$ into a contraction mapping. ∎

2.35. We are now well on our way to proving the following key

THEOREM (**Existence and uniqueness of solutions of a first-order equation**). *Given a Banach space* $\mathbf{B}$ *and a function* $\varphi(t,x)$ *defined for all* $t \in [a,b]$ *and* $x \in \mathbf{B}$, *suppose* $\varphi(t,x)$ *is continuous in both arguments and satisfies a Lipschitz condition* (5). *Then among all differentiable functions defined on* $[a,b]$ *with values in* $\mathbf{B}$, *there exists a unique function* $u(t)$ *satisfying the differential equation* (1) *and the initial condition* (2).

Proof. The validity of the theorem with $[a,b]$ replaced by $[t_0-h,\ t_0+h]$ follows at once from the Picard-Banach fixed point theorem (Sec. 2.22), with the help of Lemmas 2.33 and 2.34. There is a unique solution $u=u(t)$ of equation (1) on the interval $[t_0-h,\ t_0+h]$, subject to the initial condition (2), and we need only extend this solution onto the whole interval $[a,b]$. To this end, let $h=2/3C$, say, and apply the result just proved to the same differential equation (1), with the new initial condition

$$u^*(t_1)=u(t_1) \qquad (t_1=t_0+h),$$

where $u(t_1)$ is the value at $t=t_1$ of the solution just constructed on the interval $[t_0-h,\ t_0+h]$. This gives a new solution $u^*(t)$ defined on the interval $[t_1-h,\ t_1+h]$. But the functions $u(t)$ and $u^*(t)$ must coincide on the common part of their intervals of definition, by the uniqueness of the solution on each of these intervals, and hence there is in fact just one solution of equation (1) on the interval $[t_0-h,\ t_0+2h]$. Continuing this process (in both directions), we finally construct a solution for the whole interval $[a,b]$ after a finite number of steps. ∎

2.36. Remark. Suppose the vector function $\varphi(t,x)$ maps some fixed closed subspace $\mathbf{B}^* \subset \mathbf{B}$ into itself for every $t \in [a,b]$, and suppose we choose the initial vector u_0 in the subspace $\mathbf{B}^*$. Then *the solution of equation* (1) *satisfying the initial condition* (2) *lies in the subspace* $\mathbf{B}^*$ *for all* $t \in [a,b]$. In fact, now we need only confine ourselves from the very outset to the subspace $\mathbf{B}^*$ (instead of the whole space $\mathbf{B}$), carrying out all constructions in $\mathbf{B}^*$. This gives a solution with values in the space $\mathbf{B}^*$, and by the uniqueness theorem, there is no other solution of equation (1) taking values in the space $\mathbf{B}$, subject to the condition $u(t_0)=u_0 \in \mathbf{B}^*$.

2.37. Continuous dependence of the solution on the initial vector. Here we use $u(t;t_0,u_0)$ to denote the solution of equation (1) with the initial condition (2). This is a function which associates with every vector $u_0 \in \mathbf{B}$ a vector $u(t;t_0,u_0)$ depending continuously on t_0 and t as parameters.

THEOREM. *Under the same conditions as in Theorem 2.35, the vector $u(t;t_0,u_0)$ is a continuous function of u_0 for fixed t_0 and t.*

Proof. Consider the mappings

$$\mathbf{A}_0 x(t) = u_0 + \int_{t_0}^{t} \varphi(\tau, x(\tau))\, d\tau,$$

$$\mathbf{A}_1 x(t) = u_1 + \int_{t_0}^{t} \varphi(\tau, x(\tau))\, d\tau,$$

whose fixed points are the solutions of equation (1) with initial conditions u_0 and u_1, respectively. In the space M of vector functions $x(t)$ defined and continuous on $[t_0-h,\ t_0+h]$, where $h<1/C$, these are contraction mappings with the same value of $\theta = Ch < 1$. If $\|u_0 - u_1\| \leqslant \varepsilon$, then the contraction mappings are ε-close, and hence the distance between their fixed points does not exceed $\varepsilon/(1-\theta)$, by Lemma 2.23. It follows that

$$\max_{|t-t_0|\leqslant h} \|u(t;t_0,u_0) - u(t;t_0,u_1)\| \leqslant \frac{\varepsilon}{1-\theta}.$$

Thus two solutions differing by no more than ε for $t=t_0$ differ by no more than $\varepsilon/(1-\theta)$ on the whole interval $|t-t_0|\leqslant h$. Moving the initial point from t_0 to $t_1 = t_0 + h$ enables us, as in the proof of Theorem 2.35, to extend the solutions onto the interval $t_0 - h \leqslant t \leqslant t_0 + 2h$, and then, repeating the argument just given, we find that the two solutions differ by no more than $\varepsilon/(1-\theta)^2$. Continuing this argument, we finally get the estimate

$$\max_{a\leqslant t\leqslant b} \|u(t;t_0,u_0) - u(t;t_0,u_1)\| \leqslant \frac{\varepsilon}{(1-\theta)^m}, \tag{6}$$

where†

$$m = \left[\frac{b-a}{h}\right] + 1.$$

Thus, by making $\|u_0 - u_1\|$ sufficiently small, we can make the left-hand side of (6) as small as we please, thereby proving the required continuity of $u(t;t_0,u_0)$. ∎

2.38. The solution operator. Given an arbitrary vector $u_0 \in \mathbf{B}$, let $u(t)$ be the solution of equation (1) with the initial condition $u(t_0) = u_0$. Then the vector $u = u(t)$ is uniquely defined for $t \in [a,b]$ and depends on the instants of time t_0 and t. Suppose we write $u(t)$ in the form

$$u(t) = \Omega_{t_0}^{t} u_0,$$

† $[x]$ denotes the *integral part* of x, namely the unique integer $\leqslant x$.

where $\Omega_{t_0}^t$ is a uniquely defined operator mapping the space **B** into itself. Then we call $\Omega_{t_0}^t$ the *solution operator* of equation (1).

For example, according to Sec. 2.13, the homogeneous linear equation

$$u'(t) = \mathbf{A}u(t)$$

has the solution operator

$$\Omega_{t_0}^t = e^{(t-t_0)\mathbf{A}}.$$

a. According to Theorem 2.37, the mapping $\Omega_{t_0}^t$ is *continuous*. Thus if

$$u_0^{(n)} \qquad (n=1,2,\ldots)$$

is a sequence of vectors in the space **B** converging to the vector u_0, the corresponding sequence

$$u_1^{(n)} = \Omega_{t_0}^{t_1} u_0^{(n)} \qquad (n=1,2,\ldots)$$

converges to the vector

$$u_1 = \Omega_{t_0}^{t_1} u_0.$$

b. Obviously $\Omega_{t_0}^{t_0} u_0 = u_0$, so that $\Omega_t^t = \mathbf{E}$ is the identity operator.

c. THEOREM. *The formula*

$$\Omega_{t_0}^{t_2} = \Omega_{t_1}^{t_2} \Omega_{t_0}^{t_1} \tag{7}$$

holds for arbitrary $t_0, t_1, t_2 \in [a,b]$.

Proof. Formula (7) means that both sides give the same result when applied to an arbitrary vector $u_0 \in \mathbf{B}$. Let

$$u_1 = \Omega_{t_0}^{t_1} u_0, \; u_2 = \Omega_{t_1}^{t_2} u_1,$$

so that in particular

$$u_2 = \Omega_{t_1}^{t_2} \Omega_{t_0}^{t_1} u_0. \tag{8}$$

The vector u_1 is the value at $t=t_1$ of the solution of equation (1) which reduces to u_0 at $t=t_0$, while the vector u_2 is the value at $t=t_2$ of the solution of (1) which reduces to u_1 at $t=t_1$. But these two solutions are identical, by the uniqueness part of Theorem. 2.35. Therefore the vector u_2 is the value at $t=t_1$ of the solution of (1) which reduces to u_0 at $t=t_0$:

$$u_2 = \Omega_{t_0}^{t_2} u_0. \tag{8'}$$

Comparing (8) and (8′), we get (7). ∎

d. Setting $t_2 = t_0$ in (7), we get first

$$\Omega_{t_1}^{t_0} \Omega_{t_0}^{t_1} = \mathbf{E},$$

and then

$$\Omega_{t_0}^{t_1}\Omega_{t_1}^{t_0}=\mathbf{E},$$

after reversing the roles of t_0 and t_1. It follows that the operator $\Omega_{t_0}^{t_1}$ is *invertible*, with inverse $\Omega_{t_1}^{t_0}$.

e. The equation

$$\frac{du(t)}{dt}=\mathbf{A}(t)u(t)$$

can be written in the form

$$\frac{d(\Omega_{t_0}^{t}u_0)}{dt}=\mathbf{A}(t)\Omega_{t_0}^{t}u_0 \qquad (u_0 \in \mathbf{B}),$$

or more concisely as

$$\frac{d\Omega_{t_0}^{t}}{dt}=\mathbf{A}(t)\Omega_{t_0}^{t}.$$

2.39.a. So far we have assumed that the function $\varphi(t,x)$ is defined for all $t \in [a,b]$ and $x \in \mathbf{B}$. Suppose we make the weaker assumption that $\varphi(t,x)$ is defined for all $t \in [a,b]$ but only for x in some ball

$$V=\{x \in \mathbf{B}: \|x-u_0\| \leqslant r\}.$$

Then, provided that $\varphi(t,x)$ is continuous in both t and x and satisfies a Lipschitz condition (5) for all $t,\tilde{t} \in [a,b]$ and $x,\tilde{x} \in V$, we find that *in some neighborhood of the point* t_0 there exists a unique solution of equation (1) satisfying the initial condition (2). In fact, in this case the operator $\mathbf{A}$ defined by (4) will carry every continuous function $x(t)$ with values in V into another continuous function taking values only in V, provided that h is small enough (why?). Hence to prove the asserted existence and uniqueness, we need only replace the metric space M of *all* functions $x(t)$ continuous on $[t_0-h,\ t_0+h]$ by the metric space M_V of all functions $x(t)$ continuous on $[t_0-h,\ t_0+h]$ and *taking values only in* V. However, the resulting solution can in general no longer be extended onto the whole interval $[a,b]$.

b. A similar situation occurs in the case where the function $\varphi(t,x)$ is defined and continuous for all $t \in [a,b]$ and $x \in \mathbf{B}$, but where the constant C in the Lipschitz condition (5) depends on the distance of the points x and $\tilde{x}$ from the origin, so that (5) takes the form

$$\|\varphi(t,x)-\varphi(t,\tilde{x})\| \leqslant C(r)\|x-\tilde{x}\|$$

for all $t \in [a,b]$ and $x, \tilde{x}$ in the ball $\|x\| \leqslant r$. Once again the solution will exist and be unique in some neighborhood of the point t_0, but will in general not be extendable onto the whole interval $[a,b]$.

For example, consider the differential equation

$$x'(t) = x^2(t) \qquad (-1 \leqslant t \leqslant 1). \tag{9}$$

The right-hand side is continuous for all $x \in R_1$, and moreover

$$|x^2 - \tilde{x}^2| = |x + \tilde{x}||x - \tilde{x}| \leqslant 2r|x - \tilde{x}|$$

for all $x, \tilde{x}$ in the interval $|x| \leqslant r$. The solution of (9) satisfying the initial condition $x(0) = x_0$ is of the form

$$x(t) = \frac{x_0}{1 - x_0 t},$$

and cannot be extended onto the whole interval $-1 \leqslant t \leqslant 1$ if $|x_0| \geqslant 1$.

2.4. Systems of Equations

2.41.a. Again let **B** be a Banach space. Suppose we are given n functions

$$\varphi_1(t, x_1, \ldots, x_n), \ldots, \varphi_n(t, x_1, \ldots, x_n),$$

each depending on a real parameter $t \in [a,b]$ and on n arguments $x_1, \ldots, x_n$, where each x_k is a point in **B** and each function $\varphi_k(t, x_1, \ldots, x_n)$ takes its values in **B** as well. Consider the system of differential equations

$$\begin{aligned} u_1'(t) &= \varphi_1(t, u_1, \ldots, u_n), \\ &\cdots \\ u_n'(t) &= \varphi_n(t, u_1, \ldots, u_n), \end{aligned} \tag{1}$$

with the initial conditions

$$\begin{aligned} u_1(t_0) &= p_1 \in \mathbf{B}, \\ &\cdots \\ u_n(t_0) &= p_n \in \mathbf{B}, \end{aligned} \tag{2}$$

where $a \leqslant t_0 \leqslant b$. Then by a *solution* of the system (1) subject to the conditions (2) we mean any set of vector functions $u_1(t), \ldots, u_n(t)$ defined on $[a,b]$ satisfying both (1) and (2).

b. The function $\varphi_k(t, x_1, \ldots, x_n)$ is said to be *continuous in all its arguments* $t, x_1, \ldots, x_n$ (*jointly*) if given any $\tilde{t} \in [a,b]$, $\tilde{x}_1, \ldots, \tilde{x}_n \in \mathbf{B}$ and any $\varepsilon > 0$, there exists a $\delta = \delta(\tilde{t}, \tilde{x}_1, \ldots, \tilde{x}_n, \varepsilon) > 0$ such that $|t - \tilde{t}| < \delta$, $\|x_1 - \tilde{x}_1\| < \delta, \ldots, \|x_n - \tilde{x}_n\| < \delta$ implies $\|\varphi_k(t, x_1, \ldots, x_n) - \varphi_k(\tilde{t}, \tilde{x}_1, \ldots, \tilde{x}_n)\| < \varepsilon$ (cf. Sec. 2.33 for $n = 1$). The function $\varphi_k(t,x_1,\ldots, x_n)$ is said to satisfy a *Lipschitz condition in the arguments* $x_1,\ldots, x_n$ if there exists a constant $C_k>0$ such that the inequality

$$\|\varphi_k(t,x_1,\ldots, x_n)-\varphi_k(t,\tilde{x}_1,\ldots, \tilde{x}_n)\| \leqslant C_k \sum_{l=1}^{n} \|x_l-\tilde{x}_l\| \tag{3}$$

holds for all $t \in [a,b]$ and $x_1,\tilde{x}_1,\ldots, x_n,\tilde{x}_n \in \mathbf{B}$ (cf. Sec. 2.34 for $n=1$).

2.42. The system (1) and the initial condition (2) are easily replaced by a single differential equation and a single initial condition by going over to the space $\mathbf{B}_n$ of all vectors $X=(x_1,\ldots, x_n)$ with components $x_1,\ldots, x_n$ in the Banach space $\mathbf{B}$. We define linear operations in $\mathbf{B}_n$ in the natural way, i.e., if $X=(x_1,\ldots, x_n) \in \mathbf{B}_n$, $Y=(y_1,\ldots,y_n) \in \mathbf{B}_n$, then

$$X+Y=(x_1+y_1,\ldots, x_n+y_n),$$

while if α is a real number, then

$$\alpha X=(\alpha x_1,\ldots, \alpha x_n).$$

Thus $\mathbf{B}_n$ is a linear space, the validity of the axioms of Sec. 1.11 for $\mathbf{B}_n$ being an immediate consequence of their validity for $\mathbf{B}$. Equipping $\mathbf{B}_n$ with the norm

$$\|X\| = \sum_{k=1}^{n} \|x_k\|,$$

we make $\mathbf{B}_n$ into a *normed* linear space, the validity of the axioms of Sec. 1.31 for the norm $\|X\|$ being an immediate consequence of their validity for the norm $\|x\|$. If $X^{(m)}=(x_1^{(m)},\ldots, x_n^{(m)})$ is a sequence of vectors in $\mathbf{B}_n$, then clearly $X^{(m)}\to X$ in the norm of $\mathbf{B}_n$ if and only if $x_1^{(m)}\to x_1,\ldots, x_n^{(m)}\to x_n$ in the norm of $\mathbf{B}$. Moreover, it is easy to see that the completeness of $\mathbf{B}$ implies that of $\mathbf{B}_n$ (give the details). Thus $\mathbf{B}_n$ is a Banach space, just like $\mathbf{B}$ itself.

In terms of the space $\mathbf{B}_n$, we can now replace the system (1) and the conditions (2) by a single differential equation

$$U'(t)=\Phi(t,U(t)) \tag{1'}$$

and a single initial condition

$$U(t_0)=P, \tag{2'}$$

where $U(t)=(u_1(t),\ldots, u_n(t))$ is a vector function taking values in $\mathbf{B}_n$ for every $t \in [a,b]$, $P=(p_1,\ldots,p_n)$ is a fixed vector in $\mathbf{B}_n$, and†

† More exactly, (4) means that $\Phi(t,X)$ carries the vector $X=(x_1,\ldots, x_n) \in \mathbf{B}_n$ into the vector $Y=(y_1,\ldots,y_n) \in \mathbf{B}_n$, where $y_1=\varphi_1(t,x_1,\ldots, x_n),\ldots,y_n=\varphi_n(t,x_1,\ldots, x_n)$.

$$\Phi(t,X)=(\varphi_1(t,x_1,\dots,x_n),\dots,\varphi_n(t,x_1,\dots,x_n)) \tag{4}$$

is a mapping of $\mathbf{B}_n$ into itself depending on the parameter $t\in[a,b]$.

2.43. THEOREM (**Existence and uniqueness of solutions of a system of equations**). *Given a Banach space* **B** *and n functions*

$$\varphi_1(t,x_1,\dots,x_n),\dots,\varphi_n(t,x_1,\dots,x_n)$$

defined for all $t\in[a,b]$ *and* $x_1,\dots,x_n\in\mathbf{B}$, *suppose each* $\varphi_k(t,x_1,\dots,x_n)$ *is continuous in all its arguments and satisfies a Lipschitz condition* (3). *Then among all sets of n differentiable functions defined on* $[a,b]$ *with values in* **B**, *there exists a unique set* $u_1(t),\dots,u_n(t)$ *satisfying the system of differential equations* (1) *and the initial conditions* (2).

Proof. Since a set of n functions $u_1(t),\dots,u_n(t)$ taking values in **B** satisfies the system (1) and the conditions (2) if and only if the vector function $U(t)=(u_1(t),\dots,u_n(t))$ taking values in $\mathbf{B}_n$ satisfies the equation (1′) and the condition (2′), we need only show that the conditions imposed on the functions $\varphi_k(t,x_1,\dots,x_n)$ guarantee that the function (4) satisfies the conditions of Theorem 2.35. Thus we must show that $\Phi(t,X)$ is continuous in both its arguments and satisfies a Lipschitz condition in the argument X. To prove that $\Phi(t,X)$ is continuous in both its arguments, we argue as follows: Given any $\tilde t\in[a,b]$, $\tilde X=(\tilde x_1,\dots,\tilde x_n)\in\mathbf{B}_n$ and any $\varepsilon>0$, use the continuity of each function $\varphi_k(t,x_1,\dots,x_n)$ to choose $\delta_k>0$ such that $|t-\tilde t|<\delta_k$, $\|x_1-\tilde x_1\|<\delta_k,\dots$, $\|x_n-\tilde x_n\|<\delta_k$ implies

$$\|\varphi_k(t,x_1,\dots,x_n)-\varphi_k(\tilde t,\tilde x_1,\dots,\tilde x_n)\|<\frac{\varepsilon}{n},$$

and let $\delta=\min\{\delta_1,\dots,\delta_n\}$. Then $|t-\tilde t|<\delta$, $\|X-\tilde X\|<\delta$ implies $|t-\tilde t|<\delta$, $\|x_1-\tilde x_1\|<\delta,\dots,\|x_n-\tilde x_n\|<\delta$, since obviously

$$\|x_1-\tilde x_1\|\leqslant\|X-\tilde X\|,\dots,\|x_n-\tilde x_n\|\leqslant\|X-\tilde X\|,$$

which in turn implies

$$\|\Phi(t,X)-\Phi(\tilde t,\tilde X)\|=\sum_{k=1}^{n}\|\varphi_k(t,x_1,\dots,x_n)-\varphi_k(\tilde t,\tilde x_1,\dots,\tilde x_n)\|<n\frac{\varepsilon}{n}=\varepsilon.$$

To prove that $\Phi(t,X)$ satisfies a Lipschitz condition in the argument X, we observe that

$$\|\Phi(t,X)-\Phi(t,\tilde X)\|$$
$$=\sum_{k=1}^{n}\|\varphi_k(t,x_1,\dots,x_n)-\varphi_k(t,\tilde x_1,\dots,\tilde x_n)\|\leqslant\sum_{k=1}^{n}C_k\sum_{l=1}^{n}\|x_l-\tilde x_l\|$$

for all $t \in [a,b]$ and $X=(x_1,\ldots, x_n) \in \mathbf{B}_n$, $\tilde{X}=(\tilde{x}_1,\ldots, \tilde{x}_n) \in \mathbf{B}_n$, since each $\varphi_k(t,x_1,\ldots, x_n)$ satisfies a Lipschitz condition (3). But

$$\|X-\tilde{X}\| = \sum_{l=1}^{n} \|x_l-\tilde{x}_l\|,$$

and hence

$$\|\Phi(t,X)-\Phi(t,\tilde{X})\| \leqslant C\|X-\tilde{X}\|,$$

where $C=n \max\{C_1,\ldots, C_n\}$. ∎

2.44. Remark. Suppose there exists a closed subspace $\mathbf{B}^* \subset \mathbf{B}$ such that $\varphi_1(t,x_1,\ldots, x_n) \in \mathbf{B}^*, \ldots, \varphi_n(t,x_1,\ldots, x_n) \in \mathbf{B}^*$ for all $t \in [a,b]$ whenever $x_1, \ldots, x_n \in \mathbf{B}^*$, and suppose the initial vectors $p_1,\ldots, p_n$ also belong to $\mathbf{B}^*$. Then *the values of the solution* $u_1(t),\ldots, u_n(t)$ *of the system* (1) *satisfying the initial conditions* (2) *lie in the subspace* $\mathbf{B}^*$ *for all* $t \in [a,b]$. In fact, let $\mathbf{B}_n^*$ be the subspace of $\mathbf{B}_n$ consisting of all vectors $X=(x_1,\ldots, x_n)$ whose components $x_1,\ldots, x_n$ all belong to $\mathbf{B}^*$. Then $P=(p_1,\ldots, p_n) \in \mathbf{B}_n^*$, and the function (4) maps $\mathbf{B}_n^*$ into itself. Moreover, $\mathbf{B}_n^*$ is closed in $\mathbf{B}_n$ (why?). The italicized assertion now follows by exactly the same argument as in Sec. 2.36.

2.5. Higher-Order Equations

2.51. Once again let $\mathbf{B}$ be a Banach space, but this time suppose we are given a single function $\varphi(t,x_1,\ldots, x_n)$ depending on a real parameter $t \in [a,b]$ and on n arguments $x_1,\ldots, x_n$, where each x_k is a point in $\mathbf{B}$ and the function $\varphi(t,x_1,\ldots, x_n)$ takes its values in $\mathbf{B}$ as well. Consider the differential equation

$$u^{(n)}(t)=\varphi(t,u(t),\ldots, u^{(n-1)}(t)) \tag{1}$$

of order n, with the initial conditions

$$u(t_0)=p_1 \in \mathbf{B},\ u'(t_0)=p_2 \in \mathbf{B},\ldots, u^{(n-1)}(t_0)=p_n \in \mathbf{B}. \tag{2}$$

As before, the function $\varphi(t,x_1,\ldots, x_n)$ is said to satisfy a *Lipschitz condition in the arguments* $x_1,\ldots, x_n$ if there exists a constant $C>0$ such that the inequality

$$\|\varphi(t,x_1,\ldots, x_n)-\varphi(t,\tilde{x}_1,\ldots, \tilde{x}_n)\| \leqslant C\sum_{k=1}^{n} \|x_k-\tilde{x}_k\| \tag{3}$$

holds for all $t \in [a,b]$ and $x_1,\tilde{x}_1,\ldots, x_n,\tilde{x}_n \in \mathbf{B}$.

2.52. THEOREM (**Existence and uniqueness of solutions of a higher-order equation**). *Given a Banach space* $\mathbf{B}$ *and a function* $\varphi(t,x_1,\ldots, x_n)$ *defined for all* $t \in [a,b]$ *and* $x_1,\ldots, x_n \in \mathbf{B}$, *suppose* $\varphi(t,x_1,\ldots, x_n)$ *is continuous in all its argu-*

ments and satisfies a Lipschitz condition (3). *Then among all n-fold differentiable functions defined on* $[a,b]$ *with values in* $\mathbf{B}$, *there exists a unique function* $u(t)$ *satisfying the differential equation* (1) *and the initial conditions* (2).

Proof. Besides the system (1) and the conditions (2), consider the system of differential equations

$$\begin{aligned} u_1'(t) &= u_2(t), \\ u_2'(t) &= u_3(t), \\ &\cdots \\ u_{n-1}'(t) &= u_n(t), \\ u_n'(t) &= \varphi(t, u_1(t), \ldots, u_n(t)) \end{aligned} \tag{4}$$

and the initial conditions

$$\begin{aligned} u_1(t_0) &= p_1, \\ u_2(t_0) &= p_2, \\ &\cdots \\ u_n(t_0) &= p_n. \end{aligned} \tag{5}$$

The system (4) is a special case of the system (1), p. 169, corresponding to the special choice of functions

$$\begin{aligned} \varphi_1(t, x_1, \ldots, x_n) &\equiv x_2, \\ \varphi_2(t, x_1, \ldots, x_n) &\equiv x_3, \\ &\cdots \\ \varphi_{n-1}(t, x_1, \ldots, x_n) &\equiv x_n, \\ \varphi_n(t, x_1, \ldots, x_n) &\equiv \varphi(t, x_1, \ldots, x_n). \end{aligned} \tag{6}$$

Clearly these functions are continuous in all their arguments and satisfy Lipschitz conditions of the type (3), p. 170. (This is obvious for the first $n-1$ functions, and is explicitly assumed for the last function.) It follows from Theorem 2.43 that the system (4) has a unique solution $u_1(t), \ldots, u_n(t)$ satisfying the conditions (5). Let $u(t) \equiv u_1(t)$. Then the first equation of the system (4) implies $u'(t) = u_2(t)$, the next equation implies $u''(t) = u_2'(t) = u_3(t)$, and so on, up to $u^{(n-1)}(t) = u_{n-1}'(t) = u_n(t)$, while the last equation becomes

$$u^{(n)}(t) = u_n'(t) = \varphi(t, u(t), \ldots, u^{(n-1)}(t)),$$

so that $u(t)$ satisfies equation (1). Moreover, $u(t)$ also satisfies the conditions (2), to which the conditions (5) reduce after setting $u(t) \equiv u_1(t)$. Therefore equation (1) has a solution $u(t)$ satisfying the conditions (2).

To see that this solution $u(t)$ is unique, suppose $\tilde{u}(t)$ is another solution of (1) satisfying (2). Then the set of functions

$$\tilde{u}_1(t) \equiv \tilde{u}(t),\ \tilde{u}_2(t) \equiv \tilde{u}'(t), \ldots,\ \tilde{u}_n(t) \equiv \tilde{u}^{(n-1)}(t)$$

obviously satisfies the system (4) and the conditions (5). But the solution of (4) satisfying (5) is unique, by Theorem 2.43, and hence, in particular, $\tilde{u}(t) \equiv \tilde{u}_1(t) \equiv u_1(t) \equiv u(t)$. ∎

2.53. Remark. Suppose there exists a closed subspace $\mathbf{B}^* \subset \mathbf{B}$ such that $\varphi(t, x_1, \ldots, x_n) \in \mathbf{B}^*$ for all $t \in [a,b]$ whenever $x_1, \ldots, x_n \in \mathbf{B}^*$, and suppose the vectors $p_1, \ldots, p_n$ figuring in the initial conditions (2) also belong to $\mathbf{B}^*$. Then *the solution $u(t)$ of equation (1) satisfying the initial conditions (2) lies in the subspace $\mathbf{B}^*$ for all $t \in [a,b]$*. In fact, under these conditions, all the functions (6) take their values in $\mathbf{B}^*$ whenever $x_1, \ldots, x_n \in \mathbf{B}^*$. It follows from Sec. 2.44 that the corresponding values of the solution $u_1(t), \ldots, u_n(t)$ of the system (4) lie in $\mathbf{B}^*$ for all $t \in [a,b]$. Since $u_1(t) \equiv u(t)$, the italicized assertion follows at once.

2.6. Linear Equations and Systems

2.61. Let $\mathbf{A}(t)$ be a bounded linear operator mapping a Banach space $\mathbf{B}$ into itself and depending on a parameter $t \in [a,b]$. Such an operator $\mathbf{A}(t)$ is said to *depend continuously on the parameter t* if given any $t \in [a,b]$ and any $\varepsilon > 0$, there exists a $\delta = \delta(t, \varepsilon) > 0$ such that $|t - \tilde{t}| < \delta$ implies $\|\mathbf{A}(t) - \mathbf{A}(\tilde{t})\| < \varepsilon$, where $\|\cdots\|$ denotes the operator norm, as defined in Sec. 1.71b. It then follows (cf. Vol. 1, Sec. 5.17b) that $\mathbf{A}(t)$ is *uniformly* continuous on $[a,b]$, i.e., given any $\varepsilon > 0$, there is a $\delta = \delta(\varepsilon) > 0$ such that $|t - \tilde{t}| < \delta$ implies $\|\mathbf{A}(t) - \mathbf{A}(\tilde{t})\| < \varepsilon$ for all $t, \tilde{t} \in [a,b]$.

A function $\varphi(t,x)$ with values in a space $\mathbf{B}$ is said to be *linear in the argument $x \in \mathbf{B}$* if

$$\varphi(t,x) = \mathbf{A}(t)x + \beta(t), \tag{1}$$

where $\mathbf{A}(t)$ is a bounded linear operator depending continuously on the parameter $t \in [a,b]$ and $\beta(t)$ is a (uniformly) continuous function on $[a,b]$ with its values in the space $\mathbf{B}$.

2.62.a. LEMMA. *The function (1) is continuous in both its arguments t and x.*

Proof. The operator $\mathbf{A}(t)$ can be regarded as a continuous function of t with values in the Banach space $\mathbf{L}(\mathbf{B})$ of all bounded linear operators acting in $\mathbf{B}$ (cf. Sec. 1.73d). Such a function is bounded on the interval $a \leqslant t \leqslant b$ (cf. Vol. 1, Secs. 5.15b, 5.16b), and hence

$$C = \sup_{a \leqslant t \leqslant b} \|\mathbf{A}(t)\| < \infty. \tag{2}$$

Given any $\tilde{t} \in [a,b]$, $\tilde{x} \in \mathbf{B}$ and any $\varepsilon > 0$, use the continuity of $\mathbf{A}(t)$ to choose $\delta_1 > 0$ such that $|t - \tilde{t}| < \delta_1$ implies

$$\|\tilde{x}\| \, \|\mathbf{A}(t) - \mathbf{A}(\tilde{t})\| < \frac{\varepsilon}{3}$$

and the continuity of $\beta(t)$ to choose $\delta_2 > 0$ such that $|t - \tilde{t}| < \delta_2$ implies

$$\|\beta(t) - \beta(\tilde{t})\| < \frac{\varepsilon}{3}.$$

But then, choosing

$$\delta = \min \left\{ \delta_1, \delta_2, \frac{\varepsilon}{3C} \right\},$$

we find that $|t - \tilde{t}| < \delta$, $\|x - \tilde{x}\| < \delta$ implies

$$\begin{aligned} \|\varphi(t,x) - \varphi(\tilde{t},\tilde{x})\| &= \|\mathbf{A}(t)x - \mathbf{A}(t)\tilde{x} + \mathbf{A}(t)\tilde{x} - \mathbf{A}(\tilde{t})\tilde{x} + \beta(t) - \beta(\tilde{t})\| \\ &\leqslant \|\mathbf{A}(t)\| \, \|x - \tilde{x}\| + \|\mathbf{A}(t) - \mathbf{A}(\tilde{t})\| \, \|\tilde{x}\| + \|\beta(t) - \beta(\tilde{t})\| \\ &< C\frac{\varepsilon}{3C} + \frac{\varepsilon}{3} + \frac{\varepsilon}{3} = \varepsilon. \quad \blacksquare \end{aligned}$$

b. LEMMA. *The function* (1) *satisfies a Lipschitz condition in the argument x with constant* (2).

Proof. By the definition of the norm of an operator,

$$\|\varphi(t,x) - \varphi(t,\tilde{x})\| = \|\mathbf{A}(t)x - \mathbf{A}(t)\tilde{x}\| \leqslant \|\mathbf{A}(t)\| \, \|x - \tilde{x}\| \leqslant C\|x - \tilde{x}\|. \quad \blacksquare$$

c. THEOREM (**Existence and uniqueness of solutions of a first-order linear equation**). *Let* $\mathbf{A}(t)$ *be a bounded linear operator acting in a Banach space* $\mathbf{B}$ *and depending continuously on a parameter* $t \in [a,b]$, *and let* $\beta(t)$ *be a continuous function on* $[a,b]$ *with values in* $\mathbf{B}$. *Then among all differentiable functions defined on* $[a,b]$ *with values in* $\mathbf{B}$, *there exists a unique function* $u(t)$ *satisfying the linear differential equation*

$$u'(t) = \mathbf{A}(t)u(t) + \beta(t)$$

and the initial condition

$$u(t_0) = u_0 \in \mathbf{B}.$$

Proof. An immediate consequence of Theorem 2.35 and the preceding lemmas. $\blacksquare$

2.63. Systems of linear equations. Next consider the system of linear equations

$$\begin{aligned} u_1'(t) &= \mathbf{A}_{11}(t)u_1(t) + \cdots + \mathbf{A}_{1n}(t)u_n(t) + \beta_1(t), \\ &\cdots \\ u_n'(t) &= \mathbf{A}_{n1}(t)u_1(t) + \cdots + \mathbf{A}_{nn}(t)u_n(t) + \beta_n(t), \end{aligned} \tag{3}$$

with the initial conditions

$$\begin{aligned} u_1(t_0) &= p_1 \in \mathbf{B}, \\ &\cdots \\ u_n(t_0) &= p_n \in \mathbf{B}, \end{aligned} \tag{4}$$

where the $\mathbf{A}_{jk}(t)$ $(j,k=1,\ldots, n)$ are all bounded linear operators acting in a space $\mathbf{B}$ and depending continuously on a parameter $t \in [a,b]$, while the $\beta_j(t)$ $(j=1,\ldots, n)$ are continuous vector functions defined on $[a,b]$ and taking values in $\mathbf{B}$.

THEOREM (**Existence and uniqueness of solutions of a system of linear equations**). *Let $\mathbf{A}_{jk}(t)$ $(j,k=1,\ldots,n)$ be n^2 bounded linear operators acting in a Banach space $\mathbf{B}$ and depending continuously on a parameter $t \in [a,b]$, and let $\beta_j(t)$ $(j=1,\ldots, n)$ be n continuous functions defined on $[a,b]$ with values in $\mathbf{B}$. Then among all sets of n differentiable functions defined on $[a,b]$ with values in $\mathbf{B}$, there exists a unique set $u_1(t),\ldots, u_n(t)$ satisfying the system of linear differential equations* (3) *and the initial conditions* (4).

Proof. The system (3) is a special case of the system

$$\begin{aligned} u_1'(t) &= \varphi_1(t,u_1,\ldots, u_n), \\ &\cdots \\ u_n'(t) &= \varphi_n(t,u_1,\ldots, u_n), \end{aligned}$$

corresponding to the choice

$$\varphi_j(t,x_1,\ldots, x_n) = \mathbf{A}_{j1}(t)x_1 + \cdots + \mathbf{A}_{jn}(t)x_n + \beta_j(t) \qquad (j=1,\ldots, n). \tag{5}$$

As already shown, each term $\mathbf{A}_{jk}(t)x_k$ in (5) is continuous in both t and x_k, and satisfies a Lipschitz condition in x_k. But then the sum (5) is continuous in all the arguments $t,x_1,\ldots, x_n$, and satisfies a Lipschitz condition in $x_1,\ldots, x_n$ (why?). The proof now follows at once by applying Theorem 2.43. ∎

2.64. Remark. Suppose that for all $t \in [a,b]$ the operators $\mathbf{A}_{jk}(t)$ $(j,k=1,\ldots,n)$ map a fixed subspace $\mathbf{B}^* \subset \mathbf{B}$ into itself and the functions $\beta_j(t)$ $(j=1,\ldots, n)$ take all their values in $\mathbf{B}^*$, while the initial vectors $p_1,\ldots, p_n$ also belong to $\mathbf{B}^*$. Then *the values of the solution $u_1(t),\ldots, u_n(t)$ of the system* (3) *satisfying the initial* *conditions* (4) *lie in the subspace* $\mathbf{B}^*$ *for all* $t \in [a,b]$. In fact, under these conditions, $\varphi_1(t,x_1,\ldots, x_n) \in \mathbf{B}^*,\ldots, \varphi_n(t,x_1,\ldots, x_n) \in \mathbf{B}^*$ for all $t \in [a,b]$ whenever $x_1,\ldots, x_n \in \mathbf{B}^*$, and we can apply Remark 2.44.

2.65. THEOREM (**Existence and uniqueness of solutions of a higher-order linear equation**). *Let* $\mathbf{A}_1(t),\ldots, \mathbf{A}_n(t)$ *be bounded linear operators acting in a Banach space* $\mathbf{B}$ *and depending continuously on a parameter* $t \in [a,b]$, *and let* $\beta(t)$ *be a continuous function on* $[a,b]$ *with values in* $\mathbf{B}$. *Then among all n-fold differentiable functions defined on* $[a,b]$ *with values in* $\mathbf{B}$, *there exists a unique function* $u(t)$ *satisfying the linear differential equation*

$$u^{(n)}(t) = \mathbf{A}_1(t)u(t) + \cdots + \mathbf{A}_n(t)u^{(n-1)}(t) + \beta(t) \tag{6}$$

of order n and the initial conditions

$$u(t_0) = p_1 \in \mathbf{B},\ u'(t_0) = p_2 \in \mathbf{B},\ldots,\ u^{(n-1)}(t_0) = p_n \in \mathbf{B}. \tag{7}$$

Proof. Equation (6) is a special case of the equation

$$u^{(n)}(t) = \varphi(t,u(t),\ldots, u^{(n-1)}(t)),$$

corresponding to the choice

$$\varphi(t,x_1,\ldots, x_n) = \mathbf{A}_1(t)x_1 + \cdots + \mathbf{A}_n(t)x_n + \beta(t), \tag{8}$$

where, just as in the proof of Theorem 2.63, the function (8) is continuous in all the arguments $t,x_1,\ldots, x_n$ and satisfies a Lipschitz condition in $x_1,\ldots, x_n$. The proof now follows at once by applying Theorem 2.52. ∎

2.66. Remark. Suppose that for all $t \in [a,b]$ the operators $\mathbf{A}_1(t),\ldots, \mathbf{A}_n(t)$ map a fixed subspace $\mathbf{B}^* \subset \mathbf{B}$ into itself and the function $\beta(t)$ takes all its values in $\mathbf{B}^*$, while the initial vectors $p_1,\ldots, p_n$ also belong to $\mathbf{B}^*$. Then *the solution* $u(t)$ *of equation* (6) *satisfying the initial conditions* (7) *lies in the subspace* $\mathbf{B}^*$ *for all* $t \in [a,b]$. In fact, under these conditions, $\varphi(t, x_1, \ldots, x_n) \in \mathbf{B}^*$ for all $t \in [a,b]$ whenever $x_1,\ldots, x_n \in \mathbf{B}^*$, and we can apply Remark 2.53.

2.7. The Homogeneous Linear Equation

2.71. Choosing $\beta(t) \equiv 0$ in the general linear equation

$$u'(t) = \mathbf{A}(t)u(t) + \beta(t) \qquad (a \leqslant t \leqslant b),$$

we get the *homogeneous* linear equation

$$u'(t) = \mathbf{A}(t)u(t) \qquad (a \leqslant t \leqslant b). \tag{1}$$

Equation (1) has the obvious solution $u(t) \equiv 0$. The other solutions of (1) cannot vanish for any $t \in [a,b]$, because of the uniqueness part of Theorem 2.62c.

The set L of solutions of equation (1) is closed under linear operations, i.e., if $u_1(t)$ and $u_2(t)$ are solutions of (1), then so is $\alpha_1 u_1(t) + \alpha_2 u_2(t)$ for arbitrary real numbers α_1 and α_2. This follows at once from the observation that

$$\begin{aligned}[\alpha_1 u_1(t) + \alpha_2 u_2(t)]' &= \alpha_1 u_1'(t) + \alpha_2 u_2'(t) \\ &= \alpha_1 \mathbf{A}(t) u_1(t) + \alpha_2 \mathbf{A}(t) u_2(t) = \mathbf{A}(t)[\alpha_1 u_1(t) + \alpha_2 u_2(t)].\end{aligned}$$

It is easy to see that L satisfies all the axioms of a linear space (Sec. 1.11).

Let $\Omega_{t_0}^t$ be the solution operator (Sec. 2.38) of the homogeneous linear equation (1). Then $\Omega_{t_0}^t$ is a linear operator, i.e.,

$$\Omega_{t_0}^t(\alpha_1 u_1 + \alpha_2 u_2) = \alpha_1 \Omega_{t_0}^t u_1 + \alpha_2 \Omega_{t_0}^t u_2 \tag{2}$$

for arbitrary initial vectors u_1, u_2 and real numbers α_1, α_2. In fact, the right-hand side of (2) is a linear combination of solutions of equation (1), equal to $\alpha_1 u_1 + \alpha_2 u_2$ at $t = t_0$, and this linear combination is itself a solution of (1), as just proved. On the other hand, by the very definition of the solution operator, the left-hand side of (2) is also a solution of (1) equal to $\alpha_1 u_1 + \alpha_2 u_2$ at $t = t_0$. But then the two sides of (2) must coincide for all $t \in [a,b]$, again by the uniqueness part of Theorem 2.62c.

Thus the solution operator $\Omega_{t_0}^t$ of the homogeneous linear equation (1) is a linear operator. Moreover, it will be recalled that $\Omega_{t_0}^t$ is continuous (Sec. 2.38a) and invertible (Sec. 2.38d).

2.72. We now analyze the structure of the solution operator $\Omega_{t_0}^t$ of the homogeneous linear equation (1) for the case where the underlying Banach space $\mathbf{B}$ is the n-dimensional space R_n of vectors $x = (\xi_1, \ldots, \xi_n)$. Let $f_1, \ldots, f_n$ be any n linearly independent vectors in R_n. Then the vector equation (1), involving the unknown vector function

$$u(t) = \sum_{k=1}^{n} u_k(t) f_k,$$

gives rise to a system of n scalar equations

$$\begin{aligned} u_1'(t) &= a_{11}(t) u_1(t) + \cdots + a_{1n} u_n(t), \\ &\cdots \\ u_n'(t) &= a_{n1}(t) u_1(t) + \cdots + a_{nn} u_n(t). \end{aligned} \tag{3}$$

According to Sec. 1.17a, the solution operator $\Omega_{t_0}^t$ is associated with a matrix whose kth column is made up of the components of the vector

$$f_k(t) \equiv \Omega_{t_0}^t f_k \qquad (k=1,\dots, n).$$

In other words, with the solution operator $\Omega_{t_0}^t$ we associate a matrix

$$W_{t_0}^t = \begin{Vmatrix} f_{11}(t) & f_{12}(t) & \cdots & f_{1n}(t) \\ f_{21}(t) & f_{22}(t) & \cdots & f_{2n}(t) \\ \cdot & \cdot & \cdots & \cdot \\ f_{n1}(t) & f_{n2}(t) & \cdots & f_{nn}(t) \end{Vmatrix}, \tag{4}$$

where $f_{1k}(t),\dots,f_{nk}(t)$ are the components of the solution $f_k(t)$ which reduces to the vector f_k for $t=t_0$. The matrix (4) is called the *Wronskian matrix* of the system (3), and its determinant is called the *Wronskian determinant* (or simply the *Wronskian*) of (3). In Sec. 2.15 we in effect calculated the Wronskian matrix for the case of a constant operator $\mathbf{A}(t) \equiv \mathbf{A}$.

Since the operator $\Omega_{t_0}^t$ is invertible, the matrix (4) is nonsingular for all $t \in [a,b]$, i.e., the Wronskian is nonvanishing for all $t \in [a,b]$. Hence the solutions $f_1(t),\dots, f_n(t)$, which are linearly independent at $t=t_0$, remain linearly independent for all $t \in [a,b]$.

The solution of equation (1) with an arbitrary initial vector

$$u = \sum_{k=1}^{n} u_k f_k$$

(at $t=t_0$) is constructed by using the general formula

$$u(t) = \Omega_{t_0}^t u = \Omega_{t_0}^t \sum_{k=1}^{n} u_k f_k = \sum_{k=1}^{n} u_k \Omega_{t_0}^t f_k = \sum_{k=1}^{n} u_k f_k(t).$$

Thus every solution of the system (3) is a linear combination of the n particular solutions $f_1(t),\dots, f_n(t)$. In other words, in the case where $\mathbf{B}$ is the n-dimensional space R_n, the linear space of all solutions of the system (3) is also n-dimensional, with the vector functions

$$f_1(t),\dots,f_n(t) \tag{5}$$

as a basis. The set of functions (5) is called a *fundamental system of solutions* of (3).

2.73. Calculation of the Wronskian. The Wronskian

$$[f_1(t), f_2(t),\dots, f_n(t)] \equiv \begin{vmatrix} f_{11}(t) & f_{12}(t) & \cdots & f_{1n}(t) \\ \cdot & \cdot & \cdots & \cdot \\ f_{n1}(t) & f_{n2}(t) & \cdots & f_{nn}(t) \end{vmatrix} \tag{6}$$

can be regarded as a kind of "mixed product" of the vectors $f_1(t)$, $f_2(t)$,..., $f_n(t)$, by analogy with vector algebra. Having differentiable functions as its elements, the Wronskian is itself differentiable, with derivative

$$\begin{aligned}\frac{d}{dt}[f_1(t),f_2(t),\dots,f_n(t)] &= [f'_1(t),f_2(t),\dots,f_n(t)]+[f_1(t),f'_2(t),\dots,f_n(t)]\\ &\quad+\cdots+[f_1(t),f_2(t),\dots,f'_n(t)]\\ &= [\mathbf{A}(t)f_1(t),f_2(t),\dots,f_n(t)]+[f_1(t),\,\mathbf{A}(t)f_2(t),\dots,f_n(t)]\\ &\quad+\cdots+[f_1(t),f_2(t),\dots,\,\mathbf{A}f_n(t)] \qquad (7)\end{aligned}$$

(cf. Vol. 1, Sec. 7.14e). In the kth term of the sum on the right, the only relevant component of the vector $\mathbf{A}(t)f_k(t)$ is the component in the direction of the vector $f_k(t)$, namely $a_{kk}(t)f_k(t)$, since every other component in the mixed product leads to a determinant which has two identical columns and hence vanishes. Thus (7) reduces to

$$\begin{aligned}\frac{d}{dt}[f_1(t),\dots,f_n(t)] &= [a_{11}(t)+\cdots+a_{nn}(t)][f_1(t)+\cdots+f_n(t)]\\ &= [f_{11}(t),\dots,f_{nn}(t)]\ \mathrm{tr}\ \mathbf{A}(t), \qquad (8)\end{aligned}$$

in terms of the *trace* of the operator $\mathbf{A}(t)$,† a quantity which is independent of the choice of the basis $f_1,\dots,\ f_n$ (cf. Lin. Alg., Sec. 5.53). Integrating (8), we find that

$$[f_1(t),\dots,f_n(t)]=[f_1(t_0),\dots,f_n(t_0)]\ \exp\left\{\int_{t_0}^{t}\mathrm{tr}\ \mathbf{A}(\tau)\ d\tau\right\}. \qquad (9)$$

2.74. The nth-order equation. As we saw in Sec. 2.51, the equation

$$u^{(n)}(t)=a_1(t)u(t)+\cdots+a_n(t)u^{(n-1)}(t)+\beta(t) \qquad (10)$$

is equivalent to the system

$$\begin{aligned}&u'_1(t)=u_2(t),\\ &u'_2(t)=u_3(t),\\ &\cdots \qquad\qquad\qquad (11)\\ &u'_n(t)=a_1(t)u_1(t)+\cdots+a_n(t)u_n(t)+\beta(t),\end{aligned}$$

where

$$u_1(t)=u(t),\ u_2(t)=u'(t),\dots,\ u_n(t)=u^{(n-1)}(t),$$

so that every solution $u_1(t),\ u_2(t),\dots,\ u_n(t)$ of (11) gives rise to a solution $u(t)\equiv u_1(t)$ of (10). According to Sec. 2.72, every solution $u(t)$ of the homo-

† By the *trace* of the operator $\mathbf{A}$, denoted by tr $\mathbf{A}$, is meant the sum of the diagonal elements of the matrix of $\mathbf{A}$ (in any basis).

geneous equation

$$u^{(n)}(t)=a_1(t)u(t)+\cdots+a_n(t)u^{(n-1)}(t), \qquad (10')$$

obtained from (10) by setting $\beta(t)\equiv 0$, can be represented in the form

$$u(t)=c_1u_1(t)+\cdots+c_nu_n(t),$$

where the set of functions $u_1(t),\ldots, u_n(t)$ is the solution of the system (11) with $\beta(t)\equiv 0$, corresponding to some nonsingular "initial data matrix"

$$[u_1(t_0),\ldots, u_n(t_0)]=\left\| \begin{matrix} u_1(t_0) & \cdots & u_n(t_0) \\ \cdot & \cdots & \cdot \\ u_1^{(n-1)}(t_0) & \cdots & u_n^{(n-1)}(t_0) \end{matrix} \right\|.$$

The corresponding solution (or Wronskian) matrix

$$[u_1(t),\ldots, u_n(t)]=\left\| \begin{matrix} u_1(t) & \cdots & u_n(t) \\ \cdot & \cdots & \cdot \\ u_1^{(n-1)}(t) & \cdots & u_n^{(n-1)}(t) \end{matrix} \right\|$$

remains nonsingular for all $t\in[a,b]$. It follows from formula (9) after taking account of the special form of the system (11) that

$$\det[u_1(t),\ldots, u_n(t)]=\det[u_1(t_0),\ldots,u_n(t_0)]\exp\left\{\int_{t_0}^{t} a_n(\tau)\,d\tau\right\}.$$

2.8. The Nonhomogeneous Linear Equation

2.81. We now return to the general linear equation

$$u'(t)=\mathbf{A}(t)u(t)+\beta(t) \qquad (a\leqslant t\leqslant b), \qquad (1)$$

where the function $\beta(t)$ is not identically zero, a fact emphasized by calling (1) the *nonhomogeneous* linear equation. If $v_1(t)$ and $v_2(t)$ are two solutions of the nonhomogeneous equation (1), then their difference $v_1(t)-v_2(t)$ is obviously a solution of the homogeneous equation (1), p. 177. Therefore, starting from the solution operator $\Omega_{t_0}^t$ of the homogeneous equation and any particular solution $v_1(t)$ of the nonhomogeneous equation (for example, the solution equal to 0 at $t=t_0$), we can get an arbitrary solution of the nonhomogeneous equation by using the formula

$$v(t)=v_1(t)+\Omega_{t_0}^t v_0. \qquad (2)$$

The operator $\Omega_{t_0}^t$ can also be used to construct the particular solution $v_1(t)$ (vanishing for $t=t_0$), by using the method of "variation of constants," which goes as follows: Suppose we write $v_1(t)$ in the form

$$v_1(t) = \Omega^t_{t_0} C(t), \tag{3}$$

where $C(t)$ is a variable vector to be determined in a moment. The vector $C(t)$ must depend on t (hence the designation "variation of constants"), since if $C(t)$ were constant, (3) would give a solution of the *homogeneous* linear equation. Moreover, since $v_1(t_0)=0$, we must have $C(t_0)=0$. By Lemma 2.13 and Sec. 2.38e,

$$v_1'(t) = (\Omega^t_{t_0})' C(t) + \Omega^t_{t_0} C'(t) = \mathbf{A}(t)\Omega^t_{t_0} C(t) + \Omega^t_{t_0} C'(t), \tag{4}$$

while, on the other hand,

$$\mathbf{A}(t) v_1(t) + \beta(t) = \mathbf{A}(t)\Omega^t_{t_0} C(t) + \beta(t). \tag{5}$$

Equating the right-hand sides of (4) and (5), we find that

$$\Omega^t_{t_0} C'(t) = \beta(t). \tag{6}$$

Then applying the operator $\Omega^{t_0}_t$, which is the inverse of $\Omega^t_{t_0}$ (Sec. 2.38d), to both sides of (6), we get

$$C'(t) = \Omega^{t_0}_t \beta(t),$$

which implies

$$C(t) = \int_{t_0}^t \Omega^{t_0}_\tau \beta(\tau)\, d\tau, \tag{7}$$

where there is no constant of integration since (7) automatically gives $C(t_0)=0$.

Thus, finally, we get the formula

$$v(t) = \Omega^t_{t_0} \int_{t_0}^t \Omega^{t_0}_\tau \beta(\tau)\, d\tau + \Omega^t_{t_0} v_0 = \Omega^t_{t_0} v_0 + \int_{t_0}^t \Omega^t_\tau \beta(\tau)\, d\tau, \tag{8}$$

where in the last step we use Theorem 2.38c. We can confirm the validity of (8) by direct substitution and differentiation, using the familiar rule for differentiation of a parameter-dependent integral with a variable upper limit (cf. Vol. 1, Sec. 9.116), which is easily generalized to the case of vector functions.

2.82. If the underlying Banach space $\mathbf{B}$ is the real line R_1, and if $\mathbf{A}(t)$ reduces to a numerical function $A(t)$, then

$$\Omega^t_{t_0} = \exp\left\{ \int_{t_0}^t A(\tau)\, d\tau \right\},$$

as shown in Sec. 2.12, and hence

$$v(t)=\exp\left\{\int_{t_0}^{t} A(\tau)\,d\tau\right\}v_0+\int_{t_0}^{t}\exp\left\{\int_{\tau}^{t} A(\theta)\,d\theta\right\}\beta(\tau)\,d\tau. \tag{9}$$

In particular, for constant $A(t)\equiv A$ we have

$$v(t)=e^{(t-t_0)A}v_0+\int_{t_0}^{t} e^{(t-\tau)A}\beta(\tau)\,d\tau. \tag{10}$$

2.83. Formulas (9) and (10) continue to hold in the general case where $u(t)$ is a vector function with values in a Banach space $\mathbf{B}$ and $\mathbf{A}(t)$ is a linear operator acting in $\mathbf{B}$. The exponentials appearing in these formulas must then be interpreted in the sense of Sec. 2.13 if $\mathbf{A}(t)\equiv\mathbf{A}$ is constant and in the sense of Sec. 2.18 if $\mathbf{A}(t)$ is variable.

In particular, suppose $\mathbf{A}(t)\equiv\mathbf{A}$ and the function $\beta(t)$ is of the special form

$$\beta(t)=\sum_{k=1}^{n} P_k(t)e^{\mathbf{Q}_k(t)}\beta_k,$$

where the $P_k(t)$ are polynomials, the $\mathbf{Q}_k$ are constant operators commuting with $\mathbf{A}$, and the β_k are fixed vectors in the space $\mathbf{B}$. Then the integral (10) can be calculated explicitly, with a result whose structure can be written down in advance except for the values of certain coefficients. Hence in this case the solution can be found by using the "method of undetermined coefficients."†

2.84. In the case of an n-dimensional space $\mathbf{B}=R_n$, the solution operator $\Omega_{t_0}^{t}$ can be specified by a Wronskian matrix

$$W_{t_0}^{t}=\begin{Vmatrix} f_{11}(t) & f_{12}(t) & \cdots & f_{1n}(t)\\ f_{21}(t) & f_{22}(t) & \cdots & f_{2n}(t)\\ \cdot & \cdot & \cdots & \cdot\\ f_{n1}(t) & f_{n2}(t) & \cdots & f_{nn}(t)\end{Vmatrix}$$

(Sec. 2.72). Suppose we look for a solution of the form (3), which in this case can be written as

$$v_{1j}(t)=\sum_{k=1}^{n} f_{jk}(t)C_k(t)\qquad (j=1,\ldots,n), \tag{11}$$

in terms of the unknown functions $C_1(t),\ldots,C_n(t)$. Equation (6) then becomes a system of equations

$$\sum_{k=1}^{n} f_{jk}(t)C_k'(t)=\beta_k(t)\qquad (j=1,\ldots,n). \tag{12}$$

† W. W. Stepanow, *Lehrbuch der Differentialgleichungen*, VEB Deutscher Verlag der Wissenschaften, Berlin (1956), p. 216 ff.

Solving (12) for the $C_k'(t)$ and integrating the result from t_0 to t, we get the functions $C_k(t)$ figuring in the solution (11).

2.85. Finally consider the nonhomogeneous linear equation

$$u^{(n)}(t)=a_1(t)u(t)+\cdots+a_n(t)u^{(n-1)}(t)+\beta(t) \tag{13}$$

of order n, and the corresponding homogeneous equation

$$u^{(n)}(t)=a_1u(t)+\cdots+a_n(t)u^{(n-1)}(t). \tag{13'}$$

In this case, the solution operator of the system

$$\begin{aligned} &u_1'(t)=u_2(t),\\ &u_2'(t)=u_3(t),\\ &\cdots\\ &u_n'(t)=a_1(t)u_1(t)+\cdots+a_n(t)u_n(t) \end{aligned} \tag{14}$$

equivalent to (13′), where

$$u_1(t)=u(t),\ u_2(t)=u'(t),\ldots,\ u_n(t)=u^{(n-1)}(t),$$

has a matrix of the form

$$\left\| \begin{matrix} u_1(t) & \cdots & u_n(t) \\ \cdot & \cdots & \cdot \\ u_1^{(n-1)}(t) & \cdots & u_n^{(n-1)}(t) \end{matrix} \right\|,$$

where, as in Sec. 2.74, $u_1(t),\ldots, u_n(t)$ is a solution of (14) corresponding to a nonsingular initial data matrix. Suppose we look for a solution of the form (3), which (for the first row) can be written as

$$v(t)=\sum_{k=1}^{n} u_k(t)C_k(t). \tag{15}$$

Equation (6) then becomes a system of equations

$$\begin{aligned} &\sum_{k=1}^{n} u_k(t)C_k'(t)=0,\\ &\sum_{k=1}^{n} u_k'(t)C_k'(t)=0,\\ &\cdots\\ &\sum_{k=1}^{n} u^{(n-1)}(t)C_k'(t)=\beta(t). \end{aligned} \tag{16}$$

Solving (16) for the $C_k'(t)$ and integrating the result from t_0 to t, we get the functions $C_k(t)$ figuring in the solution (15).

Problems

1. The differential equation $y' = 3y^{2/3}$ has two distinct solutions $y \equiv 0$ and $y = x^3$, both satisfying the initial condition $y(0) = 0$. Why doesn't this contradict Theorem 2.35?

2. A heavy point P slides without friction along a curve. Find the form of the curve if the projection of P moves uniformly

(a) Along a horizontal line;
(b) Along a vertical line.

3. Starting from a particular solution $u_1(t)$ of the second-order homogeneous linear equation

$$u''(t) + a(t)u'(t) + b(t)u(t) = 0,$$

how does one find a second solution which is linearly independent of the first?

4. According to Sec. 2.17, a linear equation of order n with constant coefficients is equivalent to a first-order system whose matrix has a minimal annihilating polynomial of degree n. Show that every system of n first-order equations with this property is equivalent to a single equation of order n.

5. Given a vector equation $u'(t) = \mathbf{A}(t)u(t)$ where $u(t) \in R_n$, suppose $\mathbf{A}(t)$ is a periodic operator with period T, so that $\mathbf{A}(t+T) \equiv \mathbf{A}(t)$. Show that $\Omega_0^{T+t} = \mathbf{C}\Omega_0^t$, where $\mathbf{C}$ is a constant operator.

6. Let $u_1(t), \ldots, u_n(t)$ be $n \leqslant N$ linearly independent vector functions taking values in the space R_N, which are differentiable for every $t \in [a,b]$. Show that there exists an equation $u'(t) = \mathbf{A}(t)u(t)$, with a continuous operator $\mathbf{A}(t)$ mapping R_N into itself, which has $u_1(t), \ldots, u_n(t)$ as its solutions.

7. Let $y_1(t), \ldots, y_n(t)$ be n linearly independent scalar functions with derivatives up to order n (inclusive). Can the Wronskian of these functions vanish identically?

8. Let $y_1(t), \ldots, y_n(t)$ be n linearly independent scalar functions with derivatives up to order n and a nonzero Wronskian. Construct an nth-order equation with $y_1(t), \ldots, y_n(t)$ as its solutions.

9. Suppose the functions $\mathbf{A}(t)$ and $\beta(t)$ have continuous derivatives up to order n. Prove that the solution of the linear equation $u'(t) = \mathbf{A}(t)u(t) + \beta(t)$ has continuous derivatives up to order $n+1$.

10. Prove that the inequality

$$y'(t) - ky(t) \leqslant \varphi(t) \qquad (0 \leqslant t \leqslant T, y(0) = 0)$$

implies the inequality

$$y(t) \leqslant \int_0^t e^{k(t-s)} \varphi(s)\, ds \qquad (0 \leqslant t \leqslant T).$$

11 (*Continuation*). Prove that the inequality

$$w(t) \leqslant \varphi(t) + k \int_0^t w(s)\, ds \qquad (0 \leqslant t \leqslant T)$$

implies the inequality

$$w(t) \leqslant \varphi(t) + k \int_0^t e^{k(t-s)} \varphi(s)\, ds \qquad (0 \leqslant t \leqslant T).$$

12 (*Continuation*). A function $y(t) \in \mathbf{X}$ defined on $[0, T]$ is called an *ε-almost-solution* of the equation $u'(t) = f(t,u)$ if

$$\| y'(t) - f(t, y(t)) \| \leqslant \varepsilon \qquad (0 \leqslant t \leqslant T).$$

Prove that the inequality

$$\| u(t) - y(t) \| \leqslant \| u(0) - y(0) \| e^{kt} + \frac{\varepsilon}{k}(e^{kt} - 1) \qquad (0 \leqslant t \leqslant T)$$

holds for any solution $u(t)$ and any ε-almost-solution $y(t)$, where k is the constant in the Lipschitz condition for the function $f(t,u)$.

13 (*Continuation*). Given an equation

$$u'(t) = \mathbf{A}(t) u(t) \qquad (u(0) = u_0,\ 0 \leqslant t \leqslant T) \tag{1}$$

with a continuous operator $\mathbf{A}(t)$, let

$$\Pi = \{0 = t_0 < t_1 < \cdots < t_n = T\}$$

and define a continuous vector function $y_\Pi(t)$ as follows:

$$y_\Pi(0) = 0,$$
$$y_\Pi(t_{k+1}) = y_\Pi(t_k) + \mathbf{A}(t_k) y_\Pi(t_k) \Delta t_k \qquad (k = 0, 1, \ldots, n-1),$$
$$y_\Pi(t) \text{ is linear for } t_k \leqslant t \leqslant t_{k+1} \qquad (k = 0, 1, \ldots, n-1).$$

Given any $\varepsilon > 0$, find a partition Π such that the function $y_\Pi(t)$ becomes an ε-almost-solution of equation (1).

14 (*Continuation*). Show that the solution of equation (1) at $t = T$ is given by the expression

$$u(T) = \lim_{d(\Pi) \to 0} y(t) = \lim_{d(\Pi) \to 0} \left\{ \prod_{k=n-1}^{0} [\mathbf{E} + \mathbf{A}(t_k) \Delta t_k] \right\} u_0$$

(the order of the factors is essential!).

15. Under the conditions of Problem 13, define a continuous vector function $z_\Pi(t)$ as follows:

$$z_\Pi(0) = y_0,$$

$$z_\Pi(t_{k+1}) = \left\{ \prod_{j=k}^{0} e^{\mathbf{A}(t_j)\Delta t_j} \right\} y_0 \qquad (k=0,1,\ldots, n-1),$$

$$z_\Pi(t) = e^{\mathbf{A}(t_k)(t-t_k)} z_\Pi(t_k) \qquad (t_k \leqslant t \leqslant t_{k+1}).$$

Given any $\varepsilon > 0$, find a partition Π such that the function $z_\Pi(t)$ becomes an ε-almost-solution of equation (1).

16. Show that the solution of equation (1) at $t = T$ is given by the expression

$$u(T) = \lim_{d(\Pi)\to 0} z_\Pi(t) = \lim_{d(\Pi)\to 0} \left\{ \prod_{k=n-1}^{0} e^{\mathbf{A}(t_k)\Delta t_k} \right\} u_0$$

(the order of the factors is essential!).

3 Space Curves

3.1. Basic Concepts

3.11. By a *curve* in the n-dimensional real space R_n we mean the locus of all points $x=(\xi_1,\ldots,\xi_n)\in R_n$ satisfying a system of parametric equations

$$\xi_1=\xi_1(t),\ldots,\xi_n=\xi_n(t)\qquad(a\leqslant t\leqslant b),\tag{1}$$

or, more concisely, a single vector equation

$$x=x(t)\qquad(a\leqslant t\leqslant b).\tag{2}$$

The vector $x(t)$ is called the *position vector* of the curve.

The same curve (i.e., the same geometric set of points) can be specified by many different systems of equations of the form (1). For example, the two systems

$$x=r\cos t,\quad y=r\sin t\quad(-\infty<t<\infty)$$

and

$$x=r\cos t^3,\quad y=r\sin t^3\quad(-\infty<t<\infty)$$

specify the same set of points in the xy-plane, namely a circle of radius r with its center at the origin. As we will see later, in many cases the choice of a suitable parametric representation from among various possibilities helps to clarify the geometric properties of a given curve.

It will be assumed that the functions $\xi_k(t)$ are continuous and satisfy certain smoothness conditions,† to be made precise later. Mere continuity of the functions $\xi_k(t)$ will not suffice to give the locus of (1) the usual appearance of a curve. In fact, there exist systems (1) with continuous right-hand sides such that the corresponding locus is the whole space R_n (see Problem 5); however, it can be shown that this cannot happen for $n>1$ if the functions $\xi_k(t)$ have continuous derivatives.

3.12. More generally, the vector function $x(t)$ depending on the parameter t can take its values in a metric space or a normed linear space, and equation (2) then specifies a curve in the space in question. If $x(t)$ takes its values in a normed linear space $\mathbf{X}$, then, just as in Sec. 1.61a, we can define the *derivative* of $x(t)$ at the point $t=c$ as the limit

$$x'(c)=\lim_{\Delta t\to 0}\frac{x(c+\Delta t)-x(c)}{\Delta t},\tag{3}$$

provided the limit on the right exists in the metric of $\mathbf{X}$. If (3) exists, we say

† Let n be the order of the highest derivative of a function $f(t)$. Then the larger n, the "smoother" we call $f(t)$.

that $x(t)$ is *differentiable at the point* $t=c$. If the limit (3) exists for all $c \in [a,b]$, we say that $x(t)$ is *differentiable on the interval* $[a,b]$. In what follows, we will be primarily concerned with differentiable functions taking values in the n-dimensional real space R_n, but many of the results will also be valid for functions taking values in an infinite-dimensional normed linear space.

It should be noted that since passage to the limit in R_n is equivalent to taking the limit of each component (Corollary 1.35f), differentiability at $t=c$ of the vector function $x(t)=(\xi_1(t),\ldots, \xi_n(t)) \in R_n$ in the sense of (3) is equivalent to differentiability of all n numerical functions $\xi_1(t),\ldots, \xi_n(t)$ at $t=c$, and moreover

$$x'(c)=(\xi_1'(c),\ldots, \xi_n'(c)) \in R_n.$$

3.13. Next we discuss the geometric meaning of differentiability of a vector function, a topic already touched upon in Sec. 2.11.

The definition (3) of a derivative can also be written in the equivalent form

$$\Delta x \equiv x(t+\Delta t)-x(t)=x'(c)\Delta t+\varepsilon(t)\Delta t, \tag{4}$$

where the vector $\varepsilon(t)$ approaches zero as $\Delta t \to 0$. According to (4), the principal linear part of the increment of the function $x(t)$ as t changes from c to $c+\Delta t$ is just $x'(c)\Delta t$. A point $P=x(c)$ of the curve with equation (2) is said to be *ordinary* if $x'(c) \neq 0$ and *singular* if $x'(c)=0$ (cf. Vol. 1, Sec. 9.74).

The geometric object corresponding to the linear equation

$$x=x(c)+x'(c)(t-c)$$

at an ordinary point is a straight line Λ, going through the point $P=x(c)$ in the direction of the vector $x'(c)$ (see Figure 14). Thus the deviation of a point of the curve L from the corresponding point of the line Λ (i.e., from the point of the line with the same value of t) is of a higher order of smallness than Δt. For this reason we call Λ the *tangent to the curve L at the point P*. To summarize, the existence of a nonzero derivative $x'(c)$ is equivalent to the existence of a tangent to the curve L at the point $P=x(c)$, and the direction of the tangent is that of the vector $x'(c)$.

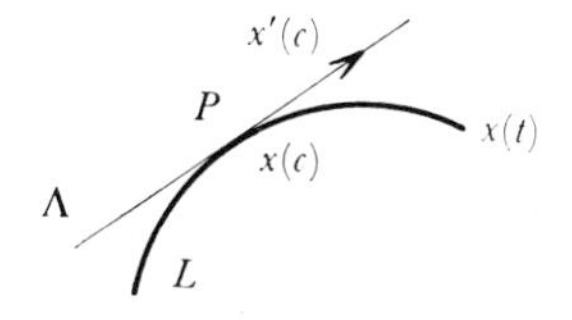

Figure 14

3.14. Suppose that in the equation of the curve L we go over to a new parameter τ, by introducing a differentiable function $t=t(\tau)$. Let $c=t(\gamma)$, where $t'(\gamma)\neq 0$, and let $y(\tau)\equiv x(t(\tau))$. Then, by the rule for differentiation of a composite function (Sec. 1.61d), we have $y'(\tau)=x'(t)t'(\tau)$. Thus the vector $y'(\gamma)$ specifying the tangent at P has the same direction as the old vector $x'(c)$ but differs from it by a numerical factor of $t'(\gamma)$, i.e., the length of the vector specifying the tangent has no direct geometric meaning. As we saw in Sec. 2.11d, the vector $x'(t)$ has the following kinematic interpretation: If t is regarded as the time, then $x'(t)$ is the velocity of motion of the point $x=x(t)$ along the curve L at the time $t=c$.

3.15. Again let $x(t)$ be a vector function with values in R_n, and suppose the argument t changes from c to $c+dt$. Then by the *differential* of $x(t)$ at $t=c$ we mean the vector

$$dx=x'(c)\,dt,$$

just as in Sec. 1.61e. Thus the differential dx is just the principal linear part of the increment of $x(t)$ corresponding to the increment dt of the independent variable t. The differential of $x(t)$ does not depend on whether its argument t is the independent variable or a function $t=t(\tau)$ of some other independent variable τ, for exactly the same reason as in Sec. 1.61e.

3.16.a. The *integral* of a vector function $x(t)$ defined on an interval $[a,b]$ and taking values in R_m is defined as the limit

$$\int_a^b x(t)\,dt=\lim_{d(\Pi)\to 0}\sum_{k=0}^{n-1} x(\tau_k)\Delta t_k, \tag{5}$$

where Π and $d(\Pi)$ have the same meaning as in Sec. 1.62a. The existence of (5) for a piecewise continuous function $x(t)$ has already been established (see Theorem 1.62c). The integral has the same properties as in Sec. 1.62e.

b. Let $u(t)$ be a differentiable function on $[a,b]$ with values in R_n (or, more generally, in a Banach space $\mathbf{X}$), and let $v(t)$ be a differentiable numerical function on $[a,b]$. Then we have the following formula for integration by parts:

$$\int_a^b u(t)dv(t)=u(t)v(t)\Big|_a^b-\int_a^b v(t)\,du(t). \tag{6}$$

The proof of (6) is the exact analogue of the proof for numerical functions (Vol. 1, Sec. 9.51a).

3.17. Arc length in a normed linear space

a. The familiar formula for arc length in a finite-dimensional space (Vol. 1, Sec. 9.71) is easily generalized to the case of an arbitrary normed linear space **X**. We again define arc length as the limiting length of a polygonal line inscribed in the given curve as the (maximum) length of the segments making up the line approaches zero. More exactly, let L be a curve in **X** with parametric equation

$$x = x(t) \qquad (a \leqslant t \leqslant b), \tag{7}$$

and let

$$\Pi = \{a = t_0 < t_1 < \cdots < t_n = b\}$$

be a partition of the interval $[a,b]$ on which the vector function $x(t)$ is defined, with parameter

$$d(\Pi) = \max\{\Delta t_0, \Delta t_1, \ldots, \Delta t_{n-1}\} \qquad (\Delta t_k = t_{k+1} - t_k).$$

Then every point t_k corresponds to a point $P_k = x(t_k)$ on the curve L. Joining the points P_k by "line segments" in **X**, we get a "polygonal line" $L_\Pi = P_0 P_1 \ldots P_n$ "inscribed" in L, of "length"

$$\sum_{k=0}^{n-1} |\Delta x_k| \qquad (\Delta x_k = x_{k+1} - x_k).$$

Suppose $x(t)$ is continuously differentiable on $[a,b]$. Then

$$\Delta x_k = \int_{t_k}^{t_{k+1}} x'(t)\, dt = x'(t_k)\Delta t_k + \varepsilon_k \Delta t_k,$$

where

$$\varepsilon_k = \frac{1}{\Delta t_k} \int_{t_k}^{t_{k+1}} [x'(t) - x'(t_k)]\, dt$$

and

$$|\varepsilon_k| \leqslant \varepsilon_\Pi = \max_{|t - \tilde{t}| \leqslant d(\Pi)} |x'(t) - x'(\tilde{t})|.$$

Therefore we have the estimate

$$\left| \sum_{k=0}^{n-1} |\Delta x_k| - \sum_{k=0}^{n-1} |x'(t_k)|\Delta t_k \right| \leqslant \sum_{k=0}^{n-1} |\varepsilon_k| \Delta t_k \leqslant \varepsilon_\Pi (b-a).$$

But $\varepsilon_\Pi \to 0$ as $d(\Pi) \to 0$, by the uniform continuity of $x'(t)$ on $[a,b]$, while the sum

$$\sum_{k=0}^{n-1} |x'(t_k)|\Delta t_k$$

approaches the limit

$$s=\int_a^b |x'(t)|\,dt \tag{8}$$

as $d(\Pi)\to 0$, since the continuity of $x'(t)$ implies that of $|x'(t)|$. Hence, as $d(\Pi)\to 0$, the length of the inscribed polygonal line L_Π exists and equals (8). Note that if $\mathbf{X}=R_m$, so that $x(t)=(x_1(t),\ldots, x_m(t))$, where each $x_j(t)$ is a numerical function on $[a,b]$, then

$$|x'(t)|=\sqrt{\sum_{j=1}^{m} [x_j'(t)]^2}\,,$$

and (8) reduces to the familiar formula

$$s=\int_a^b \sqrt{\sum_{j=1}^{m} [x_j'(t)]^2}\,dt \tag{8'}$$

for arc length in a finite-dimensional space (Vol. 1, Sec. 9.71). However, unlike (8′), formula (8) works in an arbitrary normed linear space.

b. Replacing t by τ and b by t in (8), we get an expression for the length $s(t)$ of the arc of the curve L joining the initial point of L to the variable point $P=x(t)$:

$$s(t)=\int_a^t |x'(\tau)|\,d\tau \tag{9}$$

(cf. Vol. 1, Sec. 9.74). Thus $s(t)$ is a continuously differentiable function of t, with derivative

$$s'(t)=|x'(t)|.$$

Suppose L has no singular points, so that $x'(t)$ is nonvanishing on $[a,b]$. Then, just as in the finite-dimensional case (Vol. 1, Sec. 9.74), $s=s(t)$ is increasing and has a continuously differentiable increasing inverse $t=t(s)$, with derivative

$$t'(s)=\frac{1}{s'(t)}$$

on the interval $[0,l]$, where l is the length of L. In this case, we can choose the arc length s as the original parameter t, so that L now has the parametric equation

$$x = x(s) \qquad (0 \leqslant s \leqslant l), \tag{7'}$$

where the function $x(s)$ is continuously differentiable. The variable arc length s is often called the *natural parameter* of L. In terms of this parameter, (9) becomes

$$s(s) = \int_a^s |x'(\tau)| \, d\tau,$$

which implies

$$|x'(s)| = s'(s) = 1.$$

Thus the vector $x'(s)$ is of unit length at every nonsingular point of L. The kinematic meaning of this is clear: If the parameter s is both the length of the traversed path and the time spent in traversing the path, then the velocity of motion along L must equal 1.

3.2. Higher Derivatives

3.21. Let $x(t)$ be a vector function defined on $[a,b]$ taking values in R_n. Then it will be recalled from Sec. 1.64a that the derivative of order m of $x(t)$ is the (first) derivative of the derivative of $x(t)$ of order $m-1$, provided that the latter is a differentiable function on $[a,b]$. We will henceforth tacitly assume that all higher derivatives under discussion actually exist.

3.22.a. Suppose we go over from t to a new independent variable τ, by introducing a sufficiently smooth function $t = t(\tau)$. Then, as noted in Sec. 3.14, the first derivatives of $x(t)$ and $x(t(\tau))$ with respect to t and τ, denoted by x_t and x_τ, differ by only a numerical factor:

$$x_\tau = x_t t_\tau.$$

Differentiating with respect to τ, we get

$$x_{\tau\tau} = (x_\tau)_\tau = (x_t t_\tau)_\tau = (x_t)_\tau t_\tau + x_t t_{\tau\tau} = x_{tt} t_\tau^2 + x_t t_{\tau\tau},$$

after again using the formula for differentiation of a composite function. Hence the vector $x_{\tau\tau}$, which is in general not collinear with the vector x_{tt}, lies in the same plane as the vectors x_t and x_{tt}. Thus the plane determined by the vectors x_t and x_{tt} is independent of the choice of parameter, although the position of the vector x_{tt} in the plane changes in going over to a new parameter.

b. More generally, we have the following

THEOREM. *Given any positive integer m, the vector* $x_\tau^{(m)}$ *lies in the linear manifold spanned by the vectors* $x_t, x_{tt}, \ldots, x_t^{(m)}$.†

Proof. We use induction, assuming that

$$x_\tau^{(m)} = \sum_{k=1}^{m} x_t^{(k)} \varphi_k(\tau). \tag{1}$$

Differentiating (1) once more with respect to τ, we get

$$x_\tau^{(m)} = \sum_{k=1}^{m} x_t^{(k+1)} t_\tau \varphi_k(\tau) + \sum_{k=1}^{m} x_t^{(k)} \varphi_k'(\tau).$$

Therefore $x_\tau^{(m)}$ is a linear combination of the vectors $x_t, x_{tt}, \ldots, x_t^{(m)}$. This completes the induction, since (1) obviously holds for $m=1$. ∎

Thus, in a certain sense, it is the linear manifolds spanned by the vectors $x_t, x_{tt}, \ldots, x_t^{(m)}, \ldots$ rather than the vectors themselves that are geometrically meaningful. These linear manifolds are called the *osculating (sub)spaces* of the curve $x=x(t)$, of dimensions 1, 2, ..., m, ... respectively (it is assumed that $x_t, x_{tt}, \ldots, x_t^{(m)}, \ldots$ are linearly independent).

3.23. Suppose $x(t)$ has continuous derivatives up to order $m+1$ inclusive on the interval $[a,b]$. Then we have Taylor's formula

$$\Delta x(t) \equiv x(t+\Delta t) - x(t) = x'(t)\Delta t + \frac{x''(t)}{2!}(\Delta t)^2 + \cdots + \frac{x^{(m)}(t)}{m!}(\Delta t)^m + Q_m.$$

Setting $m=1,2,\ldots$ and using the estimate of the remainder Q_m given in Sec. 1.64b, we get a sequence of successively more exact formulas

$$\Delta x(t) = x'(t)\Delta t + \varepsilon_1(t)\Delta t, \tag{2}$$

$$\Delta x(t) = x'(t)\Delta t + \frac{x''(t)}{2}(\Delta t)^2 + \varepsilon_2(t)(\Delta t)^2, \tag{3}$$

$$\Delta x(t) = x'(t)\Delta t + \frac{x''(t)}{2}(\Delta t)^2 + \frac{x'''(t)}{6}(\Delta t)^3 + \varepsilon_3(t)(\Delta t)^3, \tag{4}$$

...

where

$$\|\varepsilon_m(t)\| \leqslant \frac{\Delta t}{(m+1)!} \max_{t \leqslant \tau \leqslant t+\Delta t} \|x^{(m+1)}(\tau)\| \qquad (m=1,2,\ldots)$$

and hence $\varepsilon_m(t) \to 0$ as $\Delta t \to 0$.

† I.e., in the set of all (finite) linear combinations $\alpha_1 x_t + \alpha_2 x_{tt} + \cdots + \alpha_m x_t^{(m)}$ with real coefficients $\alpha_1, \alpha_2, \ldots, \alpha_m$.

3.24. Structure of a curve near an ordinary point. According to (2), the curve L with equation $x=x(t)$ is approximated by its tangent up to terms of order 1, provided that $x'(t)\neq 0$. Similarly, according to (3), L lies in the plane determined by the vectors $x'(t)$ and $x''(t)$, with an accuracy up to terms of order 2. In keeping with Sec. 3.23, this plane is called the *osculating plane* of L at the variable point $P=x(t)$ (in the case where $x'(t)$ and $x''(t)$ are linearly independent). Suppose that for the coordinates ξ and η in the osculating plane we choose the components of $\Delta x(t)$ with respect to the basis $x'(t)$, $x''(t)/2$. Then (3) leads to the following parametric representation of the curve L, with an accuracy up to terms of order 2:

$$\xi=\Delta t, \qquad \eta=(\Delta t)^2.$$

Thus, to the given accuracy, L is a parabola in the osculating plane, with equation $\eta=\xi^2$.

The space determined by the vectors $x'(t)$, $x''(t)/2$, and $x'''(t)/6$ (provided these vectors are linearly independent) is called the *three-dimensional osculating (sub)space* of L at the variable point $P=x(t)$. It follows from (4) that L lies in this space with an accuracy up to terms of order 3. Suppose that for the coordinates ξ, η, and ζ in the three-dimensional osculating space we choose the components of $\Delta x(t)$ with respect to the basis $x'(t)$, $x''(t)/2$, $x'''(t)/6$. Then (4) leads to the following parametric representation of the curve L, with an accuracy up to terms of order 3:

$$\xi=\Delta t, \quad \eta=(\Delta t)^2, \quad \zeta=(\Delta t)^3.$$

This gives the space curve shown in Figure 15, whose projection onto the $\xi\eta$-plane is the parabola $\eta=\xi^2$ already considered. The projection of the curve onto the $\xi\zeta$-plane is the third-order curve $\zeta=\xi^3$ (the appearance of L

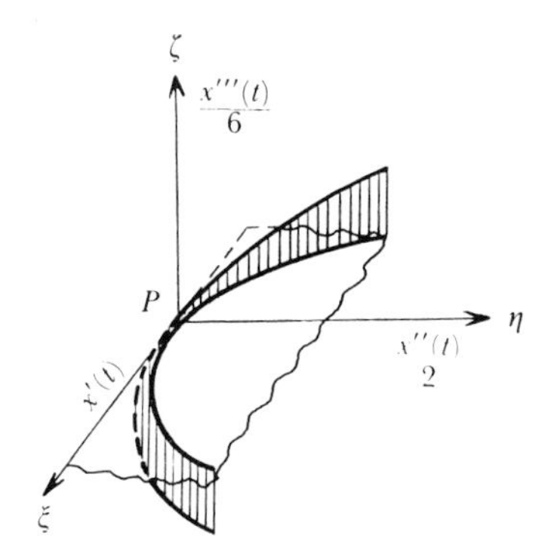

Figure 15

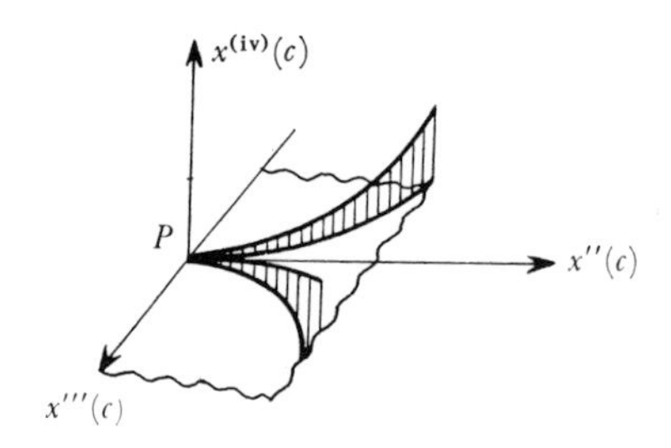

Figure 16

from the end of the vector $x''(t)/2$), while its projection onto the $\eta\zeta$-plane is the semicubical parabola $\zeta=\eta^{3/2}$ (the appearance of L from the end of the vector $x'(t)$).

3.25. Structure of a curve near a singular point. Let $P=x(c)$ be a singular point of the curve L with equation $x=x(t)$, so that $x'(c)=0$ (Sec. 3.13). Provided that $x''(c)\neq 0$, $x'''(c)\neq 0$, formula (4) then implies

$$\Delta x(c)=\frac{x''(c)}{2}(\Delta t)^2+\frac{x'''(c)}{6}(\Delta t)^3+\varepsilon_3(t)(\Delta t)^3. \tag{5}$$

Thus, with an accuracy up to terms of order 3, the curve L lies in the plane determined by the vectors $x''(c)/2$ and $x'''(c)/6$, where it has the equation $\zeta=\eta^{3/2}$ corresponding to a *cusp* at P (see Figure 16). Suppose we include fourth-order terms by replacing the last term in the right-hand side of (5) by

$$\frac{x^{(iv)}(c)}{24}(\Delta t)^4+\varepsilon_4(t)(\Delta t)^4.$$

Then, if $x^{(iv)}(c)\neq 0$, the effect of the terms involving $(\Delta t)^4$ is to cause the curve to deviate from the plane of $x''(c)$ and $x'''(c)$ in the direction of the vector $x^{(iv)}(c)$, as shown in the figure. Moreover, because of the constant sign of this term (for sufficiently small Δt), both branches of the cusp are bent away from the plane in the same direction.

3.3. Curvature

3.31. In what follows we will deal with angles between vectors as well as with lengths of vectors. Hence the natural context for our subsequent considerations is a Hilbert space (Sec. 1.4) rather than a normed linear space.

LEMMA. *Let $x(t), y(t)$ $(a\leqslant t\leqslant b)$ be two differentiable functions with values in a real*

Hilbert space **H**. *Then the scalar product*

$$\varphi(t) = (x(t), y(t))$$

(a numerical function) is also differentiable, with derivative

$$\varphi'(t) = (x'(t), y(t)) + (x(t), y'(t)). \tag{1}$$

Proof. Clearly

$$\begin{aligned}\Delta\varphi &= (x(t+\Delta t), y(t+\Delta t)) - (x(t), y(t)) \\ &= (x(t) + x'(t)\Delta t + \varepsilon_1 \Delta t, y(t) + y'(t)\Delta t + \varepsilon_2 \Delta t) - (x(t), y(t)),\end{aligned}$$

where $\varepsilon_1 = \varepsilon_1(t,\Delta t) \to 0$, $\varepsilon_2 = \varepsilon_2(t,\Delta t) \to 0$ as $\Delta t \to 0$ (cf. Sec. 3.13). It follows that

$$\Delta\varphi = [(x'(t), y(t)) + (x(t), y'(t))]\Delta t + \varepsilon_3 \Delta t,$$

where $\varepsilon_3 = \varepsilon_3(t,\Delta t) \to 0$ as $\Delta t \to 0$ (why?). But then the expression in brackets, equal to (1), is the principal linear part of $\Delta\varphi$, or equivalently the derivative of $\varphi(t)$. ∎

COROLLARY. *If the length of the vector $x(t)$ does not vary with t, then the vector $x'(t)$ is orthogonal to $x(t)$.*

Proof. Using (1) to differentiate the function $(x(t), x(t))$, which is independent of t by hypothesis, we get

$$0 = (x(t), x(t))' = 2(x(t), x'(t)). \quad ∎$$

3.32. Consider a curve L with parametric equation $x = x(s)$, where s is the arc length of L measured from some fixed point. Then the vector $e_1(s) = x'(s)$ has constant length 1, as noted in Sec. 3.29. If $x'(s)$ and $x''(s)$ are linearly independent, then L has an osculating plane (Sec. 3.24). The vector $e_1(s) = x''(s)$ lying in the osculating plane, which, according to the corollary, is orthogonal to the vector $e_1(s)$, is called the *curvature vector* of L at the point s.† Let

$$e_1'(s) = \kappa(s) e_2(s), \tag{2}$$

where $e_2(s)$ is a unit vector orthogonal to $e_1(s)$ and the coefficiently $\kappa(s)$ is positive. Then the number $\kappa(s)$, called the *curvature* of L at the point s, is

† For simplicity, we often say "at (the point) s," meaning "at the point (of L) with parameter value s."

given by

$$\kappa(s)=|e_1'(s)|=\lim_{\Delta s\to 0}\left|\frac{e_1(s+\Delta s)-e_1(s)}{\Delta s}\right|. \tag{3}$$

The absolute value of the difference between the vectors $e_1(s+\Delta s)$ and $e_1(s)$ is a chord of the unit circle, and is a small quantity differing only by terms of higher order from the angle between the tangents to the curve L at the points with values of arc length s and $s+\Delta s$. Thus the coefficient $\kappa(s)$ is just the velocity of rotation of the tangent per unit arc length; the faster the rotation of the tangent per unit arc length, the larger the number $\kappa(s)$.

It should be noted that formula (2) gives a somewhat more general definition of curvature than formula (1), since (2) does not require that $e_1'(s)\neq 0$, but only that $e_1'(s)$ exist. If $e_1'(s)=0$, then, according to (2), the curvature vanishes at the corresponding point of the curve.

3.33. Introducing the curvature $\kappa(s)$ into Taylor's formula (3), p. 194 (for $e_1'(s)\neq 0$), we get

$$\begin{aligned}\Delta x(s)&=x'(s)\Delta s+\frac{x''(s)}{2}(\Delta s)^2+\varepsilon_2(s)(\Delta s)^2\\ &=e_1(s)\Delta s+\frac{\kappa(s)}{2}e_2(s)(\Delta s)^2+\varepsilon_2(s)(\Delta s)^2,\end{aligned} \tag{4}$$

where $\varepsilon_2(s)\to 0$ as $\Delta s\to 0$. We can also write $\Delta x(s)$ in the form

$$\Delta x(s)=e_1(s)\Delta s+\frac{\kappa(s)}{2}e_2(s)(\Delta s)^2+\alpha_2(s)(\Delta s)^3, \tag{4'}$$

where $\alpha_2(s)$ satisfies the estimate

$$|\alpha_2(s)|\leqslant\frac{1}{6}\max_s|x'''(s)|$$

(cf. Sec. 1.64b).

3.34.a. We now derive the formula for curvature in the case where the curve L has an equation of the form $x=x(t)$ with an arbitrary parameter t. Since

$$s_t=|x_t|=\sqrt{(x_t,x_t)}$$

(Sec. 3.29), then

$$s_{tt}=\frac{(x_t,x_{tt})}{\sqrt{(x_t,x_t)}}=\frac{(x_t,x_{tt})}{|x_t|},$$

by Lemma 3.31. Moreover

$$x_s = x_t t_s,$$

$$x_{ss} = x_{tt} t_s^2 + x_t t_{ss} = \frac{x_{tt}}{s_t^2} + x_t \left(\frac{1}{s_t}\right)_t \frac{1}{s_t}$$

$$= \frac{x_{tt}}{|x_t|^2} - \frac{x_t}{s_t^2} \frac{s_{tt}}{s_t} = \frac{x_{tt}}{|x_t|^2} - \frac{x_t(x_t, x_{tt})}{|x_t|^4},$$

and hence, finally,

$$\kappa(s) = |x_{ss}| = \left| \frac{x_{tt}}{|x_t|^2} - \frac{x_t(x_t, x_{tt})}{|x_t|^4} \right|. \tag{5}$$

In deriving (5), we used the definition (3). Therefore (5) holds for both $\kappa(s) \neq 0$ and for $\kappa(s) = 0$.

b. For example, let L be the circle in the plane with parametric equation

$$x = (R \cos t, R \sin t) \quad (0 \leqslant t \leqslant 2\pi). \tag{6}$$

Differentiating (6), we get

$$x_t = (-R \sin t, R \cos t), \qquad |x_t| = R,$$

$$x_{tt} = (-R \cos t, -R \sin t), \qquad (x_t, x_{tt}) = 0,$$

and hence

$$\kappa(s) = \frac{|x_{tt}|}{|x_t|^2} = \frac{1}{R}, \tag{7}$$

i.e., *the curvature of a circle equals the reciprocal of its radius.*

3.35. The osculating circle. Let L be a space curve with equation $x = x(s)$ in terms of the natural parameter s. Suppose we draw a circle Q of radius R tangent to L at s and lying in the osculating plane of L at s. Then the deviation of a given point of L from the corresponding point of Q will in general be a term of order 2 in the increment Δs. However, the radius of Q can be adjusted in such a way as to make the deviation of the third order rather than of the second order. To this end, let $x = x^*(s)$ be the equation of Q with natural parameter s and the same direction of the curvature vector as the curve L itself (see Figure 17). Then, by (4′) and (7),

$$\Delta x(s) = e_1(s)\Delta s + \frac{\kappa(s)}{2} e_2(s)(\Delta s)^2 + \alpha_2(\Delta s)^3,$$

$$\Delta x^*(s) = e_1(s)\Delta s + \frac{1}{2R} e_2(s)(\Delta s)^2 + \alpha_2^*(\Delta s)^3.$$

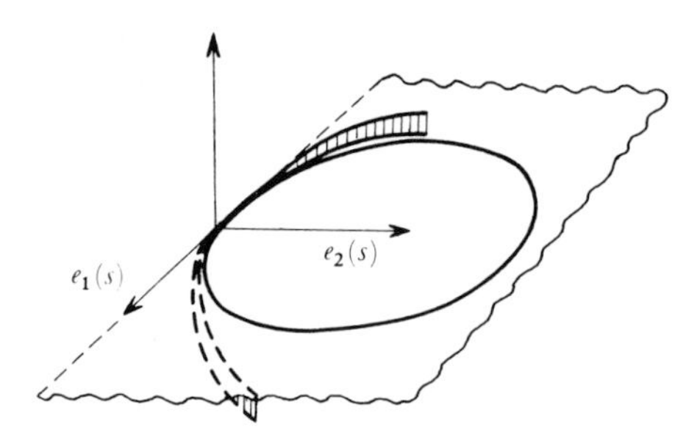

Figure 17

Hence, to make $|\Delta x(s) - \Delta x^*(s)|$ of the third order of smallness in Δs, we need only choose

$$R = \frac{1}{\kappa(s)}. \tag{8}$$

The circle Q corresponding to this choice of the radius R (and of the vector $e_2(s)$) is called the *osculating circle* of L at s; by the same token, the center of Q is called the *center of curvature* of L at s, and the number (8) is called the *radius of curvature* of L at s. If the curve L is a circle, then, because of (7) and (8), its radius of curvature coincides with its ordinary radius.

3.4. The Moving Basis

3.41. Again let L be a curve in the space R_n with equation $x = x(s)$ in terms of the natural parameter, and suppose the vectors $x'(s), x''(s), \ldots, x^{(n)}(s)$ exist and are linearly independent at the point s. Then the osculating subspaces $E_1 \subset E_2 \subset \cdots \subset E_n$ of dimensions $1, 2, \ldots, n$, respectively (Sec. 3.22b), all exist at s. We now construct an orthonormal basis (Sec. 1.43i) in the space E_n as follows: For the first two vectors we choose the vectors

$$e_1(s) = x'(s), \qquad e_2(s) = \frac{x''(s)}{|x''(s)|}$$

already constructed, while for the third vector $e_3(s)$ in the subspace E_3 we choose a unit vector orthogonal to the plane E_2 and pointing into the same half-space as the vector $x'''(s)$. Similarly, once having constructed the first $m-1$ vectors $e_1(s), \ldots, e_{m-1}(s)$ in the subspace E_m, we choose the vector $e_m(s)$ in E_m to be a unit vector orthogonal to the subspace E_{m-1} and pointing into the same half-space as the vector $x^{(m)}(s)$.† These conditions uniquely deter-

† A vector $x \in R_n$ is said to be *orthogonal* to a subspace $E \subset R_n$ if $(x, y) = 0$ for every $y \in E$ (cf. Sec. 1.43g).

mine an orthonormal basis $e_1(s), \ldots, e_n(s)$, called the *moving basis* of the curve L at the point s, whose position in space varies with the point $P=x(t)$. By construction, every vector $e_m(s)$ is a linear combination

$$e_m(s)=\varphi_1(s)x'(s)+\cdots+\varphi_m(s)x^{(m)}(s) \tag{1}$$

of the vectors $x'(s), \ldots, x^{(m)}(s)$, where $\varphi_m(s)>0$ automatically.

3.42. The Frenet-Serret formulas. Next we calculate the derivatives of the vectors of the moving basis of a curve $L\subset R_n$ with respect to the parameter s. Differentiating (1), we get

$$e'_m(s)=\sum_{j=1}^{m}\varphi'_j(s)x^{(j)}(s)+\sum_{j=1}^{m}\varphi_j(s)x^{(j+1)}(s) \tag{2}$$

(cf. Sec. 1.61d). It follows that the vector $e'_m(s)$ lies in the subspace E_{m+1} if $m<n$, and hence we can write

$$e'_m(s)=a_{m1}(s)e_1(s)+\cdots+a_{mm}(s)e_m(s)+a_{m,m+1}(s)e_{m+1}(s). \tag{3}$$

Formula (3) remains valid for $m=n$ if we set $e_{n+1}=0$. Comparing (1), (2), and (3), we see at once that $a_{m,m+1}=\varphi_m(s)>0$. The other coefficients in (3) are also easily calculated. First we note that $a_{mm}(s)\equiv 0$, since the derivative of the unit vector $e_m(s)$ is orthogonal to $e_m(s)$, by Lemma 3.31. Then, differentiating the obvious formula

$$(e_j(s),e_m(s))=0 \qquad (j<m),$$

we get

$$(e'_j(s),e_m(s))+(e_j(s),e'_m(s))=0,$$

which implies

$$a_{mj}=(e'_m(s),e_j(s))=-(e'_j(s),e_m(s)). \tag{4}$$

But the right-hand side of (4) vanishes for $j<m-1$, since $e'_j(s)\in E_{j+1}$. Therefore (3) reduces to the form

$$e'_m(s)=a_{m,m-1}(s)e_{m-1}(s)+a_{m,m+1}(s)e_{m+1}(s). \tag{5}$$

Moreover, setting $j=m-1$ in (4), we get

$$a_{m,m-1}=(e'_m(s),e_{m-1}(s))=-(e'_{m-1}(s),e_m(s))=-a_{m-1,m}.$$

Now let

$$\kappa_1=\kappa_1(s)=(e'_1(s),e_2(s))=a_{12}$$

(this is the quantity previously denoted by $\kappa(s)$), and let

$$\kappa_2 = \kappa_2(s) = (e_2', e_3),$$
$$\kappa_3 = \kappa_3(s) = (e_3', e_4),$$
$$\ldots$$

Then we get the sequence of formulas

$$
\begin{aligned}
e_1'(s) &= \kappa_1 e_2(s),\\
e_2(s) &= -\kappa_1 e_1(s) + \kappa_2 e_3(s),\\
&\ldots\\
e_{n-1}'(s) &= -\kappa_{n-2} e_{n-2}(s) + \kappa_{n-1} e_n(s),\\
e_n'(s) &= -\kappa_{n-1} e_{n-1}(s),
\end{aligned}
\tag{6}
$$

called the *Frenet-Serret formulas* in R_n. The quantity κ_m ($m=2, \ldots, n-1$) is called the *mth curvature* of the curve L at the point s. The second curvature κ_2 is also called the *torsion* of L at s.

To interpret the coefficients $\kappa_2, \ldots, \kappa_{n-1}$ geometrically, we observe that according to the formula

$$e_m'(s) = -\kappa_{m-1} e_{m-1}(s) + \kappa_m e_{m+1}(s),$$

the velocity of rotation of the vector $e_m(s)$ is the sum of two components: The first component, in the direction of the vector $e_{m-1}(s)$, represents a rotation of the subspace E_m "into itself," while the second component, in the direction of the vector $e_{m+1}(s)$, represents a rotation of the subspace E_m in a direction orthogonal to E_m. The coefficient κ_m is just the velocity of the latter rotation per unit arc length. Thus the torsion is the velocity of rotation of the osculating plane in the direction from the vector e_2 to the vector e_3.

In other words, the number κ_m can be interpreted as the velocity of rotation of the subspace E_m in the direction orthogonal to E_m, in keeping with the geometric definition of the curvature κ_1 as the velocity of rotation of the tangent (Sec. 3.32). Just as in the latter case, the geometric definition of κ_m is of a somewhat more general character than the definition associated with the Frenet-Serret formulas, which requires that the space E_{m+1} be "nondegenerate" (i.e., that it actually be $(m+1)$-dimensional). In fact, the geometric definition requires only that E_m be nondegenerate and that the derivative $x^{(m+1)}(s)$ exist. If $x^{(m+1)}(s)=0$, then the velocity of rotation of the space E_m vanishes at the given point, and the geometric definition assigns the value 0 to the quantity κ_m.

3.43. Calculation of higher curvatures. Let $a_1, a_2, \ldots, a_m$ be m linearly independent vectors in the n-dimensional Euclidean space R_n ($m \leqslant n$). Then by the *mixed product* of $a_1, a_2, \ldots, a_m$, denoted by $[a_1, a_2, \ldots, a_m]$, we mean the volume of the m-dimensional parallelepiped constructed on the vectors a_1, $a_2, \ldots, a_m$. Let

$$a_1 = (a_{11}, a_{21}, \ldots, a_{n1}),$$
$$a_2 = (a_{12}, a_{22}, \ldots, a_{n2}),$$
$$\ldots$$
$$a_m = (a_{1m}, a_{2m}, \ldots, a_{nm}),$$

and consider the $n \times m$ matrix

$$\left\| \begin{matrix} a_{11} & a_{12} & \cdots & a_{1m} \\ a_{21} & a_{22} & \cdots & a_{2m} \\ \cdot & \cdot & \cdots & \cdot \\ a_{n1} & a_{n2} & \cdots & a_{nm} \end{matrix} \right\|, \qquad (7)$$

with the components of the vector a_1 as its first column, the components of a_2 as its second column, and so on. Then it can be shown (see Lin. Alg., Sec. 8.73) that $[a_1, a_2, \ldots, a_n]^2$ equals the sum of the squares of all the minors of (7) of order m. In particular, if $m = n$, then $[a_1, a_2, \ldots, a_m]$ is just the absolute value of the determinant

$$\left| \begin{matrix} a_{11} & a_{12} & \cdots & a_{1n} \\ a_{21} & a_{22} & \cdots & a_{2n} \\ \cdot & \cdot & \cdots & \cdot \\ a_{n1} & a_{n2} & \cdots & a_{nn} \end{matrix} \right|$$

of order n (Lin. Alg., Sec. 8.74).†

We now calculate the mixed product of the vectors $x_s, x_{ss}, \ldots, x_s^{(n)}$. To this end, we first write the formulas

$$x_s = x_t t_s,$$
$$x_{ss} = \cdots + x_{tt} t_s^2,$$
$$\ldots$$
$$x_s^{(n)} = \cdots + x_t^{(n)} t_s^n$$

(cf. Sec. 3.22), expressing the derivatives with respect to s in terms of the derivatives with respect to any other parameter t, where the dots indicate vectors which are linearly dependent on vectors already written in the preceding rows. It follows from these formulas and the properties of determinants that

$$[x_s, \ldots, x_s^{(n)}] = [x_t, \ldots, x_t^{(n)}] t^{1+\cdots+n}. \qquad (8)$$

† Apart from a possible sign difference, this is in keeping with the notation of Sec. 2.73.

On the other hand, we have

$$x_s = e_1,$$
$$x_{ss} = \kappa_1 e_2,$$
$$x_{sss} = (\kappa_1 e_2)_s = \cdots + \kappa_1 \kappa_2 e_3,$$
$$\cdots$$
$$x_s^{(n)} = \cdots + \kappa_1 \kappa_2 \cdots \kappa_{n-1} e_n,$$

with the help of the Frenet-Serret formulas, where the dots again indicate vectors which are linearly dependent on vectors already written in preceding rows. It follows that

$$[x_s, \ldots, x_s^{(n)}] = \kappa_1^{n-1} \kappa_2^{n-2} \cdots \kappa_{n-1} [e_1, \ldots, e_n] = \kappa_1^{n-1} \kappa_2^{n-2} \cdots \kappa_{n-1}. \tag{9}$$

Comparing (8) and (9), we find that

$$\kappa_1^{n-1} \kappa_2^{n-2} \cdots \kappa_{n-1} = [x_t, \ldots, x_t^{(n)}] t_s^{n(n+1)/2} = \frac{[x_t, \ldots, x_t^{(n)}]}{|x_t|^{n(n+1)/2}}, \tag{10}$$

which allows us to determine κ_{n-1} from a knowledge of $\kappa_1, \ldots, \kappa_{n-2}$.

For $n=2$ we get a formula

$$\kappa_1 = \frac{[x_t, x_{tt}]}{|x_t|^3} \tag{11}$$

for the curvature, which looks somewhat simpler than formula (5), p. 199. However, the new formula (11) is actually more suitable than the old formula only in the case of a plane curve, where the quantity $[x_t, x_{tt}]$ reduces to the absolute value of a determinant of order 2.

For $n>2$ we divide both sides of equation (10) by the corresponding sides of the analogous formula

$$\kappa_1^{n-2} \kappa_2^{n-3} \cdots \kappa_{n-2} = \frac{[x_t, \ldots, x_t^{(n-1)}]}{|x_t|^{(n-1)n/2}},$$

obtaining

$$\kappa_1 \kappa_2 \cdots \kappa_{n-1} = \frac{[x_t, \ldots, x_t^{(n)}]}{[x_t, \ldots, x_t^{(n-1)}]} \frac{1}{|x_t|^n}. \tag{12}$$

Then, dividing both sides of this formula by the corresponding sides of the analogous formula

$$\kappa_1 \kappa_2 \cdots \kappa_{n-2} = \frac{[x_t, \ldots, x_t^{(n-1)}]}{[x_t, \ldots, x_t^{(n-2)}]} \frac{1}{|x_t|^{n-1}},$$

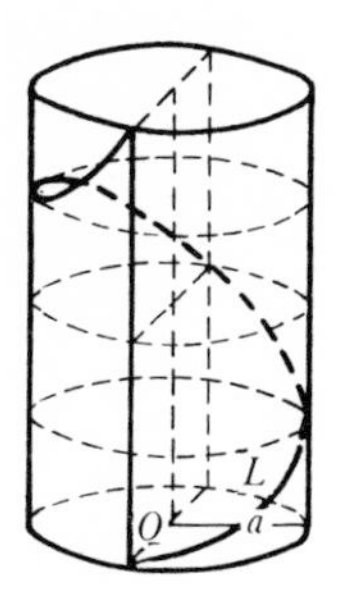

Figure 18

we finally get

$$\kappa_{n-1} = \frac{[x_t, \ldots, x_t^{(n)}][x_t, \ldots, x_t^{(n-2)}]}{[x_t, \ldots, x_t^{(n-1)}]^2 |x_t|}.$$

Geometrically this means that, apart from a factor of $1/|x_t|$, the curvature κ_{n-1} equals the ratio of the altitude of the n-dimensional parallelepiped constructed on the vectors $x_t, \ldots, x_t^{(n)}$ to the altitude of the $(n-1)$-dimensional parallelepiped constructed on the vectors $x_t, \ldots, x_t^{(n-1)}$.

3.44. Example. To illustrate these considerations, we now calculate the curvature and torsion of the helix L shown in Figure 18, with equation

$$x = x(t) = (a \cos t, a \sin t, bt). \tag{13}$$

Differentiating (13), we get

$$x_t = (-a \sin t, a \cos t, b), \quad |x_t| = \sqrt{a^2 + b^2},$$
$$x_{tt} = (-a \cos t, -a \sin t, 0), \quad (x_t, x_{tt}) = 0,$$
$$x_{ttt} = (a \sin t, -a \cos t, 0), \quad [x_t, x_{tt}, x_{ttt}] = a^2 b.$$

Using (11) and the fact that $(x_t, x_{tt}) = 0$, we find that the curvature equals†

$$\kappa_1 = \frac{|x_{tt}|}{|x_t|^2} = \frac{a}{a^2 + b^2}.$$

Then (10) gives

$$\kappa_1^2 \kappa_2 = \frac{[x_t, x_{tt}, x_{ttt}]}{|x_t|^6} = \frac{a^2 b}{(a^2 + b^2)^3},$$

so that the torsion equals

$$\kappa_2 = \frac{b}{a^2 + b^2}.$$

† Note that the same result follows at once from formula (5), p. 199.

3.45. Degeneracy of the moving basis

a. In Sec. 3.22b we defined the osculating subspaces $E_1 \subset E_2 \subset \cdots \subset E_n$ of a curve $x = x(t)$ in the case where the vectors $x'(t), x''(t), \ldots, x^{(n)}(t)$ exist and are linearly independent. We now consider the case where the latter condition fails to hold.

Suppose the vectors $x'(t), \ldots, x^{(n)}(t)$ become linearly dependent at some point t, while the vectors $x'(t), \ldots, x^{(n-1)}(t)$ are still linearly independent. Then the space E_n no longer exists, although E_{n-1} continues to exist; by the same token, the curvature $\kappa_{n-1}(t)$ and the vector $e_n(t)$ become meaningless. We might try to construct the vector $e_n(t)$ by continuity, taking the limit as $\tilde{t} \to t$ of the well-defined vector $e_n(\tilde{t})$,† but then it may well turn out that the one-sided limits

$$\lim_{\tilde{t} \nearrow t} e_n(\tilde{t}), \qquad \lim_{\tilde{t} \searrow t} e_n(\tilde{t})$$

(Vol. 1, Sec. 5.24) exist and are unequal. For example, in the case of the plane curve shown in Figure 19, the vector $e_2(t)$ changes its direction by a jump of 180° in passing through the inflection point C. Guided by the considerations at the end of Sec. 3.42, we complete the definition of the curvatures $\kappa_1(t), \ldots, \kappa_n(t)$ by the convention of setting them equal to zero at the points where they fail to exist.

b. Example. A straight line L has an equation of the form

$$x(t) = x_0 + t x_1,$$

where x_0 and x_1 are constant vectors. Therefore

$$x'(t) = x_1, \; x''(t) = 0.$$

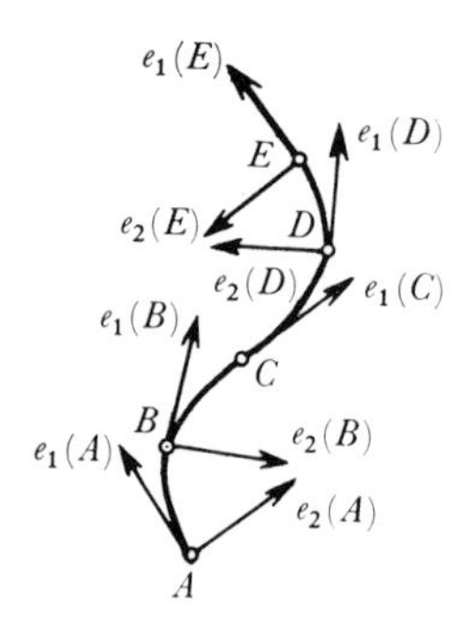

Figure 19

† Under the assumption that the vectors $x'(\tilde{t}), \ldots, x^{(n)}(\tilde{t})$ are linearly independent.

But then L has no osculating plane, and hence, by our convention, the curvature of the straight line L equals zero.

Conversely, suppose the curvature of a curve L with equation

$$x = x(t) \qquad (a \leqslant t \leqslant b)$$

vanishes identically, i.e., suppose the vectors $x'(t) \neq 0$ and $x''(t)$ are linearly dependent for all $t \in [a,b]$. Then L must be a straight line. In fact, linear dependence of $x'(t)$ and $x''(t)$ means that $x''(s) = 0$, since $x''(s)$ is a linear combination of $x'(t)$ and $x''(t)$, and is also orthogonal to $x'(t)$. But then $x'(s)$ is a constant vector, by Theorem 1.61f, where this constant vector, which we denote by x_1, must be of unit length, like every vector of the form $x'(s)$. Integrating the equation

$$x'(s) = x_1,$$

and using the uniqueness of the solution (also a consequence of Theorem 1.61f), we get

$$x(s) = x_0 + sx_1,$$

where $x_0 = x(0)$, i.e., L is a straight line, as asserted.

c. A somewhat more complicated situation occurs when the curve $x = x(t)$ is entirely contained in an m-dimensional hyperplane (Lin. Alg., Sec. 2.6). Then the hyperplane contains all the vectors $x'(t), x''(t), \ldots$, and hence the vectors $x'(t), x''(t), \ldots, x^{(m+1)}(t)$ are linearly dependent. It follows that $\kappa_m(t)$ and all higher curvatures vanish. As we now show, the converse is also true:

THEOREM. *Let L be a curve in Hilbert space with equation*

$$x = x(t) \qquad (a \leqslant t \leqslant b),$$

and suppose the vectors $x'(t), \ldots, x^{(m)}(t)$ are linearly independent for all $t \in [a,b]$, while the vectors $x'(t), \ldots, x^{(m)}(t), x^{(m+1)}(t)$ are linearly dependent for all $t \in [a,b]$. Then L lies in the hyperplane going through the point $x(a)$ and determined by the vectors $x'(a), \ldots, x^{(m)}(a)$.

Proof. Being linearly dependent for all $t \in [a,b]$, the vectors $x'(t), \ldots, x^{(m)}(t)$, $x^{(m+1)}(t)$ satisfy a linear relation of the form

$$x^{(m+1)}(t) = a_0(t)x'(t) + \cdots + a_{m-1}(t)x^{(m)}(t) \qquad (a \leqslant t \leqslant b). \tag{14}$$

If $x^{(m+1)}(t)$ is continuous, then so are all the coefficients $a_k(t)$ in (14). In fact, taking the scalar product of (14) with the functions $x'(t), \ldots, x^{(m)}(t)$ in turn, we get the following system of linear equations in the functions $a_k(t)$:

$$(x^{(m+1)}(t),x'(t))=a_0(t)(x_t,x_t)+\cdots+a_{m-1}(t)(x_t^{(m)},x_t),$$

$$\ldots$$

$$(x^{(m+1)}(t),x^{(m)}(t))=a_0(t)(x_t,x_t^{(m)})+\cdots+a_{m-1}(t)(x_t^{(m)},x_t^{(m)}).$$

All the coefficients $(x_t^{(j)},x_t^{(k)})$ are continuous, being scalar products of continuous functions; moreover, the determinant of the system is nonzero, being the Gram determinant of the linearly independent vectors $x_t,\ldots,x_t^{(m)}$ (Lin. Alg., Sec. 8.7). Hence, by Cramer's rule (Lin. Alg., Sec. 1.7), the solutions $a_0(t),\ldots,a_{m-1}(t)$ of the system are themselves all continuous on $[a,b]$.

We now observe that according to Theorem 2.65 and the subsequent remark, there exists a unique solution of the equation

$$u^{(m)}(t)=a_0(t)u(t)+\cdots+a_{m-1}(t)u^{(m-1)}(t) \tag{15}$$

satisfying the initial conditions

$$u(a)=u_0=x'(a),\ldots,u^{(m-1)}(a)=u_{m-1}=x^{(m)}(a),$$

and this solution lies entirely in the subspace determined by the initial vectors $u_0,\ldots,u_{m-1}$. But the vector function $x'(t)$ satisfies equation (15) (which reduces to (14) under the substitution $u(t)=x'(t)$) and the indicated initial conditions. It follows that for all $t\in[a,b]$ the vector $x'(t)$ stays in the subspace determined by the vectors $x'(a)$, …, $x^{(m)}(a)$. By the same token, for all $t\in[a,b]$ the vector

$$x(t)=x(a)+\int_a^t x'(\tau)\,d\tau \qquad (a\leqslant t\leqslant b)$$

lies in the hyperplane going through the point $x(a)$ and parallel to the given subspace. ∎

In particular, if the torsion $\kappa_2(t)$ of a curve L vanishes identically, while its curvature is nonvanishing, then L is a plane curve.

3.5. The Natural Equations

3.51. On a curve L written in terms of the natural parameter s all the curvatures can be regarded as functions of s:

$$\kappa_1=\kappa_1(s),\ \kappa_2=\kappa_2(s),\ldots \qquad (0\leqslant s\leqslant l) \tag{1}$$

(l is the length of L). Here we allow curvatures to vanish, interpreting this as the kind of degeneracy described in Sec. 3.45.

If a curve L^* is obtained from L by an isometric linear transformation, i.e., by a transformation preserving all distances in the underlying space, then, since all the functions (1) are determined only by the metric, they are the same for the curve L^* as for the curve L. As we now show, the converse is true for finite-dimensional curves:

THEOREM. *Given two curves L and L^* in the n-dimensional Euclidean space R_n, each described by an n-fold differentiable vector function, suppose both L and L^* have the same set of continuous positive functions $\kappa_1(s),\ldots,\kappa_{n-1}(s)$ for their curvatures (in terms of the natural parameter). Then there exists an isometric linear transformation (a rigid motion, possibly accompanied by a reflection) of the space R_n into itself, which carries L into complete coincidence with L^*.*

Proof. Let $e_1(s),\ldots, e_n(s)$ be the moving basis of L and $e_1^*(s),\ldots,e_n^*(s)$ the moving basis of L^*. Consider the rigid motion (possibly accompanied by a reflection) of R_n into itself which carries the initial configuration $e_1(0),\ldots,e_n(0)$ of the moving basis of the curve L into the initial configuration $e_1^*(0),\ldots,e_n^*(0)$ of the moving basis of the curve L^*, so that $e_1(0)$ goes into $e_1^*(0)$, $e_2(0)$ goes into $e_2^*(0)$, and so on. Then, as we now show, this transformation carries L itself into complete coincidence with L^*.

The functions $\kappa_1(s),\ldots,\kappa_{n-1}(s)$ are continuous, by hypothesis, and hence, by Theorem 2.63, there exists a unique solution of the system

$$\begin{aligned}
&u_1'(s) = \kappa_1(s)u_2(s),\\
&u_2'(s) = -\kappa_1(s)u_1(s) + \kappa_2(s)u_3(s),\\
&\cdots\\
&u_n'(s) = -\kappa_{n-1}(s)u_{n-1}(s)
\end{aligned} \tag{2}$$

satisfying the initial conditions

$$u_1(0) = e_1^*(0),\ldots,u_n(0) = e_n^*(0). \tag{3}$$

But, according to the Frenet-Serret formulas (Sec. 3.42), both sets of vectors $e_1(s),\ldots,e_n(s)$ and $e_1^*(s),\ldots,e_n^*(s)$ satisfy the system (2) and the initial conditions (3), after carrying out the indicated transformation. It follows from Theorem 2.63 that

$$e_1(s) = e_1^*(s),\ldots,e_n(s) = e_n^*(s).$$

Let $x(s)$ be the position vector of the curve L (after the transformation) and $x^*(s)$ that of the curve L^*. Then, since L and L^* start from the same point $x^*(0)$, we have

$$x(s) = x^*(0) + \int_0^s e_1(\sigma)\, d\sigma = x^*(0) + \int_0^s e_1^*(\sigma)\, d\sigma = x^*(s),$$

i.e., L and L^* coincide completely. ∎

The equations

$$\kappa_1 = \kappa_1(s), \ldots, \kappa_{n-1} = \kappa_{n-1}(s) \qquad (0 \leqslant s \leqslant l)$$

are called the *natural equations* of the curve $L \subset R_n$. As just shown, they determine L to within an isometric linear transformation of R_n.

3.52. Curves with preassigned curvatures

a. Let $\mathbf{A}$ be a linear operator acting in the Euclidean space R_n. Then by the *adjoint* of $\mathbf{A}$ we mean the unique linear operator $\mathbf{A}'$ acting in R_n such that $(\mathbf{A}x, y) = (x, \mathbf{A}'y)$ for arbitrary $x, y \in R_n$ (Lin. Alg., Sec. 8.91).† If $\|a_{jk}\|$ is the matrix of $\mathbf{A}$ in some orthonormal basis, then the matrix of $\mathbf{A}'$ is just the *transpose* of that of $\mathbf{A}$, namely the matrix

$$\|a_{jk}\|' = \|a_{kj}\|$$

obtained from $\|a_{jk}\|$ by interchanging rows and columns (Lin. Alg., loc. cit.). We say that $\mathbf{A}$ is *antisymmetric* if $\mathbf{A}' = -\mathbf{A}$, or, in terms of matrix elements, if $a_{kj} = -a_{jk}$ $(j, k = 1, \ldots, n)$. We say that $\mathbf{A}$ is *orthogonal* if $\mathbf{A}' = \mathbf{A}^{-1}$ (provided $\mathbf{A}$ is invertible, with inverse $\mathbf{A}^{-1}$). Clearly

$$(\mathbf{AB})' = \mathbf{B}'\mathbf{A}'$$

for any two linear operators $\mathbf{A}$ and $\mathbf{B}$ acting in R_n (why?).

b. LEMMA. *Let* $\mathbf{A}(t)$ *be an antisymmetric operator acting in* R_n. *Then the solution operator* $\Omega_{t_0}^t$ *of the differential equation*

$$\frac{du(t)}{dt} = \mathbf{A}(t)u(t)$$

is orthogonal.

Proof. The operator $\Omega_{t_0}^t$ satisfies the equation

$$\frac{d\Omega_{t_0}^t}{dt} = \mathbf{A}(t)\Omega_{t_0}^t \tag{4}$$

and the initial condition $\Omega_{t_0}^{t_0} = \mathbf{E}$ (Secs. 2.38b,e). Moreover, we have

$$\Omega_{t_0}^t \Omega_t^{t_0} = \mathbf{E} \tag{5}$$

(Sec. 2.38d). Differentiating (5) with respect to t, we find that

$$\frac{d\Omega_{t_0}^t}{dt}\Omega_t^{t_0} + \Omega_{t_0}^t \frac{d\Omega_t^{t_0}}{dt} = 0,$$

† Here we temporarily use the prime to denote the operation of taking the adjoint. In Sec. 3.52c the prime will again denote differentiation (as usual).

or

$$\mathbf{A}(t)\Omega_{t_0}^{t}\Omega_{t}^{t_0}+\Omega_{t_0}^{t}\frac{d\Omega_{t}^{t_0}}{dt}=0,$$

which implies

$$\frac{d\Omega_{t}^{t_0}}{dt}=-\Omega_{t}^{t_0}\mathbf{A}(t).$$

Going over to adjoint operators, we get

$$\frac{d(\Omega_{t}^{t_0})'}{dt}=-\mathbf{A}'(t)(\Omega_{t}^{t_0})'. \tag{6}$$

We now use the condition $\mathbf{A}'(t)=-\mathbf{A}(t)$, obtaining

$$\frac{d(\Omega_{t}^{t_0})'}{dt}=\mathbf{A}(t)(\Omega_{t}^{t_0})'. \tag{7}$$

Comparing (4) and (7), and noting that the operator $(\Omega_{t}^{t_0})'$ satisfies the same initial condition $(\Omega_{t_0}^{t_0})'=\mathbf{E}'=\mathbf{E}$ as the operator $\Omega_{t_0}^{t}$, we find that

$$(\Omega_{t}^{t_0})'=\Omega_{t_0}^{t}, \tag{8}$$

by the uniqueness of the solution of the differential equation. But

$$\Omega_{t}^{t_0}=(\Omega_{t_0}^{t})^{-1}, \tag{9}$$

and hence, comparing (8) and (9), we get

$$(\Omega_{t_0}^{t})'=(\Omega_{t_0}^{t})^{-1},$$

i.e., the operator $\Omega_{t_0}^{t}$ is orthogonal. ∎

c. THEOREM. *Given any $n-1$ continuous positive functions*

$$\varphi_1(s),\ldots,\varphi_{n-1}(s) \qquad (0\leqslant s\leqslant l), \tag{10}$$

there exists an n-fold differentiable vector function $x(s)$ with values in R_n such that the curve $L\subset R_n$ with equation

$$x=x(s) \qquad (0\leqslant s\leqslant l)$$

has the functions (10) as its curvatures as a function of the arc length s:

$$\kappa_1(s)=\varphi_1(s),\ldots,\kappa_{n-1}(s)=\varphi_{n-1}(s) \qquad (0\leqslant s\leqslant l). \tag{11}$$

Proof. Consider the system of vector equations

$$\begin{aligned}
u_1'(s)&=\varphi_1(s)u_2(s),\\
u_2'(s)&=-\varphi_1(s)u_1(s)+\varphi_2(s)u_3(s),\\
&\cdots\\
u_n'(s)&=-\varphi_{n-1}(s)u_{n-1}(s)
\end{aligned} \tag{12}$$

with initial conditions

$$u_1(0)=e_1,\ldots,u_n(0)=e_n, \tag{13}$$

where $e_1,\ldots,e_n$ is any set of n orthonormal vectors in R_n. Then the unique solution $u_1(s),\ldots,u_n(s)$ of (12) and (13), guaranteed by Theorem 2.63, is itself a set of orthonormal vectors in R_n for all $s\in[0,l]$. In fact, by the lemma (what is $\mathbf{A}(t)$ here?), the matrix $\|\omega_{jk}(s)\|$ of the solution operator Ω_0^s of the system (12) is orthogonal, and hence

$$\sum_{k=1}^{n}\omega_{ik}(s)\omega_{jk}(s)=\begin{cases}1 & \text{if } i=j,\\ 0 & \text{if } i\neq j\end{cases}$$

(Lin. Alg., Sec. 8.93). But

$$u_i(s)=\sum_{k=1}^{n}\omega_{ik}(s)e_k \qquad (i=1,\ldots,n),$$

and hence

$$(u_i(s),u_j(s))=\sum_{k=1}^{n}\omega_{ik}(s)\omega_{jk}(s)=\begin{cases}1 & \text{if } i=j,\\ 0 & \text{if } i\neq j,\end{cases}$$

since $e_1,\ldots,e_n$ is an orthonormal basis. Thus the vectors $u_1(s),\ldots,u_n(s)$ are orthonormal for all $s\in[0,l]$, as asserted.

Now let

$$x(s)=\int_0^s u_1(\sigma)\,d\sigma,$$

and let L be the curve in R_n with equation $x=x(s)$. Then $|x'(s)|=|u_1(s)|=1$, so that the parameter s is just the arc length of L. As already proved, the vectors $u_1(s),\ldots,u_m(s)$ are orthonormal for all $m=1,2,\ldots,n$, and moreover, because of (12),

$$x^{(m)}(s)=y_1^{(m-1)}(s)$$

is a linear combination of $u_1(s),\ldots,u_m(s)$, where the coefficient of $u_m(s)$, namely $\varphi_{m-1}(s)$, is positive. Hence the vectors $u_1(s),\ldots,u_n(s)$ are the vectors of the moving basis of L for every s. But the vectors of the moving basis satisfy the Frenet-Serret formulas

$$u_1'(s)=\kappa_1(s)u_2(s),$$
$$u_2'(s)=-\kappa_1(s)u_1(s)+\kappa_2(s)u_3(s),$$
$$\cdots$$
$$u_{n-1}'(s)=-\kappa_{n-1}(s)u_n(s).$$

Comparing these formulas with (12), we successively verify the formulas (11). ∎

3.6. Helices

3.61.a. By a *helix* we mean a curve all of whose curvatures are constant. An obvious example of a helix is a straight line (all of whose curvatures vanish). The circle Q of radius R is also a helix in the plane, since Q has constant curvature $\kappa = 1/R$ (Sec. 3.34b), and all its higher curvatures vanish. Besides straight lines and circles, there are no other helices in the plane. In fact, if L is a plane curve of constant curvature $\kappa > 0$, then besides L we can consider the curve Q of radius $R = 1/\kappa$, which also has constant curvature κ. But then, according to Theorem 3.51, there is a rigid motion (possibly accompanied by a reflection) carrying L into Q, so that L is itself a circle.

b. In three-dimensional space we have the "classical" helix Q with parametric equations

$$x_1 = a \cos t, \qquad x_2 = a \sin t, \qquad x_3 = bt. \tag{1}$$

As shown in Sec. 3.44, the curvature and torsion of Q have the constant values

$$\kappa_1 = \frac{a}{a^2 + b^2}, \qquad \kappa_2 = \frac{b}{a^2 + b^2}. \tag{2}$$

Therefore Q is also a helix in the sense of the above definition. Every other helix in R_3 is either a straight line ($\kappa_1 = \kappa_2 = 0$), a circle ($\kappa_1 > 0$, $\kappa_2 = 0$), or a curve congruent to a curve Q with suitable values of a and b. In fact, given any helix L in R_3 with $\kappa_1 > 0$, $\kappa_2 > 0$, we determine the parameters a and b from (2), i.e., we choose

$$a = \frac{\kappa_1}{\kappa_1^2 + \kappa_2^2}, \qquad b = \frac{\kappa_2}{\kappa_1^2 + \kappa_2^2}.$$

Using these values of a and b to construct a helix with parametric equations (1), we see that Q has constant curvature κ_1 and constant torsion κ_2. But then, again by Theorem 3.51, there is a rigid motion carrying L into Q, i.e., L is congruent to Q.

***3.62. Helices in n-dimensional space.** A helix in R_n is a curve L such that $\kappa_1(s) = \kappa_1, \ldots, \kappa_{n-1}(s) = \kappa_{n-1}$, where $\kappa_1, \ldots, \kappa_{n-1}$ are positive constants. The vectors $e_1(s), \ldots, e_n(s)$ of the moving basis of L satisfy the Frenet-Serret formulas

$$\begin{aligned}
&e_1'(s)=\kappa_1 e_2(s),\\
&e_2'(s)=-\kappa_1 e_1(s)+\kappa_2 e_3(s),\\
&\cdots\\
&e_{n-1}'(s)=-\kappa_{n-2}e_{n-2}(s)+\kappa_{n-1}e_n(s),\\
&e_n'(s)=-\kappa_{n-1}e_{n-1}(s),
\end{aligned} \tag{3}$$

with constant $n\times n$ coefficient matrix of the form

$$K=\left\|\begin{matrix}
0 & \kappa_1 & 0 & & & & \\
-\kappa_1 & 0 & \kappa_2 & & & & \\
0 & -\kappa_2 & 0 & & & & \\
 & & & \ddots & & & \\
 & & & & 0 & \kappa_{n-2} & 0\\
 & & & & -\kappa_{n-2} & 0 & \kappa_{n-1}\\
 & & & & 0 & -\kappa_{n-1} & 0
\end{matrix}\right\|. \tag{4}$$

To calculate the rank of K, we delete the row and column containing the element κ_1, thereby lowering the rank of K by 1 and obtaining a new matrix K^*. Then from K^* we delete the row and column containing the element $-\kappa_1$, thereby lowering the rank of K^* by 1 and obtaining a new matrix

$$K^{**}=\left\|\begin{matrix}
0 & \kappa_3 & 0 & & & & \\
-\kappa_3 & 0 & \kappa_4 & & & & \\
0 & -\kappa_4 & 0 & & & & \\
 & & & \ddots & & & \\
 & & & & 0 & \kappa_{n-2} & 0\\
 & & & & -\kappa_{n-2} & 0 & \kappa_{n-1}\\
 & & & & 0 & -\kappa_{n-1} & 0
\end{matrix}\right\|$$

with the same structure as that of K, but with a rank which is 2 lower. If $n=2m$, then after $2m-2$ steps we arrive at the matrix

$$\left\|\begin{matrix} 0 & \kappa_{n-1}\\ -\kappa_{n-1} & 0\end{matrix}\right\|$$

which is obviously of rank 2, so that the rank of K equals $2m-2+2=2m=n$. On the other hand, if $n=2m+1$, then after $2m$ steps we arrive at the matrix $\|0\|$, which is obviously of rank 0, so that the rank of K equals $2m+0=2m=n-1$.

The operator $\mathbf{K}$ with matrix (4) is obviously antisymmetric, since the transpose of K is just $-K$. To analyze $\mathbf{K}$ further, we make use of a known result on the structure of antisymmetric operators (Lin. Alg., Sec. 9.46) to deduce that if $n=2m$, then there exists an orthonormal "canonical basis" $x_1,y_1,\ldots,x_m,y_m$ in R_n such that

$$\mathbf{K}x_1=-\tau_1 x_1,\quad \mathbf{K}x_2=-\tau_2 x_2,\quad \ldots,\quad \mathbf{K}x_m=-\tau_m x_m,$$
$$\mathbf{K}y_1=\tau_1 y_1,\quad \mathbf{K}y_2=\tau_2 y_2,\quad \ldots,\quad \mathbf{K}y_m=\tau_m y_m.$$

If $n=2m+1$, there is another basis vector z_n such that $\mathbf{K}z_n=0$. Since the rank of the matrix K equals $2m$ in both cases, the numbers $\tau_1,\ldots,\tau_m$ are all nonzero.

Besides (3), we also consider the system of scalar differential equations

$$\begin{aligned} &u_1'(s)=\kappa_1 u_2(s),\\ &u_2'(s)=-\kappa_1 u_1(s)+\kappa_2 u_3(s),\\ &\cdots\\ &u_n'(s)=-\kappa_{n-1}u_{n-1}(s). \end{aligned} \tag{5}$$

According to Sec. 2.72, this system has n linearly independent vector solutions directed along one of the vectors $x_1,y_1,\ldots,x_m,y_m(,z_n)$ for $s=0$.† Moreover, according to Sec. 2.14, the solution operator e^{tK} of the system (5) has a matrix of the form

$$e^{tK}=\begin{Vmatrix} \cos\tau_1 t & \sin\tau_1 t & & & \\ -\sin\tau_1 t & \cos\tau_1 t & & & \\ & & \ddots & & \\ & & & \cos\tau_m t & \sin\tau_m t \\ & & & -\sin\tau_m t & \cos\tau_m t \end{Vmatrix}$$

if $n=2m$. If $n=2m+1$, there is an extra element 1 in the lower right-hand corner of the matrix.

Suppose we write

$$x_j=(x_{j1},\ldots,x_{jm}),\quad x_j(s)=(x_{j1}(s),\ldots,x_{jm}(s)),$$
$$y_j=(y_{j1},\ldots,y_{jm}),\quad y_j(s)=(y_{j1}(s),\ldots,y_{jm}(s)),$$

where $j=1,\ldots,m$, and

$$z_n=(z_{n1},\ldots,z_{nn}),\ z_n(s)=(z_{n1}(s),\ldots,z_{nn}(s)).$$

Then, according to Sec. 2.14,

$$\begin{aligned} &x_{jk}(s)=x_{jk}\cos\tau_j s-y_{jk}\sin\tau_j s, \\ &y_{jk}(s)=x_{jk}\sin\tau_j s+y_{jk}\cos\tau_j s, \end{aligned}\qquad (j=1,\ldots,m;\ k=1,\ldots,n),$$
$$z_{nk}(s)=z_{nk}\qquad (n=2m+1).$$

† The notation indicates that z_n may or may not be included, depending on whether n is odd or even.

Now let $e_1(s),\ldots,e_n(s)$ be any solution of the system (3), and write

$$e_j(s)=(e_{j1}(s),\ldots,e_{jn}(s)) \quad (j=1,\ldots,n). \tag{6}$$

For fixed k, the functions $e_{jk}(s)$ satisfy the system (5). Suppose the solutions (6) satisfy the initial conditions

$$e_{11}(0)=x_{11},\ldots,e_{n1}(0)=x_{1n},$$
$$e_{12}(0)=y_{11},\ldots,e_{n2}(0)=y_{1n},$$
$$e_{13}(0)=x_{21},\ldots,e_{n3}(0)=x_{2n},$$
$$e_{14}(0)=y_{21},\ldots,e_{n4}(0)=y_{2n},$$
$$\ldots$$
$$e_{1n}(0)=z_{n1},\ldots,e_{nn}(0)=z_{nn},$$

where the last equation is written only if $n=2m+1$. Then the vectors $e_1(0),\ldots, e_n(0)$ are orthonormal, as required to reconstruct a curve from given curvatures (see the proof of Theorem 3.52c). In particular, we have

$$e_1(s)=(e_{11}(s),\ldots,e_{1n}(s))=(x_{11}(s),y_{11}(s),x_{21}(s),y_{21}(s),\ldots(,z_{n1}(s)))$$
$$=(x_{11}\cos\tau_1 s-y_{11}\sin\tau_1 s,\ x_{11}\sin\tau_1 s+y_{11}\cos\tau_1 s,\ldots(,z_{n1})),$$

where again the quantity z_{n1} appears only if $n=2m+1$. Integrating with respect to s and setting constants of integration (corresponding to shifts along the axes) equal to zero, we get the following expression for the position vector of our helix L:

$$r(s)=\left(x_{11}\frac{\sin\tau_1 s}{\tau_1}+y_{11}\frac{\cos\tau_1 s}{\tau_1},\ -x_{11}\frac{\cos\tau_1 s}{\tau_1}+y_{11}\frac{\sin\tau_1 s}{\tau_1},\ldots(,z_{n1}s)\right).$$

This can be written more concisely as

$$x(s)=(A_1\cos\tau_1 S_1,\ A_1\sin\tau_1 S_1,\ A_2\cos\tau_2 S_2,\ A_2\sin\tau_2 S_2,\ldots(,B_n s)), \tag{7}$$

where $S_1=s-s_1$, $S_2=s-s_2$, ... For even $n=2m$, the whole curve L lies on the sphere

$$x_1^2+x_2^2+x_3^2+x_4^2+\cdots+x_{2m-1}^2+x_{2m}^2=A_1^2+A_2^2+\cdots+A_m^2.$$

The curve L is closed if the numbers $\tau_1,\ldots,\tau_m$ are all commensurable,† but fails to be closed if at least one pair of numbers τ_j,τ_k are incommensurable. For odd $n=2m+1$, the curve L is unbounded, since as $t\to+\infty$ it approaches infinity along the coordinate x_{2m+1}.

*3.63. Helices in infinite-dimensional space

a. Self-congruent curves. Let L be a curve in R_n, and suppose that for a given (initial) point $A\in L$ and any other point $P\in L$ there exists a rigid

† I.e., if $\tau_1,\ldots,\tau_m$ are all rational multiples of some τ_k.

motion of R_n (possibly accompanied by a reflection) carrying L into itself and the point A into the point P, while preserving the direction of increase of the parameter. Then, since the motion preserves the metric, the values of the curvatures of L at A (under the assumption that L is sufficiently smooth) coincide with their values at P. But then, since P is arbitrary, all the curvatures must be constant along L, i.e., the curve L must be a helix. Conversely, by Theorem 3.51, given any two points A and P of a helix $L \subset R_n$, there exists a rigid motion of R_n (possibly accompanied by a reflection) carrying L into itself and the point A into the point P, while preserving the direction of increase of the parameter.

These considerations suggest another way of defining a helix, namely as a curve L which can be carried into itself by a rigid motion of space (possibly accompanied by a reflection), while at the same time carrying a given point $A \in L$ into another given point $P \in L$ and preserving the direction of increase of the parameter. Note that this new definition imposes no smoothness requirements whatsoever on L. A curve L with the indicated property will be called *self-congruent*. It can be shown that the class of helices in R_n (as previously defined) coincides with the class of self-congruent curves in R_n (see Problem 2).

In infinite-dimensional spaces it turns out that there exist self-congruent curves which are continuous but have no tangent. Here we confine ourselves to a single (but very characteristic) example of a self-congruent curve in Hilbert space.

b. The Wiener spiral. Consider the space $\mathbf{H}_2(0,\infty)$, consisting of all piecewise continuous real functions $x(\tau)$ defined on the half-line $0 \leqslant \tau < \infty$, each vanishing outside some closed interval (which varies from function to function). We equip $\mathbf{H}_2(0,\infty)$ with a scalar product

$$(x(\tau),y(\tau)) = \int_0^\infty x(\tau)y(\tau)\,d\tau$$

and a corresponding norm

$$\|x(\tau)\| = \sqrt{\int_0^\infty x^2(\tau)\,d\tau},$$

thereby making $\mathbf{H}_2(0,\infty)$ into a Hilbert space.† Given any $t \in [0,\infty)$, let

$$Z(t) \equiv z(t,\tau) = \begin{cases} 1 & \text{if } 0 < \tau < t, \\ 0 & \text{otherwise.} \end{cases}$$

† Although the integrals have upper limit ∞, they reduce to integrals over finite intervals. Actually $\mathbf{H}_2(0,\infty)$ is a pre-Hilbert space, but can be made into a Hilbert space by the procedure described in Sec. 1.48 (give the details).

Then, as t varies from 0 to ∞, the point $Z(t)$ describes a curve L in the space $\mathbf{H}_2(0,\infty)$. This curve, called the *Wiener spiral*, is continuous, in fact uniformly continuous, since

$$\|Z(t)-Z(\tilde{t})\|^2=\left|\int_{\tilde{t}}^{t} 1^2\,dt\right|=|t-\tilde{t}|.$$

THEOREM. *The Wiener spiral L is self-congruent.*

Proof. Given any $t_0>0$, consider the linear transformation $\mathbf{U}$ of the space $\mathbf{H}_2(0,\infty)$ into itself which carries the function $x(\tau)$ into the function

$$\mathbf{U}x(\tau)=z(t_0,\tau)+x(\tau-t_0),$$

where we set $x(\tau-t_0)=0$ if $\tau<t_0$. Then $\mathbf{U}$ is isometric, since

$$\|\mathbf{U}x(\tau)-\mathbf{U}y(\tau)\|^2=\|x(\tau-t_0)-y(\tau-t_0)\|^2=\int_{t_0}^{\infty}[x(\tau-t_0)-y(\tau-t_0)]^2\,d\tau$$
$$=\int_0^{\infty}[x(\tau)-y(\tau)]^2\,d\tau=\|x(\tau)-y(\tau)\|^2.$$

The initial point $Z(0)$ of the curve L is just the zero element of the space $\mathbf{H}_2(0,\infty)$, which $\mathbf{U}$ carries into the point $Z(t_0)\in L$. Moreover, $\mathbf{U}$ carries the point $Z(t)\in L$ into the point $Z(t+t_0)\in L$. Therefore L is self-congruent. ∎

Problems

1. Prove that the locus of the centers of curvature of a helix L in R_3 is also a helix L^* with the same axis. Prove that the locus of the centers of curvature of L^* is just the original helix L.

2. Prove that every self-congruent curve in R_n is a helix (without assuming smoothness).

3. Suppose there is a one-to-one correspondence between two curves L and L^* in R_n such that corresponding vectors of the moving bases are parallel at corresponding points of the curves. If κ_j,κ_j^* $(j=1,\ldots,n)$ are the curvatures of the two curves, show that

$$\frac{\kappa_1}{\kappa_1^*}=\frac{\kappa_2}{\kappa_2^*}=\cdots=\frac{\kappa_{n-1}}{\kappa_{n-1}^*}.$$

4. By the *osculating m-sphere S_m* of a space curve L is meant the sphere in the $(m+1)$-dimensional space determined by the first $m+1$ vectors of the moving basis of L such that the deviation of a variable point of L from S_m is of order $(\Delta s)^{m+2}$. Let r_m be the radius of S_m. Prove that $r_1\leqslant r_2\leqslant\cdots\leqslant r_m\leqslant\cdots$.

5 (*Peano's curve*). Let t be a number in $[0,1)$ with ternary representation

$$t=0.t_1t_2\ldots t_{2n-1}t_{2n}\cdots$$

(cf. Vol. 1, Sec. 1.78) consisting only of the digits 0 and 2. Show that the functions $x(t)$ and $y(t)$, whose values have the binary representations

$$x(t)=0.\frac{t_1}{2}\frac{t_3}{2}\cdots\frac{t_{2n-1}}{2}\cdots, \qquad y(t)=0.\frac{t_2}{2}\frac{t_4}{2}\cdots\frac{t_{2n}}{2}\cdots,$$

are uniquely defined on the set of all such t, and can be extended continuously from their domain of definition onto the whole interval $[0,1]$. Show that the curve $r(t)=(x(t),y(t))$ passes through every point of the square $0\leqslant x\leqslant 1, 0\leqslant y\leqslant 1$.

6. Prove that the function $Z(t)$ defining the Wiener spiral (Sec. 3.63b) has no derivative in $\mathbf{H}_2(0,\infty)$.

7. Prove the following property of the Wiener spiral L: Any two chords of L corresponding to nonoverlapping intervals of the parameter values are orthogonal.

8. Construct smooth analogues of the Wiener spiral and find the corresponding curvatures.

4 Orthogonal Expansions

4.1. Orthogonal Expansions in Hilbert Space

4.11. Statement of the problem. We now return to the problem of approximation of functions. This problem can be approached in various ways. In Sec. 1.5 we considered *uniform* approximation in the space $C^s(Q)$ of continuous complex functions on a compactum Q. This kind of approximation corresponds to the norm

$$\|f\| = \max_x |f(x)|. \tag{1}$$

The problem of approximation *in the integral mean*, corresponding to the norm

$$\|f\|_p = \sqrt[p]{\int_Q |f(x)|^p \, dx} \tag{2}$$

(Sec. 1.39c), is also important in analysis, especially the problem of approximation *in the mean square*, corresponding to the norm

$$\|f\|_2 = \sqrt{\int_Q |f(x)|^2 \, dx}. \tag{3}$$

In this chapter, we will be mainly concerned with the problem of mean square approximation, as a rule choosing Q to be a closed interval of the real line.

Our basic problem can be stated as follows: Let $C(Q)$ be the set of all piecewise continuous complex functions defined on Q (the piecewise continuity guarantees the existence of the relevant integrals). Then, under what conditions on a set (linear manifold) $B(Q) \subset C(Q)$ can one find a sequence $\varphi_1(x), \varphi_2(x), \ldots$ of functions in $B(Q)$ such that

$$\|f - \varphi_n\| = \sqrt{\int_Q |f(x) - \varphi_n(x)|^2 \, dx} \to 0 \tag{4}$$

as $n \to \infty$?†

If the sequence $\varphi_n(x)$ converges uniformly to $f(x)$ on Q, then obviously (4) holds. However, (4) does not imply uniform convergence of $\varphi_n(x)$ to $f(x)$, and in fact (4) does not even imply *pointwise* convergence of $\varphi_n(x)$ to $f(x)$, i.e., convergence at the separate points of Q. Hence we can say that the problem of mean square approximation is "easier" than that of uniform approximation. Moreover, the construction of mean square approximations can be cast into a transparent geometric form, thanks to the fact that the

† Of course, the sequence $\varphi_n(x)$ depends on the particular function $f(x) \in C(Q)$.

appropriate space of functions, with the norm (3), can be taken to be a Hilbert space, in which we can not only measure lengths of vectors (as in any normed linear space), but also make use of the property of orthogonality.

4.12. Approximation in Hilbert space. Let $\mathbf{H}$ be a real or complex Hilbert space, with scalar product (x,y) and norm $\|x\|=\sqrt{(x,x)}$, and let $e_1,\ldots,e_n$ be an orthonormal system of elements of $\mathbf{H}$ (Sec. 1.43i), so that

$$(e_j,e_k)=\begin{cases}1 & \text{if } j=k,\\ 0 & \text{if } j\neq k.\end{cases}$$

Let $\mathbf{B}\subset\mathbf{H}$ be the n-dimensional subspace of $\mathbf{H}$ consisting of all linear combinations of the form $\alpha_1 e_1+\cdots+\alpha_n e_n$. Then we pose the following problem: Given any vector $f\in\mathbf{H}$, find the vector

$$g=\sum_{k=1}^{n} c_k e_k\in\mathbf{B}$$

such that the quantity $\|f-g\|$ is the smallest.

In solving this problem, we assume that the space $\mathbf{H}$ is complex, since the real case differs only by a few notational simplifications. Given any vector

$$x=\sum_{k=1}^{n}\xi_k e_k,$$

we have

$$\begin{aligned}\|f-x\|^2&=\left(f-\sum_{k=1}^{n}\xi_k e_k,f-\sum_{k=1}^{n}\xi_k e_k\right)\\
&=(f,f)-\sum_{k=1}^{n}\xi_k\overline{(f,e_k)}-\sum_{k=1}^{n}\bar{\xi}_k(f,e_k)+\sum_{k=1}^{n}|\xi_k|^2\\
&=(f,f)+\sum_{k=1}^{n}[|(f,e_k)|^2-\xi_k\overline{(f,e_k)}-\bar{\xi}_k(f,e_k)+|\xi_k|^2]-\sum_{k=1}^{n}|(f,e_k)|^2\\
&=(f,f)+\sum_{k=1}^{n}[(f,e_k)-\xi_k]\overline{[(f,e_k)-\xi_k]}-\sum_{k=1}^{n}|(f,e_k)|^2\\
&=(f,f)+\sum_{k=1}^{n}|(f,e_k)-\xi_k|^2-\sum_{k=1}^{n}|(f,e_k)|^2.\end{aligned}\tag{5}$$

Obviously, the last expression on the right achieves its minimum when the components ξ_k have the values

$$\xi_k=(f,e_k)\qquad(k=1,\ldots,n).$$

The corresponding vector

$$g=\sum_{k=1}^{n}(f,e_k)e_k$$

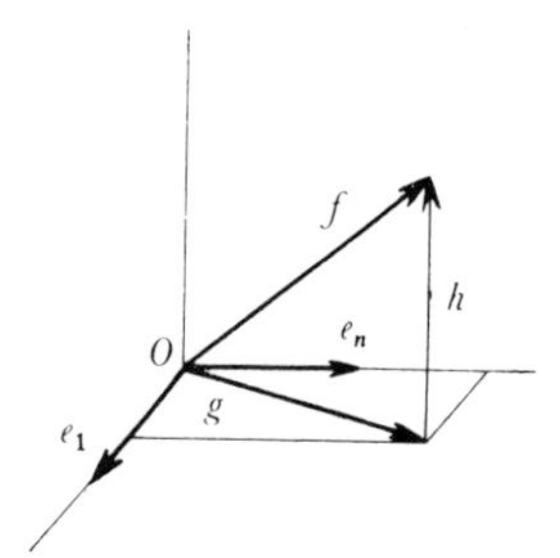

Figure 20

is called the *projection of f onto the subspace* **B**, while the vector $h = f - g$ is called the *perpendicular dropped from the end of f onto the subspace* **B**. These definitions are in keeping with the corresponding geometric concepts in the real case (see Figure 20). It follows from (5) that the square of the length of h equals

$$\|h\|^2 = (f,f) - \sum_{k=1}^{n} |(f,e_k)|^2.$$

In particular, this implies *Bessel's inequality*

$$\sum_{k=1}^{n} |(f,e_k)|^2 \leqslant \|f\|^2, \tag{6}$$

valid for any vector $f \in \mathbf{H}$ and any orthonormal system $e_1,\ldots,e_n$.

Thus the best "Hilbert approximation" of the vector f by a vector of the subspace **B** is obtained by choosing the approximating vector g to be the projection of f onto the subspace **B**.

4.13.a. Next suppose we have an infinite orthonormal system $e_1,\ldots,e_n,\ldots$ Then, carrying out the above construction for every finite subsystem $e_1,\ldots,e_n$, we get a corresponding best Hilbert approximation

$$g_n = \sum_{k=1}^{n} (f,e_k)e_k.$$

Note that the coefficients (f,e_k) of this best approximation do not depend on the index $n \geqslant k$, while the deviation of g_n from f, which we denote by h_n, has norm

$$\|h_n\| = \sqrt{(f,f) - \sum_{k=1}^{n} |(f,e_k)|^2}.$$

It is now natural to ask whether $\|h_n\|$ can be made arbitrarily small by choosing n sufficiently large. In general, this is not possible. For example, suppose the system $e_1,e_2,\ldots$ is "incomplete," in the sense that there exists a vector f different from zero which is orthogonal to all the vectors $e_1,e_2,\ldots$ Then all the numbers (f,e_k) vanish, and all the numbers $\|h_n\|$ equal $\|f\|$.

b. However, under certain circumstances we can still assert that the numbers $\|h_n\|$ approach 0 as $n\to\infty$. This happens if we know from other considerations, for example, from results like Stone's theorem (Sec. 1.53b), that there is a sequence of linear combinations of the vectors $e_1,e_2,\ldots$ which converges to f (in the Hilbert norm). In fact, suppose that given any $\varepsilon>0$, we can find a linear combination

$$\sum_{k=1}^{n} \xi_k e_k$$

such that

$$\left\| f-\sum_{k=1}^{n} \xi_k e_k \right\| < \varepsilon.$$

Then, for the best Hilbert approximation

$$\sum_{k=1}^{n} (f,e_k)e_k,$$

we certainly have

$$\|h_n\| = \sqrt{\|f\|^2 - \sum_{k=1}^{n} |(f,e_k)|^2} = \left\| f-\sum_{k=1}^{n} (f,e_k)e_k \right\| < \varepsilon. \tag{7}$$

Letting $\varepsilon\to 0$, we get in the limit

$$f=\lim_{n\to\infty} \sum_{k=1}^{n} (f,e_k)e_k = \sum_{k=1}^{\infty} (f,e_k)e_k. \tag{8}$$

The series on the right in (8) is called the *Fourier series of f with respect to the orthonormal system $e_1,e_2,\ldots$*, and the numbers (f,e_k) are called the *Fourier coefficients of f with respect to the system $e_1,e_2,\ldots$* This terminology is independent of whether or not the series (8) actually converges. The Fourier series of f can still be considered *formally*, even if it converges.

c. If the series (8) converges to the vector f, then, taking the limit of (7) as $n\to\infty$, we get

$$\|f\|^2 = \sum_{k=1}^{\infty} |(f,e_k)|^2. \tag{9}$$

This infinite-dimensional analogue of the Pythagorean theorem is known as *Parseval's theorem*. Even if nothing is known about the convergence of the

Fourier series (8) to the vector f, we can still take the limit as $n\to\infty$ of Bessel's inequality (6), obtaining the inequality

$$\sum_{k=1}^{\infty}|(f,e_k)|^2\leqslant\|f\|^2, \tag{10}$$

also called *Bessel's inequality*.

4.14. If the Fourier series

$$\sum_{k=1}^{\infty}(f,e_k)e_k$$

converges to f, then every rearrangement

$$\sum_{k=1}^{\infty}(f,e_{j_k})e_{j_k}$$

of the Fourier series still converges to f. In fact, by (7),

$$\left\|f-\sum_{k=1}^{n}(f,e_{j_k})e_{j_k}\right\|^2=\|f\|^2-\sum_{k=1}^{n}|(f,e_{j_k})|^2,$$

where the right-hand side approaches zero as $n\to\infty$ because of (9) and the fact that a convergent nonnegative series can be rearranged arbitrarily without changing its sum (Vol. 1, Sec. 6.35).

4.15. The fact that Fourier series necessarily appear in approximation problems is also shown by the following

LEMMA. *Given any vector $f\in\mathbf{H}$ and any orthonormal system $e_1,e_2,\ldots$ in $\mathbf{H}$, suppose it is known that*

$$f=\sum_{k=1}^{\infty}c_ke_k \tag{11}$$

in the sense that

$$\lim_{p\to\infty}\left\|f-\sum_{k=1}^{n_p}c_ke_k\right\|=0$$

for some increasing sequence of positive integers $n_1,\ldots,n_p,\ldots$ Then every coefficient c_m $(m=1,2,\ldots)$ coincides with the corresponding Fourier coefficient (f,e_k), and the series (11) *converges in the norm of* $\mathbf{H}$ *in the usual sense.*

Proof. Taking the scalar product of (11) with e_m and using the continuity of the scalar product (Sec. 1.43d) and the orthonormality of the system e_1, $e_2,\ldots$, we get

$$(f,e_m)=\left(\lim_{p\to\infty}\sum_{k=1}^{n_p}c_ke_k,e_m\right)=\lim_{p\to\infty}\left(\sum_{k=1}^{n_p}c_ke_k,e_m\right)=c_m. \quad \blacksquare$$

4.16. LEMMA. *Given an orthonormal system* $e_1, e_2, \ldots$ *in* $\mathbf{H}$, *let*

$$f = \sum_{k=1}^{\infty} a_k e_k, \qquad g = \sum_{k=1}^{\infty} b_k e_k$$

in the same sense as in the preceding lemma. Then

$$(f,g) = \sum_{k=1}^{\infty} a_k \bar{b}_k, \tag{12}$$

where the series on the right is absolutely convergent.

Proof. The absolute convergence of the series

$$\sum_{k=1}^{\infty} a_k b_k$$

follows from the inequality

$$|a_k b_k| \leqslant \tfrac{1}{2}(|a_k|^2 + |b_k|^2)$$

and Bessel's inequality (10). Moreover, if

$$f = \lim_{p\to\infty} \sum_{k=1}^{m_p} a_k e_k, \qquad g = \lim_{p\to\infty} \sum_{k=1}^{n_p} b_k e_k$$

in the norm of $\mathbf{H}$, then, setting $l_p = \min\{m_p, n_p\}$, we have

$$\begin{aligned}(f,g) = \left(\lim_{p\to\infty} \sum_{k=1}^{m_p} a_k e_k, \lim_{p\to\infty} \sum_{k=1}^{n_p} b_k e_k\right) &= \lim_{p\to\infty}\left(\sum_{k=1}^{m_p} a_k e_k, \sum_{k=1}^{n_p} b_k e_k\right) \\ &= \lim_{p\to\infty} \sum_{k=1}^{l_p} a_k \bar{b}_k = \sum_{k=1}^{\infty} a_k \bar{b}_k. \quad \blacksquare\end{aligned}$$

4.17. Later we will need to write Fourier series and Parseval's theorem for a system of vectors $g_1, g_2, \ldots$ which is orthogonal but not orthonormal (so that the vectors $g_1, g_2, \ldots$ are not all unit vectors). Suppose the vectors $g_1, g_2, \ldots$ are all nonzero. Then the system

$$e_1 = \frac{g_1}{\|g_1\|}, \qquad e_2 = \frac{g_2}{\|g_2\|}, \qquad \ldots \tag{13}$$

is orthonormal, and the Fourier series of f with respect to (13) can be written in the form

$$\sum_{k=1}^{\infty} (f,e_k)e_k = \sum_{k=1}^{\infty} \left(f, \frac{g_k}{\|g_k\|}\right)\frac{g_k}{\|g_k\|} = \sum_{k=1}^{\infty} \frac{(f,g_k)}{\|g_k\|^2} g_k = \sum_{k=1}^{\infty} \alpha_k g_k, \tag{14}$$

where

$$\alpha_k = \frac{(f,g_k)}{\|g_k\|^2}.$$

Just as before, the right-hand side of (14) is called the *Fourier series of f with respect to the orthogonal system* $g_1, g_2, \ldots$, and the numbers $\alpha_1, \alpha_2, \ldots$ are called the *Fourier coefficients of f with respect to the system* $g_1, g_2, \ldots$ If the series (14) converges to f, then

$$\|f\|^2 = \sum_{k=1}^{\infty} |(f,e_k)|^2 = \sum_{k=1}^{\infty} \left|\left(f, \frac{g_k}{\|g_k\|^2}\right)\right|^2 = \sum_{k=1}^{\infty} \frac{|(f,g_k)|^2}{\|g_k\|^2} = \sum_{k=1}^{\infty} \|g_k\|^2 |\alpha_k|^2. \quad (15)$$

This is just *Parseval's theorem* for the system $g_1, g_2, \ldots$ By the same token, if

$$f = \sum_{k=1}^{\infty} \alpha_k g_k, \qquad g = \sum_{k=1}^{\infty} \beta_k g_k,$$

then we get

$$(f,g) = \sum_{k=1}^{\infty} \alpha_k \bar{\beta}_k \|g_k\|^2 \quad (16)$$

instead of (12).

4.2. Trigonometric Fourier Series

4.21. The trigonometric functions

$$1, \cos t, \sin t, \cos 2t, \sin 2t, \ldots, \cos kt, \sin kt, \ldots \quad (1)$$

form an infinite orthogonal system in the real Hilbert space $\mathbf{H}_R(-\pi,\pi)$ of all piecewise continuous real functions $f(t)$ defined on the interval $-\pi \leqslant t \leqslant \pi$ (or equivalently on the unit circle $Q = \{(x,y) : x^2 + y^2 = 1\}$).† The orthogonality of the system (1) is easily verified by calculating the integrals of the functions

$$\cos kt \cos mt, \ \cos kt \sin mt, \ \sin kt \sin mt$$

over the interval $[-\pi,\pi]$. Moreover, we see at once that

$$\|1\|^2 = \int_{-\pi}^{\pi} 1^2\, dt = 2\pi,$$

$$\|\cos kt\|^2 = \int_{-\pi}^{\pi} \cos^2 kt\, dt = \int_{-\pi}^{\pi} \frac{1 + \cos 2kt}{2}\, dt = \pi,$$

$$\|\sin kt\|^2 = \int_{-\pi}^{\pi} \sin^2 kt\, dt = \int_{-\pi}^{\pi} \frac{1 - \cos 2kt}{2}\, dt = \pi.$$

† As in Sec. 1.48d, $\mathbf{H}_R(-\pi,\pi)$ is actually a factor space generated by the given set of functions.

Hence, by Sec. 4.17, given any function $f(t) \in \mathbf{H}_R(-\pi,\pi)$, we can write the Fourier series of $f(t)$ with respect to the system (1) as

$$\frac{1}{2\pi}\int_{-\pi}^{\pi} f(\tau)\,d\tau + \cos t\cdot\frac{1}{\pi}\int_{-\pi}^{\pi} f(\tau)\cos\tau\,d\tau + \sin t\cdot\frac{1}{\pi}\int_{-\pi}^{\pi} f(\tau)\sin\tau\,d\tau + \cdots$$

$$= \frac{a_0}{2} + \sum_{k=1}^{\infty}(a_k\cos kt + b_k\sin kt), \tag{2}$$

where

$$a_k = \frac{1}{\pi}\int_{-\pi}^{\pi} f(\tau)\cos k\tau\,d\tau \qquad (k=0,1,\ldots),$$
$$b_k = \frac{1}{\pi}\int_{-\pi}^{\pi} f(\tau)\sin k\tau\,d\tau \qquad (k=1,2,\ldots). \tag{3}$$

The numbers a_k and b_k are called the *Fourier coefficients of $f(t)$ with respect to the trigonometric system* (1). The mean square convergence of the Fourier series (2) will be proved in Theorem 4.24.

Fourier series with respect to the trigonometric system (1) are called "trigonometric Fourier series," or simply "Fourier series" when it is clear from the context that the underlying orthogonal system is the trigonometric system.

4.22. Next we consider the complex form of a trigonometric Fourier series. Let $\mathbf{H}_C(-\pi,\pi)$ be the complex Hilbert space consisting of all piecewise continuous complex functions $f(t)$ defined on the interval $-\pi\leqslant t\leqslant\pi$ (Sec. 1.48e). In this space we have an infinite orthogonal system consisting of the functions

$$e^{ikt} \qquad (k=0,\pm1,\pm2,\ldots). \tag{1'}$$

To prove the orthogonality of the functions (1′), we need only note that

$$(e^{ikt},e^{imt}) = \int_{-\pi}^{\pi} e^{ikt}e^{-imt}\,dt = \int_{-\pi}^{\pi} e^{i(k-m)t}\,dt = \left.\frac{e^{i(k-m)t}}{i(k-m)}\right|_{-\pi}^{\pi} = 0$$

if $k\neq m$, while

$$\|e^{ikt}\|^2 = \int_{-\pi}^{\pi} |e^{ikt}|^2\,dt = \int_{-\pi}^{\pi} 1^2\,dt = 2\pi.$$

Hence, by Sec. 4.17, given any function $f(t)\in\mathbf{H}_C(-\pi,\pi)$, we can write the Fourier series of $f(t)$ with respect to the system (1′) in the form of the two-sided series

$$\sum_{k=-\infty}^{\infty} c_k e^{ikt} \tag{2'}$$

(Vol. 1, Sec. 6.48), where the numbers

$$c_k = \frac{1}{2\pi}\int_{-\pi}^{\pi} f(\tau)e^{-ikt}\,d\tau \qquad (k=0,1,\ldots) \tag{3'}$$

are the Fourier coefficients of $f(t)$ with respect to the system (1′).

4.23. Using Euler's formula

$$e^{ikt} = \cos kt + i \sin kt,$$

we can transform the series (2′) into the form (2) (so far we have written (2) only for real functions). Thus let

$$c_0 = \frac{1}{2\pi}\int_{-\pi}^{\pi} f(\tau)\,d\tau = \frac{a_0}{2},$$

$$\begin{aligned} c_k e^{ikt} + c_{-k}e^{-ikt} &= (c_k + c_{-k}) \cos kt + i(c_k - c_{-k}) \sin kt \\ &= \frac{\cos kt}{\pi}\int_{-\pi}^{\pi} f(\tau) \cos k\tau\,d\tau + \frac{\sin kt}{\pi}\int_{-\pi}^{\pi} f(\tau) \sin k\tau\,d\tau \\ &= a_k \cos kt + b_k \sin kt, \end{aligned}$$

where the coefficients a_k and b_k are still given by (3), but $f(t)$ can now be complex.

For any $n=1,2,\ldots$ the partial sum

$$\frac{a_0}{2} + \sum_{k=1}^{n} (a_k \cos kt + b_k \sin kt)$$

of the series (2) and the symmetric sum

$$\sum_{k=-n}^{n} c_k e^{ikt}$$

of the series (2′) coincide. Hence the series (2) converges if and only if the series (2′) is symmetrically summable (Vol. 1, Sec. 6.49b). It should be noted that the series (2′) can be symmetrically summable for all real t, without being convergent (in the sense that the limit

$$\lim_{\substack{m\to\infty \\ n\to\infty}} \sum_{k=-m}^{n} c_k e^{ikt}$$

exists).†

† See Vol. 1, Secs. 6.49a, 6.49c.

4.24. THEOREM (**Mean square approximation of Fourier series**). *Let $\mathbf{H}_R(-\pi,\pi)=\mathbf{H}_R(Q)$ be the Hilbert space of all piecewise continuous real functions $f(t)$ defined on the interval $-\pi\leqslant t\leqslant\pi$ (or equivalently on the unit circle $Q=\{(x,y):x^2+y^2=1\}$). Then the Fourier series* (2) *of $f(t)$ converges to $f(t)$ in the mean square, i.e., in the norm of the space $\mathbf{H}_R(Q)$.*

Proof. According to Sec. 4.13b, we need only prove the existence of a sequence of trigonometric polynomials $T_n(t)$ converging to $f(t)$ in the norm of $\mathbf{H}_R(Q)$. To this end, consider the trigonometric polynomials

$$T_n(t)=\int_{-\pi}^{\pi} D_n(\tau;t)f(\tau)\,d\tau,$$

where

$$D_n(\tau;t)=\frac{\cos^{2n}\dfrac{\tau-t}{2}}{\displaystyle\int_{-\pi}^{\pi}\cos^{2n}\tau\,d\tau}\qquad (n=1,2,\dots)$$

is the delta-like sequence of Sec. 1.57a. According to Sec. 1.57c, the polynomials $T_n(t)$ converge uniformly to $f(t)$ on every closed set $E\subset Q$ of continuity points of $f(t)$. Moreover, if

$$M=\sup_{t\in Q}|f(t)|,$$

then

$$\sup_{t\in Q}|T_n(t)|\leqslant\sup_{t\in Q}|f(t)|\cdot\int_{-\pi}^{\pi}D_n(\tau;t)\,d\tau=M$$

(why?). Given any $\varepsilon>0$, we can cover the (finite) set of discontinuity points of $f(t)$ by an open set G, consisting of a finite number of nonintersecting open intervals of total length less than ε^2 (cf. Vol. 1, Sec. 3.23). Clearly $f(t)$ is continuous on the closed set $Q-G$, and hence there is an integer $N>0$ such that $|f(t)-T_n(t)|<\varepsilon$ on $Q-G$ if $n>N$. But then

$$\begin{aligned}\|f(t)-T_n(t)\|^2&=\int_Q[f(t)-T_n(t)]^2\,dt\\&=\int_G[f(t)-T_n(t)]^2\,dt+\int_{Q-G}[f(t)-T_n(t)]^2\,dt\\&\leqslant 4M^2\varepsilon^2+2\pi\varepsilon^2=(4M^2+2\pi)\varepsilon^2\end{aligned}$$

if $n>N$, and hence

$$\lim_{n\to\infty} T_n(t)=f(t)$$

in the norm of $\mathbf{H}_R(Q)$. ▮

The theorem has an obvious analogue for the space $\mathbf{H}_C(Q)=\mathbf{H}_C(-\pi,\pi)$ of all piecewise continuous *complex* functions $f(t)$ on Q, with Fourier series of the form (2′). Moreover, according to Sec. 4.14, both series (2) and (2′) can be rearranged arbitrarily without affecting their mean square convergence to $f(t)$.

4.25. Theorem 4.24 implies the validity of Parseval's theorem (formula (15), p. 226) for trigonometric Fourier series. Thus we have

$$\int_{-\pi}^{\pi} f^2(t)\,dt=\pi\frac{a_0^2}{2}+\pi\sum_{k=1}^{\infty}(a_k^2+b_k^2) \tag{4}$$

in the real case, since

$$\|1\|^2=2\pi,\ \|\cos kt\|^2=\|\sin kt\|^2=\pi$$

(Sec. 4.21), and

$$\int_{-\pi}^{\pi} |f(t)|^2\,dt=2\pi\sum_{k=-\infty}^{\infty}|c_k|^2 \tag{4′}$$

in the complex case, since

$$\|e^{ikt}\|^2=2\pi$$

(Sec. 4.22). Moreover, let $f(t)$ and $g(t)$ be two piecewise continuous functions on $[-\pi,\pi]$, with Fourier series

$$f(t)=\frac{a_0}{2}+\sum_{k=1}^{\infty}(a_k\cos kt+b_k\sin kt),$$

$$g(t)=\frac{b_0}{2}+\sum_{k=1}^{\infty}(c_k\cos kt+d_k\sin kt)$$

in the real case and

$$f(t)=\sum_{k=-\infty}^{\infty}a_ke^{ikt},$$

$$g(t)=\sum_{k=-\infty}^{\infty}b_ke^{ikt}$$

in the complex case. Then it follows from formula (16), p. 226 that

$$\int_{-\pi}^{\pi} f(t)g(t)\,dt=\pi\frac{a_0c_0}{2}+\pi\sum_{k=1}^{\infty}(a_kc_k+b_kd_k) \tag{5}$$

in the real case and

$$\int_{-\pi}^{\pi} f(t)\,\overline{g(t)}\,dt = 2\pi \sum_{k=-\infty}^{\infty} a_k \bar{b}_k \tag{5'}$$

in the complex case.

4.26. The following propositions are all immediate consequences of Parseval's theorem for trigonometric Fourier series (formulas (4) and (4′)):

a. *The Fourier coefficients a_k and b_k of a piecewise continuous real function $f(t)$ approach 0 as $k \to \infty$, while the Fourier coefficients c_k of a piecewise continuous complex function $f(t)$ approach 0 as $k \to \pm\infty$.*

b. *If all the Fourier coefficients (3) or (3′) of a piecewise continuous real or complex function $f(t)$ equal 0, then $f(t) = 0$ everywhere except possibly at a finite number of points (Vol. 1, Sec. 9.16e).*

c. *If two piecewise continuous real or complex functions $f(t)$ and $g(t)$ have the same Fourier coefficients, then $f(t) = g(t)$ everywhere except possibly at a finite number of points.*

d. *The orthogonal system of trigonometric functions (1) is complete, in the sense that a piecewise continuous real function $f(t)$ orthogonal to all the functions of the system (1) must be the zero element of the space $\mathbf{H}_R(Q)$, i.e., must vanish everywhere except possibly at a finite number of points. The same is true of the system (1′) in the complex space $\mathbf{H}_C(Q)$.*

4.27.a. LEMMA. *Suppose the trigonometric series*

$$\sum_{k=-\infty}^{\infty} c_k e^{ikt} \tag{6}$$

converges uniformly on $[-\pi, \pi]$ to a function $s(t)$ in the sense that

$$s(t) = \lim_{p\to\infty} \sum_{k=-m_p}^{n_p} c_k e^{ikt}$$

uniformly on $[-\pi, \pi]$ for arbitrary sequences $m_p \to \infty$, $n_p \to \infty$. Then the numbers c_k equal the Fourier coefficients of $s(t)$.

Proof. An immediate consequence of Lemma 4.15, since the uniform convergence implies convergence in the norm of $\mathbf{H}_C(Q)$.† ∎

b. COROLLARY. *Given a piecewise continuous complex function $f(t)$ with Fourier coefficients c_k $(k = 0, \pm 1, \pm 2, \ldots)$, suppose the series (6) converges uniformly to a func-*

† Note that $s(t)$ is continuous, being the limit of a uniformly convergent sequence of continuous functions (Vol. 1, Sec. 5.95b).

tion $s(t)$ in the sense of the lemma. Then $f(t)=s(t)$ everywhere except possibly at a finite number of points.

Proof. The continuous function $s(t)$ has the numbers c_k as its Fourier coefficients, by the lemma. But $f(t)$ also has the numbers c_k as its Fourier coefficients. The theorem is now an immediate consequence of Sec. 4.26c. ▮

4.3. Convergence of Fourier Series

4.31. Let $f(t)$ be any piecewise continuous function on an interval $[a,b]$. Then, according to Theorem 4.24, there is a Fourier series approximating $f(t)$ to arbitrary accuracy in the *mean square sense.* We now examine the question of whether $f(t)$ can be approximated to arbitrary accuracy *at any given point* $t=t_0$. In other words, does the Fourier series of $f(t)$ converge at $t=t_0$ to the number $f(t_0)$ in the ordinary sense?

In studying this problem, we will use Fourier series in complex form (Sec. 4.22). Given any $m>0$, $n>0$, the partial sum

$$s_{m,n}(t)=\sum_{k=-m}^{n} c_k e^{ikt}$$

of the complex Fourier series of $f(t)$ equals

$$s_{m,n}(t)=\frac{1}{2\pi}\sum_{k=-m}^{n}\int_{-\pi}^{\pi} f(\tau)e^{-ik\tau}\,d\tau\, e^{ikt}=\frac{1}{2\pi}\int_{-\pi}^{\pi} f(\tau)\sum_{k=-m}^{n} e^{ik(t-\tau)}\,d\tau.$$

But

$$\begin{aligned}\sum_{k=-m}^{n} e^{ik\theta}&=\frac{e^{-im\theta}-e^{i(n+1)\theta}}{1-e^{i\theta}}\\ &=\frac{e^{i(n+\frac{1}{2})\theta}-e^{-i(m+\frac{1}{2})\theta}}{e^{i\theta/2}-e^{-i\theta/2}}=\frac{e^{i(n+\frac{1}{2})\theta}-e^{-i(m+\frac{1}{2})\theta}}{2i\sin\dfrac{\theta}{2}},\end{aligned}$$

and hence†

$$\begin{aligned}s_{m,n}(t)&=\frac{1}{4\pi i}\int_{-\pi}^{\pi} f(\tau)\frac{e^{i(n+\frac{1}{2})(t-\tau)}-e^{-i(m+\frac{1}{2})(t-\tau)}}{\sin\dfrac{t-\tau}{2}}\,d\tau\\ &=\frac{1}{4\pi i}\int_{-\pi}^{\pi} f(t+h)\frac{e^{i(m+\frac{1}{2})h}-e^{-i(n+\frac{1}{2})h}}{\sin\dfrac{h}{2}}\,dh. \qquad (1)\end{aligned}$$

† Here we use the fact that the integral of a periodic function has the same value over every interval of length equal to the period (Vol. 1, Sec. 9.54b).

If $f(t)\equiv 1$, then obviously $s_{m,n}(t)\equiv 1$ for every $m>0$, $n>0$. In this case, formula (1) reduces to

$$\frac{1}{4\pi i}\int_{-\pi}^{\pi}\frac{e^{i(m+\frac{1}{2})h}-e^{-i(n+\frac{1}{2})h}}{\sin\frac{h}{2}}\,dh=1.$$

Hence the difference $s_{m,n}(t)-f(t)$ can be written in the form

$$s_{m,n}(t)-f(t)=\frac{1}{4\pi i}\int_{-\pi}^{\pi}[f(t+h)-f(t)]\frac{e^{i(m+\frac{1}{2})h}-e^{-i(n+\frac{1}{2})h}}{\sin\frac{h}{2}}\,dh. \tag{2}$$

Our aim is to find the conditions under which $s_{m,n}(t)$ converges to $f(t)$, or equivalently under which the integral (2) converges to zero.

4.32. LEMMA. *Given a complex function $\varphi(h)$ defined on the half-open interval $a<h\leqslant b$, suppose $\varphi(h)$ is bounded and piecewise continuous on every closed interval $[a+\delta,b]$, $\delta>0$, and absolutely integrable in the improper sense on $[a,b]$. Then the integrals*

$$\int_a^b \varphi(h)\cos\nu h\,dh,\qquad \int_a^b \varphi(h)\sin\nu h\,dh,\qquad \int_a^b \varphi(h)e^{i\nu h}\,dh$$

all approach zero as $\nu\to+\infty$.

Proof. Since

$$\cos\nu h=\frac{1}{2}(e^{i\nu h}-e^{-i\nu h}),\qquad \sin\nu h=\frac{1}{2i}(e^{i\nu h}-e^{-i\nu h}),$$

we need only prove the lemma for the factor $e^{i\nu h}$. First consider the case where the function $\varphi(h)$ is continuous on the closed interval. Then, given any $\nu>0$, there are only a finite number of points of the form

$$a+k\frac{2\pi}{\nu}\qquad (k=0,1,2,\ldots)$$

in the interval $[a,b]$. Denoting these points by $h_0,h_1,\ldots,h_m$, where $h_0=a<h_1<\cdots<h_m$, let $g(h)$ be the function

$$g(h)=\begin{cases}\varphi(h_0) & \text{if } h_0\leqslant h<h_1,\\ \varphi(h_1) & \text{if } h_1\leqslant h<h_2,\\ \cdots & \\ \varphi(h_m) & \text{if } h_m\leqslant h<b.\end{cases}$$

Then obviously

$$|g(h)-\varphi(h)|\leqslant\omega_\varphi\left(\frac{2\pi}{\nu}\right) \tag{3}$$

for all $h \in [a,b]$, where $\omega_\varphi(\delta)$ is the modulus of continuity of the function $\varphi(h)$ on the interval $[a,b]$. Since the interval $[h_j,h_{j+1}]$ is a period of the function e^{ivh}, we have

$$\int_{h_j}^{h_{j+1}} g(h)e^{ivh}\,dh = \varphi(h_j)\int_{h_j}^{h_{j+1}} e^{ivh}\,dh = 0.$$

Therefore

$$\left|\int_a^b g(h)e^{ivh}\,dh\right| = \left|\int_{h_m}^b g(h)e^{ivh}\,dh\right| \leqslant M\frac{2\pi}{v}, \tag{4}$$

where $M = \max|\varphi(h)|$. It follows from (3) and (4) that

$$\begin{aligned}\left|\int_a^b \varphi(h)e^{ivh}\,dh\right| &\leqslant \int_a^b |\varphi(h)-g(h)|\,dh + \left|\int_a^b g(h)e^{ivh}\,dh\right| \\ &\leqslant \omega_\varphi\left(\frac{2\pi}{v}\right)(b-a) + M\frac{2\pi}{v}.\end{aligned} \tag{5}$$

But the modulus of continuity $\omega(2\pi/v)$ of the continuous function $\varphi(h)$ approaches zero as $v\to\infty$ (Vol. 1, Sec. 5.17c), and hence (5) implies

$$\lim_{v\to\infty}\int_a^b \varphi(h)e^{ivh}\,dh = 0 \tag{6}$$

in this case.

Next let $\varphi(h)$ be any function satisfying the conditions of the lemma. Given any $\varepsilon > 0$, we choose $\delta > 0$ such that

$$\int_a^{a+\delta} |\varphi(h)|\,dh < \frac{\varepsilon}{3}. \tag{7}$$

The function $\varphi(h)$ is bounded in absolute value on the interval $[a+\delta,b]$ by a number $M = M(\varepsilon)$, say, and is piecewise continuous, so that the interval $[a+\delta,b]$ can be partitioned into a finite number $N = N(\varepsilon)$, say, of subintervals $(a_1,b_1),\ldots,(a_N,b_N)$ on which $\varphi(h)$ is continuous. Applying the estimate (5) to each of these intervals and using (7), we get

$$\begin{aligned}\left|\int_a^b \varphi(h)e^{ivh}\,dh\right| &\leqslant \int_a^{a+\delta}|\varphi(h)|\,dh + \sum_{k=1}^N \left|\int_{a_k}^{b_k}\varphi(h)e^{ivh}\,dh\right| \\ &\leqslant \frac{\varepsilon}{3} + \omega_\varphi\left(\frac{2\pi}{v}\right)\sum_{k=1}^N (b_k - a_k) + N(\varepsilon)M(\varepsilon)\frac{2\pi}{v} \\ &< \frac{\varepsilon}{3} + \omega_\varphi\left(\frac{2\pi}{v}\right)(b-a) + N(\varepsilon)M(\varepsilon)\frac{2\pi}{v}.\end{aligned} \tag{8}$$

Then, choosing $\nu > 0$ large enough to make both quantities

$$\omega_\varphi\left(\frac{2\pi}{\nu}\right)(b-a), \qquad N(\varepsilon)M(\varepsilon)\frac{2\pi}{\nu}$$

less than $\varepsilon/3$, we find that

$$\left|\int_a^b \varphi(h)e^{i\nu h}\,dh\right| < \varepsilon,$$

which proves (6) for a function satisfying the conditions of the lemma. The proof that

$$\lim_{\nu\to-\infty}\int_a^b \varphi(h)e^{i\nu h}\,dh = 0 \tag{6'}$$

is almost identical (give the details). ▮

4.33. Returning to formula (2), we can now prove the following

THEOREM (**Pointwise convergence of Fourier series**). *Given a piecewise continuous complex function $f(t)$ defined on the circle $Q=[-\pi,\pi]$,† suppose the function*

$$\frac{f(t_0+h)-f(t_0)}{h} \tag{9}$$

is absolutely integrable with respect to h in the improper sense in a neighborhood of the point $h=0$. Then the Fourier series of $f(t)$ converges to the value $f(t_0)$ at the point $t=t_0$.

Proof. If the function (9) is absolutely integrable in the improper sense in a neighborhood of the point $h=0$, then so is the function

$$\frac{f(t_0+h)-f(t_0)}{\sin\dfrac{h}{2}} = \frac{f(t_0+h)-f(t_0)}{h}\,\frac{h}{\sin\dfrac{h}{2}}.$$

Hence, writing (2) in the form

† As usual, we identify the end points $-\pi$ and π of the interval $[-\pi,\pi]$, thereby making $[-\pi,\pi]$ equivalent to (the circumference of) the unit circle $\{(x,y): x^2+y^2=1\}$.

$$s_{m,n}(t_0)-f(t_0)=\frac{1}{4\pi i}\int_{-\pi}^{0}\frac{f(t_0+h)-f(t_0)}{\sin\frac{h}{2}}\{e^{i(m+\frac{1}{2})h}-e^{-i(n+\frac{1}{2})h}\}\,dh$$

$$+\frac{1}{4\pi i}\int_{0}^{\pi}\frac{f(t_0+h)-f(t_0)}{\sin\frac{h}{2}}\{e^{i(m+\frac{1}{2})h}-e^{-i(n+\frac{1}{2})h}\}\,dh, \qquad (10)$$

we see that both terms on the right approach 0 as $m\to\infty, n\to\infty$, the second by the lemma and the first by an obvious modification of the lemma. But then the left-hand side of (10) approaches 0 as $m\to\infty, n\to\infty$, i.e., the Fourier series of $f(t)$ converges to $f(t_0)$ at the point $t=t_0$. ∎

4.34. The condition that the ratio (9) be absolutely integrable in a neighborhood of $h=0$ is called *Dini's condition* for the function $f(t)$ at $t=t_0$. This condition holds, for example, if $f(t)$ satisfies a *Lipschitz condition of order* $\alpha>0$ at $t=t_0$, i.e., if

$$|f(t_0+h)-f(t_0)|\leqslant C|h|^{\alpha} \qquad (\alpha>0).$$

In particular, if $f(t)$ has a finite derivative at $t=t_0$, then $f(t)$ satisfies a Lipschitz condition of order 1 at $t=t_0$, and hence the Fourier series of $f(t)$ converges to $f(t_0)$ at $t=t_0$.

4.35. By further analyzing the proof of Theorem 4.32, we can use the same method to deduce an analogous result on the *uniform* convergence of a Fourier series on a set E:

THEOREM (**Uniform convergence of Fourier series**). *Given a piecewise continuous function $f(t)$ defined on the circle $Q=[-\pi,\pi]$, suppose $f(t)$ satisfies Dini's condition uniformly on a subset $E\subset Q$, i.e., suppose that given any $\varepsilon>0$, there is a $\delta>0$ such that the inequality*

$$\int_{|h|<\delta}\left|\frac{f(t+h)-f(t)}{h}\right|dh<\varepsilon$$

holds for all $t\in E$. Then the Fourier series of $f(t)$ converges uniformly to $f(t)$ on E.

Proof. We want to show that the difference

$$s_{m,n}(t)-f(t)=\frac{1}{4\pi i}\int_{-\pi}^{\pi}[f(t+h)-f(t)]\frac{e^{i(m+\frac{1}{2})h}-e^{-i(n+\frac{1}{2})h}}{\sin\frac{h}{2}}\,dh$$

converges uniformly to zero on the set E. To this end, let

$$\varphi(t,h)=\frac{f(t+h)-f(t)}{\sin\frac{h}{2}}.$$

Then, by hypothesis, given any $\varepsilon > 0$, there exists a $\delta_0 > 0$, which is independent of t, such that

$$\int_{|h|<\delta_0} \left| \frac{f(t+h)-f(t)}{\sin \frac{h}{2}} \right| dh = \int_{|h|<\delta_0} \left| \frac{f(t+h)-f(t)}{h} \right| \frac{h}{\sin \frac{h}{2}} dh < \frac{\varepsilon}{3}. \tag{11}$$

Outside the interval $|h| < \delta_0$ we have

$$|\varphi(t,h)| \leqslant M(\varepsilon) = \frac{2M_f}{\sin \frac{\delta_0}{2}}, \tag{12}$$

where $M_f = \sup |f(t)|$ and the estimate is independent of t. On the other hand, the set Z_t of discontinuity points of the function $\varphi(t,h)$ depends on t, and is in fact the set of discontinuity points of $f(h)$ shifted by $-t$ (the possible discontinuity point $h=0$ lies in the excluded interval $|h| < \delta_0$). Hence for every t we can cover Z_t by the set S_t obtained when some fixed finite set of open intervals, of total length less than $\varepsilon/3M(\varepsilon)$, is shifted by t. Let F_t denote the interval $[-\pi,\pi]$ minus the set S_t. Then F_t is the union of a fixed number $N(\varepsilon)$ of closed intervals on which the function $\varphi(t,h)$ is continuous (the fact that some of these intervals may fall in the excluded interval $|h| < \delta_0$ can only strengthen the estimate to be given in a moment).

Now let

$$g(h) = \frac{1}{\sin \frac{h}{2}} \qquad (|h| \geqslant \delta_0),$$

and let $\omega_\varphi(\delta)$ be the modulus of continuity of the function $\varphi(t,h)$. Then

$$\begin{aligned} \omega_\varphi(\delta) &\leqslant \sup |f(t+h)-f(t)| \cdot \omega_g(\delta) + \omega_f(\delta) \sup |g(h)| \\ &\leqslant 2M_f \omega_g(\delta) + M_g \omega_f(\delta) \end{aligned}$$

(Vol. 1, Sec. 5.17d). It then follows from (11), (12), and the estimate (8), p. 234, that

$$\begin{aligned} &4\pi |s_{m,n}(t) - f(t)| \\ &\quad = \left| \int_{-\pi}^{\pi} \varphi(t,h) \{ e^{i(m+\frac{1}{2})h} - e^{-i(n+\frac{1}{2})h} \} \, dh \right| \\ &\quad < \frac{\varepsilon}{3} + 2M_f \omega_g \left(\frac{2\pi}{m+\frac{1}{2}} \right) + M_g \omega_f \left(\frac{2\pi}{m+\frac{1}{2}} \right) 2\pi + N(\varepsilon) M(\varepsilon) \frac{2\pi}{m+\frac{1}{2}} \\ &\qquad + \frac{\varepsilon}{3M(\varepsilon)} M(\varepsilon) + 2M_f \omega_g \left(\frac{2\pi}{n+\frac{1}{2}} \right) + M_g \omega_f \left(\frac{2\pi}{n+\frac{1}{2}} \right) 2\pi + N(\varepsilon) M(\varepsilon) \frac{2\pi}{n+\frac{1}{2}}. \end{aligned}$$

Choosing m and n large enough, we can make the right-hand side less than ε for all $t \in E$, i.e., the Fourier series of $f(t)$ converges uniformly to $f(t)$ on E. ∎

4.36. COROLLARY. *Given a piecewise continuous function $f(t)$ defined on the circle $Q = [-\pi, \pi]$, suppose $f(t)$ satisfies a Lipschitz condition of order $\alpha > 0$ uniformly on a subset $E \subset Q$, i.e., suppose that*

$$|f(t+h) - f(t)| \leqslant C|h|^{\alpha} \qquad (\alpha > 0)$$

for all $t \in E$, where the constant C is independent of the point $t \in E$. Then the Fourier series of $f(t)$ converges uniformly to $f(t)$ on E.

Proof. If $f(t)$ satisfies a Lipschitz condition of order $\alpha > 0$ uniformly on E, then $f(t)$ clearly satisfies Dini's condition uniformly on E. ∎

In particular, suppose $f(t)$ has a bounded derivative at every point of an interval $(a,b) \subset Q$. Then $f(t)$ satisfies a Lipschitz condition of order 1 uniformly on every subinterval $[a+\delta, b-\delta] \subset (a,b)$ (why?), and hence the Fourier series of $f(t)$ converges uniformly to $f(t)$ on $[a+\delta, b-\delta]$.

4.37. If $f(t)$ does not satisfy Dini's condition at the point $t = t_0$, then Theorem 4.33 no longer works and the Fourier series of $f(t)$ may well fail to converge at $t = t_0$ (we will confirm this in Sec. 4.51), in which case we can only hope for the relation

$$\lim_{\substack{m \to \infty \\ n \to \infty}} s_{m,n}(t_0) = f(t_0)$$

to hold for some generalized method of passing to the limit. It is only natural to begin by considering the *symmetric* partial sums

$$s_{n,n}(t) = \sum_{k=-n}^{n} c_k e^{ikt}.$$

According to formula (1), p. 232, the partial sum $s_{n,n}(t)$, which we henceforth denote by $s_n(t)$, is given by

$$s_n(t) = \frac{1}{4\pi i} \int_{-\pi}^{\pi} f(t+h) \frac{e^{i(n+\frac{1}{2})h} - e^{-i(n+\frac{1}{2})h}}{\sin \frac{h}{2}} dh$$

$$= \frac{1}{2\pi} \int_{-\pi}^{\pi} f(t+h) \frac{\sin (n+\frac{1}{2})h}{\sin \frac{h}{2}} dh = \int_{-\pi}^{\pi} f(t+h) D_n(h)\, dh, \tag{13}$$

in terms of the function

$$D_n(h) = \frac{1}{2\pi} \frac{\sin\,(n+\frac{1}{2})h}{\sin \dfrac{h}{2}},$$

known as the *Dirichlet kernel*.† Clearly $D_n(h)$ is an even function, i.e., $D_n(-h) \equiv D_n(h)$. Choosing $f(t) \equiv 1$ in (13) and noting that then $s_n(t) \equiv 1$ for every n, we find that

$$\int_{-\pi}^{\pi} D_n(h)\, dh = 1. \tag{14}$$

These properties will now be used to investigate the convergence behavior of the symmetric sums of the Fourier series of $f(t)$ at the discontinuity points of $f(t)$.

4.38. A point t_0 is called a *discontinuity point of the first kind* of a function $f(t)$ if $f(t)$ is discontinuous at $t=t_0$ and if both one-sided limits

$$f(t_0-0) = \lim_{t \nearrow t_0} f(t), \quad f(t_0+0) = \lim_{t \searrow t_0} f(t)$$

exist. Note that if $f(t)$ is piecewise continuous on a finite interval $[a,b]$, then $f(t)$ has only a finite number of discontinuity points in $[a,b]$, all of the first kind (Vol. 1, Sec. 9.16c).

THEOREM. *Given a piecewise continuous complex function $f(t)$ defined on the circle $Q=[-\pi,\pi]$, with a discontinuity point at $t=t_0$, suppose $f(t)$ satisfies "one-sided Dini conditions" at $t=t_0$, i.e., suppose both integrals*

$$\int_{t_0-\delta}^{t_0} \left| \frac{f(t)-f(t_0-0)}{t} \right| dt, \qquad \int_{t_0}^{t_0+\delta} \left| \frac{f(t)-f(t_0+0)}{t} \right| dt$$

converge for some $\delta>0$. Then the symmetric partial sums of the Fourier series of $f(t)$ converge to the quantity‡

$$\tfrac{1}{2}[f(t_0-0)+f(t_0+0)] \tag{15}$$

at $t=t_0$.

† If the functions $D_n(h)$ $(n=1,2,\ldots)$ formed a delta-like sequence for the point 0, we could then apply Theorem 1.55b to infer that the symmetric sums of the Fourier series of $f(t)$ converge to the value $f(t_0)$ at every continuity point t_0 of $f(t)$. But the functions $D_n(h)$ actually do not form a delta-like sequence, as we will see in Sec. 4.51.

‡ The quantity (15) is just the arithmetic mean of the one-sided limits of $f(t)$ at $t=t_0$. Note that $f(t)$ satisfies one-sided Dini conditions at $t=t_0$ if both one-sided derivatives

$$f'(t_0-0) = \lim_{h \searrow 0} \frac{f(t_0-0)-f(t_0-h)}{h}, \qquad f'(t_0+0) = \lim_{h \searrow 0} \frac{f(t_0+h)-f(t_0+0)}{h}$$

exist at $t=t_0$.

Proof. It follows from (13) and (14) that

$$\begin{aligned} s_n(t_0) &= \int_{-\pi}^{\pi} f(t_0+h)D_n(h)\,dh \\ &= \int_{-\pi}^{0} [f(t_0+h)-f(t_0-0)]D_n(h)\,dh \\ &\quad + \int_{0}^{\pi} [f(t_0+h)-f(t_0+0)]D_n(h)\,dh \\ &\quad + f(t_0-0)\int_{-\pi}^{0} D_n(h)\,dh + f(t_0+0)\int_{0}^{\pi} D_n(h)\,dh \\ &= I_1+I_2+\tfrac{1}{2}[f(t_0-0)+f(t_0+0)]\int_{-\pi}^{\pi} D_n(h)\,dh, \end{aligned}$$

where we have used the fact that the Dirichlet kernel $D_n(h)$ is even. But I_1 and I_2 approach 0 as $n\to\infty$, by Lemma 4.32, while the last term is independent of n and equals (15). ∎

4.39. Suppose we write the Fourier series of $f(t)$ in the form

$$\frac{a_0}{2}+\sum_{k=1}^{\infty}(a_k \cos kt + b_k \sin kt). \tag{16}$$

Then, as already noted in Sec. 4.23, the partial sums

$$\frac{a_0}{2}+\sum_{k=1}^{n}(a_k \cos kt + b_k \sin kt)$$

of the series (16) coincide with the *symmetric* partial sums of the Fourier series of $f(t)$ written in complex form. Hence for Fourier series of the form (16) we can combine Theorems 4.33 and 4.38, asserting that

$$\lim_{n\to\infty}\sum_{k=1}^{n}(a_k \cos kt_0 + b_k \sin kt_0) = \tfrac{1}{2}[f(t_0-0)+f(t_0+0)];$$

here it is assumed that the piecewise continuous function $f(t)$ satisfies one-sided Dini conditions at $t=t_0$ if t_0 is a discontinuity point (of the first kind) of $f(t)$, while $f(t)$ satisfies the ordinary ("two-sided") Dini condition at $t=t_0$ if t_0 is a continuity point of $f(t)$.† Note that since $a_k\to 0$, $b_k\to 0$ as $k\to\infty$ (Sec. 4.26a), the particular way the terms of (15) are grouped together, as indicated by the parentheses, has no effect on the convergence or divergence of the series (see Vol. 1, Secs. 6.33, 6.34b).

† Note that $\frac{1}{2}[f(t_0-0)+f(t_0+0)]=f(t_0)$ if t_0 is a continuity point of $f(t)$.

4.4. Computations with Fourier Series

4.41. Calculation of Fourier coefficients. Special properties of the function $f(t)$ can often be used to simplify the calculation of its Fourier coefficients.

a. If $f(t)$ is an even function, so that $f(-t) \equiv f(t)$, then

$$b_k = \frac{1}{\pi}\int_{-\pi}^{\pi} f(t) \sin kt \, dt = 0,$$

and $f(t)$ has a Fourier series of the form

$$f(t) = \sum_{k=0}^{\infty} a_k \cos kt, \tag{1}$$

involving only the functions $\cos kt$, with coefficients

$$a_k = \frac{1}{\pi}\int_{-\pi}^{\pi} f(t) \cos kt \, dt = \frac{2}{\pi}\int_{0}^{\pi} f(t) \cos kt \, dt.$$

On the other hand, if $f(t)$ is an odd function, so that $f(-t) \equiv -f(t)$, then

$$a_k = \frac{1}{\pi}\int_{-\pi}^{\pi} f(t) \cos kt \, dt = 0,$$

and $f(t)$ has a Fourier series of the form

$$f(t) = \sum_{k=1}^{\infty} b_k \sin kt, \tag{1'}$$

involving only the functions $\sin kt$, with coefficients

$$b_k = \frac{1}{\pi}\int_{-\pi}^{\pi} f(t) \sin kt \, dt = \frac{2}{\pi}\int_{0}^{\pi} f(t) \sin kt \, dt.$$

We call (1) a *Fourier cosine series* and (1′) a *Fourier sine series.*

b. Let $f(t)$ be a "piecewise polynomial function," i.e., suppose the interval $[-\pi,\pi]$ can be partitioned into a finite number of intervals $[t_j,t_{j+1}]$ ($j = 0, 1, \ldots, n-1$) without common interior points, such that $f(t)$ coincides with a polynomial

$$p_j(t) = \sum_{k=0}^{m} p_{jk} t^k$$

on $[t_j,t_{j+1}]$. Then

$$c_k = \frac{1}{2\pi}\int_{-\pi}^{\pi} f(t)e^{-ikt}\,dt = \sum_{j=0}^{n-1}\int_{t_j}^{t_{j+1}} p_j(t)e^{-ikt}\,dt.$$

Using integration by parts to transform each of these terms, we get

$$\int_{t_j}^{t_{j+1}} p_j(t)e^{-ikt}\,dt = p_j(t)\frac{e^{-ikt}}{-ik}\Big|_{t_j}^{t_{j+1}} - \int_{t_j}^{t_{j+1}} p_j'(t)\frac{e^{-ikt}}{-ik}\,dt$$
$$= \frac{1}{ik}[p_j(t_j)e^{-ikt_j} - p_j(t_{j+1})e^{-ikt_{j+1}}] - \frac{i}{k}\int_{t_j}^{t_{j+1}} p_j'(t)e^{-ikt}\,dt.$$

After a finite number of repetitions of this transformation, we get a result free of integrals, and the Fourier coefficients turn out to be certain polynomials in $1/k$ and e^{ikt_j}. Making an analogous calculation for the coefficients a_k and b_k, we find that in this case they are polynomials in $1/k$, $\cos kt_j$, and $\sin kt_j$.

It follows from the general theorems of Sec. 4.3 that the Fourier series of a piecewise polynomial function $f(t)$ converges to the value $f(t)$ at every continuity point of $f(t)$. This convergence is uniform on every closed interval which does not contain discontinuity points of $f(t)$. At every discontinuity point the symmetric partial sums of the Fourier series of $f(t)$ converge to the quantity $\frac{1}{2}[f(t+0)+f(t-0)]$.

c. Example. Consider the function

$$f(t) = \frac{\pi - t}{2} \qquad (0<t<\pi),$$

making first the odd extension of $f(t)$ onto the interval $-\pi<t<0$ and then the periodic extension (with period 2π) onto the whole real line $-\infty<t<\infty$ (see Figure 21). According to Sec. 4.41a, $f(t)$ has a Fourier series of the form (1′), where

$$\frac{\pi}{2}b_k = \int_0^{\pi}\frac{\pi - t}{2}\sin kt\,dt = \frac{\pi - t}{2}\frac{\cos kt}{k}\Big|_0^{\pi} + \frac{1}{2}\int_{\pi}^{0}\frac{\cos kt}{k}\,dt = \frac{\pi}{2k}.$$

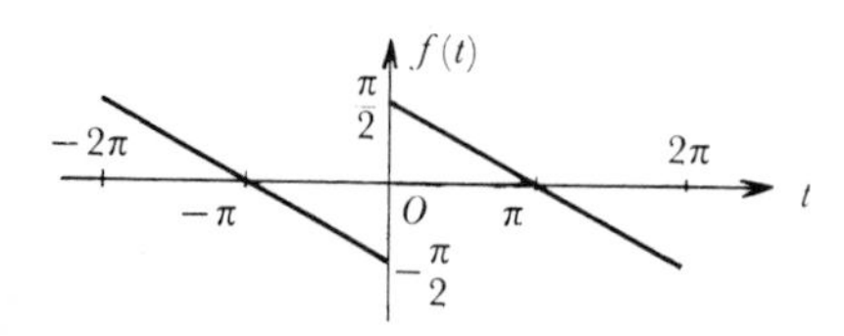

Figure 21

Therefore

$$\frac{\pi - t}{2} = \sum_{k=1}^{\infty} \frac{\sin kt}{k}$$

on the interval $(0,2\pi)$, where the series converges at every point $t \in (0,2\pi)$ and converges uniformly on every interval $[\delta,\ 2\pi - \delta] \subset (0,2\pi)$. At the points $t=0$ and $t=2\pi$, the series converges to the number 0 (see Sec. 4.39).

d. In some cases, from a knowledge of the coefficients a_k and b_k as polynomials in $1/k$, $\cos kt_j$, and $\sin kt_j$, we can sum the corresponding Fourier series, obtaining an explicit formula for some piecewise polynomial function $f(t)$.† However, even a simple series of this type, for example

$$\sum_{k=1}^{\infty} \frac{\cos kt}{k},$$

may fail to be the Fourier series of a piecewise polynomial function (see Example 4.44b).

4.42. Relation between the smoothness of a function and the rate of decrease of its Fourier coefficients.‡ So far we have assumed only that $f(t)$ is piecewise continuous. We now examine the implications of assuming that $f(t)$ has derivatives of various orders.

a. First suppose $f(t)$ is continuous on the circle $Q = [-\pi,\pi]$ and has a piecewise continuous derivative $f'(t)$ on Q. Suppose $f(t)$ has Fourier coefficients c_k (with respect to the system e^{ikt}), while $f'(t)$ has Fourier coefficients c'_k. Then

$$c_k = \frac{1}{2\pi} \int_{-\pi}^{\pi} f(t) e^{-ikt}\, dt = \frac{1}{2\pi} \frac{f(t) e^{-ikt}}{-ik} \bigg|_{-\pi}^{\pi} + \frac{1}{2\pi i k} \int_{-\pi}^{\pi} f'(t) e^{-ikt}\, dt = \frac{c'_k}{ik}, \qquad (2)$$

where the integrated term vanishes since $f(-\pi) = f(\pi)$ (a consequence of the assumption that $f(t)$ is continuous on the whole circle Q). The numbers c'_k, being Fourier coefficients of the piecewise continuous function $f'(t)$, approach 0 as $k \to \pm\infty$ (Sec. 4.26a). Thus we see that the Fourier coefficients c_k of the original function $f(t)$ approach 0 faster than $1/|k|$. The numerical series

$$\sum_{k=-\infty}^{\infty} |c'_k|^2$$

† G. M. Fichtenholz, *A Course of Differential and Integral Calculus*, Vol. III (in Russian), Gos. Izd. Fiz.-Mat. Lit., Moscow (1960), Sec. 710.
‡ See the footnote on p. 188.

converges (why?), and therefore so does the series

$$\sum_{k=-\infty}^{\infty} |c_k|, \tag{3}$$

because of the inequality

$$|c_k| = \frac{1}{|k|}|c_k'| \leqslant \frac{1}{2}\left(\frac{1}{k^2} + |c_k'|^2\right).$$

Moreover, because of Weierstrass' test (Vol. 1, Sec. 6.53), this gives another proof (independent of the considerations of Sec. 4.36) of the uniform convergence of the Fourier series of a differentiable function.

If $f(t)$ is continuous but nondifferentiable, then the series (3) will in general fail to converge (see the end of Sec. 4.51).

b. Next suppose $f(t)$ is continuous and has continuous derivatives up to order $n-1$ on Q, while $f^{(n)}(t)$ is piecewise continuous on Q. Then, using formula (2) repeatedly, we get

$$c_k = \frac{c_k'}{ik} = \frac{c_k''}{(ik)^2} = \cdots = \frac{c_k^{(n-1)}}{(ik)^{n-1}} = \frac{c_k^{(n)}}{(ik)^n} \qquad (k = \pm 1, \pm 2, \ldots), \tag{4}$$

where the function $f^{(m)}(t)$ has Fourier coefficients $c_k^{(m)}$. This time the numerical series

$$\sum_{k=-\infty}^{\infty} |c_k^{(n-1)}|$$

is convergent, and hence we can write c_k in the form

$$c_k = \frac{\varepsilon_k}{|k|^{n-1}} \qquad (k = \pm 1, \pm 2, \ldots), \tag{4'}$$

where the series

$$\sum_{k=-\infty}^{\infty} |\varepsilon_k| \tag{5}$$

is convergent.

c. These results have the following partial converse:

THEOREM. *Suppose the Fourier coefficients c_k of a continuous function $f(t)$ can be written in the form*

$$c_k = \frac{\theta_k}{|k|^n}, \qquad |\theta_k| \leqslant C \qquad (n \geqslant 2),$$

or in the form

$$c_k = \frac{\varepsilon_k}{|k|^{n-2}} \qquad (n \geqslant 2),$$

where the series (5) *is convergent. Then* $f(t)$ *has continuous derivatives up to order* $n-2$ (*inclusive*).

Proof. Under these conditions, the Fourier series

$$\sum_{k=-\infty}^{\infty} c_k e^{ikt} \equiv s_0(t) \tag{6}$$

is uniformly convergent (by Weierstrass' test), and so are the series obtained by successive formal differentiation of (6) up to order $n-2$:

$$\begin{aligned} &\sum_{k=-\infty}^{\infty} ikc_k e^{ikt} \equiv s_1(t), \\ &\sum_{k=-\infty}^{\infty} (ik)^2 c_k e^{ikt} \equiv s_2(t), \\ &\ldots \\ &\sum_{k=-\infty}^{\infty} (ik)^{n-2} c_k e^{ikt} \equiv s_{n-2}(t). \end{aligned} \tag{7}$$

In particular, the functions $s_0(t), s_1(t), \ldots, s_{n-2}(t)$ are all continuous, being sums of uniformly convergent series of continuous functions. Clearly $f(t) \equiv s_0(t)$, for the same reason as in Sec. 4.26c. Moreover, by the theorem on term-by-term differentiation of series of functions (Vol. 1, Sec. 9.107), the function $s_0(t)$ is differentiable and has derivative $s_1(t)$, the function $s_1(t)$ is differentiable and has derivative $s_2(t)$, and so on, until we get the last of the series in (7). Thus the function $f(t)$ has continuous derivatives up to order $n-2$. ▮

d. If the function $f(t)$ has continuous derivatives of all orders $n = 1, 2, \ldots$, then its Fourier coefficients satisfy the inequalities

$$|c_k| \leqslant \frac{\theta_n}{|k|^n} \qquad (n = 1, 2, \ldots), \tag{8}$$

and hence fall off more rapidly than any power of $1/|k|$. Conversely, if the Fourier coefficients of a function $f(t)$ satisfy the inequalities (8) for all $n = 1, 2, \ldots$, then, by the preceding theorem, $f(t)$ is continuous and has continuous derivatives of all orders. Thus the class of infinitely differentiable functions is completely characterized by the conditions (8) on the Fourier coefficients c_k.

***4.43. The isoperimetric problem.** The following classical problem is known as the *isoperimetric problem*: Among all closed piecewise smooth plane curves of a given length, find the curve enclosing the largest area. The solution turns out to be a circle, as we now show, using a method due to Hurwitz.

Let L be a closed piecewise smooth plane (Jordan) curve, with parametric representation

$$z = x(s) + iy(s)$$

in terms of the natural parameter s (the arc length), where we first assume that L is of length 2π, so that $z(2\pi) = z(0)$. Suppose the functions $x(s)$ and $y(s)$ have trigonometric Fourier series

$$x(s) = \frac{a_0}{2} + \sum_{k=1}^{\infty} (a_k \cos ks + b_k \sin ks), \tag{9}$$

$$y(s) = \frac{c_0}{2} + \sum_{k=1}^{\infty} (c_k \cos ks + d_k \sin ks). \tag{9'}$$

Then, by Sec. 4.42a,

$$x'(s) = \sum_{k=1}^{\infty} (-ka_k \sin ks + kb_k \cos ks),$$

$$y'(s) = \sum_{k=1}^{\infty} (-kc_k \sin ks + kd_k \cos ks).$$

Since

$$[x'(s)]^2 + [y'(s)]^2 = 1$$

(Sec. 3.17b), it follows from formula (4), p. 230, that

$$2\pi = \int_0^{2\pi} \{[x'(s)]^2 + [y'(s)]^2\} ds = \pi \sum_{k=1}^{\infty} (a_k^2 + b_k^2 + c_k^2 + d_k^2). \tag{10}$$

On the other hand, by the familiar formula expressing the area S of the domain enclosed by a closed curve L in terms of a line integral (Vol. 1, Sec. 9.124), we have

$$S = \frac{1}{2} \int_0^{2\pi} [xy'(s) - yx'(s)]\, ds = \pi \sum_{k=1}^{\infty} k(a_k d_k - b_k c_k), \tag{11}$$

with the help of formula (5), p. 230. Together (10) and (11) imply

$$2 - \frac{2S}{\pi} = \sum_{k=1}^{\infty} [k^2(a_k^2 + b_k^2 + c_k^2 + d_k^2) - 2k(a_k d_k - b_k c_k)]$$

$$= \sum_{k=1}^{\infty} [(ka_k - d_k)^2 + (kb_k + c_k)^2 + (k^2 - 1)(c_k^2 + d_k^2)] \geqslant 0. \tag{12}$$

Thus *the area S of the domain bounded by any piecewise smooth closed plane curve L of length 2π cannot exceed π.*

If formula (12) becomes an equality, then

$$ka_k - d_k = 0, \quad kb_k + c_k = 0, \quad (k^2-1)(c_k^2 + d_k^2) = 0.$$

In particular, if $k>1$, this implies $c_k = d_k = 0$ and hence $a_k = b_k = 0$ as well. Setting $k=1$, we get $a_1 = d_1$, $b_1 = -c_1$. Formula (10) then implies $a_1^2 + b_1^2 = 1$, and hence we can write $a_1 = \cos\alpha$, $b_1 = \sin\alpha$. Substituting these values into (9) and (9′), we finally obtain

$$x(s) = \frac{a_0}{2} + \cos(s-\alpha),$$

$$y(s) = \frac{c_0}{2} + \sin(s-\alpha),$$

so that L is just a circle of radius 1, with center at the point $(a_0/2, c_0/2)$.

If the length l of the curve L does not equal 2π, we make the similarity transformation

$$x^* = \frac{2\pi}{l}x, \qquad y^* = \frac{2\pi}{l}y.$$

Then L goes into a curve L^* of length 2π (Vol. 1, Sec. 9.72e), enclosing a domain of area

$$S^* = \left(\frac{2\pi}{l}\right)^2 S$$

(Vol. 1, Sec. 9.62d). But $S^* \leqslant \pi$, as just shown, and hence $S \leqslant l^2/4\pi$. In the extremal case, where L^* is a circle of radius 1, the curve L is also a circle, of radius $l/2\pi$.

4.44. Use of complex variables

a. Let the variable point of the unit circle $Q = \{(x,y) : x^2 + y^2 = 1\}$ be described by the complex variable

$$z = x + iy = e^{it} \qquad (-\pi \leqslant t \leqslant \pi),$$

and let $f(t) \equiv F(z)$. Then the Fourier series of $f(t)$ takes the form

$$f(t) \equiv F(z) = \sum_{k=-\infty}^{\infty} c_k e^{ikt} = \sum_{k=-\infty}^{\infty} c_k z^k. \tag{13}$$

Series of the form (13), involving both positive and negative powers of the variable z, are just the familiar Laurent series of complex analysis (Vol. 1,

Sec. 10.45). The coefficients c_k can easily be expressed as integrals with respect to the complex variable z. In fact,

$$c_k = \frac{1}{2\pi}\int_{-\pi}^{\pi} f(t)e^{-ikt}\,dt = \frac{1}{2\pi i}\oint_{|z|=1} F(z)z^{-k-1}\,dz,$$

since $dz = ie^{it}\,dt = iz\,dt$ and hence $dt = dz/iz$. The Laurent series (13) reduces to a Taylor series

$$F(z) = \sum_{k=1}^{\infty} c_k z^k$$

if the function $F(z)$ can be continued analytically from the circle $|z| = 1$ into the disk $|z| < 1$. In this case, the Fourier series takes the form

$$f(t) = \sum_{k=0}^{\infty} c_k e^{ikt}.$$

b. Example. Find the sum of the series

$$\sum_{k=1}^{\infty} \frac{\cos kt}{k}.$$

Solution. Writing

$$\sum_{k=1}^{\infty} \frac{\cos kt}{k} + i\sum_{k=1}^{\infty} \frac{\sin kt}{k} = \sum_{k=1}^{\infty} \frac{e^{ikt}}{k},$$

we reduce our problem to that of calculating the sum of the power series

$$\sum_{k=1}^{\infty} \frac{e^{ikt}}{k} = \sum_{k=1}^{\infty} \frac{z^k}{k} \qquad (z = e^{it}), \tag{14}$$

and afterwards taking the real part. The right-hand side of (14) is the series obtained by term-by-term integration of the geometric series

$$\sum_{k=0}^{\infty} \zeta^k = \frac{1}{1-\zeta} \tag{15}$$

along a path joining the point 0 to the point z. But

$$\int_0^z \frac{d\zeta}{1-\zeta} = -\int_1^{1-z} \frac{d\omega}{\omega} = -\ln\,(1-z)$$

(Vol. 1, Sec. 10.57); here the path of integration in the ω-plane is any piecewise smooth curve joining the points 1 and $1-z$ which does not intersect the negative real axis, and hence the path of integration in the ζ-plane ($1-\zeta = \omega$) is any piecewise smooth curve joining the points 0 and z which does not intersect the ray $\zeta \geqslant 1$ (the part of the positive real axis going from $+1$ to infinity). For our purposes, it is sufficient to integrate along the line segment joining the points 0 and z.† The series (15) converges for $|\zeta| < 1$, and hence

$$-\ln(1-z) = \sum_{k=1}^{\infty} \frac{z^k}{k!} \tag{16}$$

for $|z| < 1$. But the series (16) also converges for $|z| = 1$, $z \neq 1$ (see Vol. 1, Sec. 6.64c), and since the function $\ln(1-z)$ remains continuous at these points, it follows from Abel's theorem (Vol. 1, Sec. 6.68) that the equality (16) continues to hold for these values of z. For any $w = |w| e^{i \arg w}$ we have

$$\ln w = \ln |w| + i \arg w$$

(Vol. 1, Sec. 10.57). In particular, for $z = e^{it}$, $\operatorname{Im} z > 0$ we can easily find the modulus and argument of the quantity $1 - z$ by using Figure 22, obtaining

$$|1-z| = 2 \sin \frac{t}{2}, \qquad \arg(1-z) = \frac{t-\pi}{2}.$$

It follows that

$$\ln |1-z| = \ln\left(2 \sin \frac{t}{2}\right), \qquad \ln(1-z) = \ln\left(2 \sin \frac{t}{2}\right) + i\,\frac{t-\pi}{2},$$

and hence

$$\sum_{k=1}^{\infty} \frac{z^k}{k} = -\ln(1-z) = -\ln\left(2 \sin \frac{t}{2}\right) + i\frac{\pi - t}{2} \qquad (z = e^{it}) \tag{17}$$

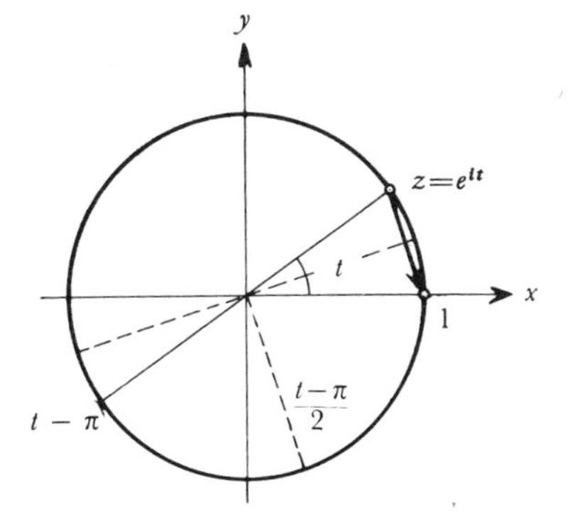

Figure 22

† Note that this guarantees single-valuedness of the function $-\ln(1-z)$.

for every $t \in (0,2\pi)$. Taking the real part of (17), we find that

$$\sum_{k=1}^{\infty} \frac{\cos kt}{k} = -\ln\left(2 \sin\frac{t}{2}\right) \qquad (0<t<2\pi),$$

thereby solving our problem. Taking the imaginary part of (17), we get the Fourier series

$$\sum_{k=1}^{\infty} \frac{\sin kt}{k} = \frac{\pi - t}{2} \qquad (0<t<2\pi),$$

already familiar from Example 4.43c.

***4.45. Periodic solutions of differential equations.** We now investigate the periodic solutions of the nth-order nonhomogeneous linear differential equation

$$a_0u^{(n)}(t) + a_1u^{(n-1)}(t) + \cdots + a_nu(t) = g(t), \tag{18}$$

with constant coefficients $a_0, a_1, \ldots, a_n$ and a periodic right-hand side $g(t)$, with period 2π. Assuming that (18) has a periodic solution $u(t)$, also with period 2π, we write $u(t)$ as a Fourier series

$$u(t) = \sum_{k=-\infty}^{\infty} u_k e^{ikt}, \tag{19}$$

involving unknown coefficients u_k. Assuming further that the series (19) can be differentiated term by term n times and introducing the polynomial

$$a_0\lambda^n + a_1\lambda^{n-1} + \cdots + a_n \equiv p(\lambda),$$

we find that

$$\sum_{k=-\infty}^{\infty} u_k p(ik) e^{ikt} = a_0u^{(n)}(t) + \cdots + a_nu(t). \tag{20}$$

On the other hand, let

$$g(t) = \sum_{k=-\infty}^{\infty} g_k e^{ikt} \tag{21}$$

be the Fourier series of the function $g(t)$. Then, comparing the orthogonal expansions (20) and (21), we get

$$g_k = u_k p(ik) \qquad (k=0, \pm 1, \pm 2, \ldots),$$

and hence

$$u_k = \frac{g_k}{p(ik)} \qquad (k=0, \pm 1, \pm 2, \ldots)$$

if $p(ik) \neq 0$, so that *formally* the solution becomes

$$u(t) = \sum_{k=-\infty}^{\infty} \frac{g_k}{p(ik)} e^{ikt}. \tag{22}$$

Having made these introductory remarks, we now prove two theorems.

a. THEOREM. *Suppose the Fourier coefficients g_k of the periodic function* (21), *with period 2π, are such that the series*

$$\sum_{k=-\infty}^{\infty} |g_k| \tag{23}$$

converges,† *and suppose $p(ik)$ is nonvanishing for all $k=0, \pm 1, \pm 2, \ldots$ Then equation* (18) *has a unique periodic solution, with period 2π.*

Proof. The polynomial $p(ik)$, of degree n, satisfies the estimates

$$|p(0)| \geqslant c > 0, \qquad |p(ik)| \geqslant ck^n \qquad (k = \pm 1, \pm 2, \ldots)$$

(why?), and hence

$$\left| \frac{g_k}{p(ik)} \right| \leqslant \frac{|g_k|}{c|k|^n}.$$

It follows from Theorem 4.42c that the function $u(t)$ defined by (22) has continuous derivatives up to order n, where the derivatives can be obtained by term-by-term differentiation of the series (22). Substituting these derivatives into equation (18), we obviously get an identity. Moreover, if there were another periodic solution $u^*(t)$ of equation (18), then the difference $v(t) = u(t) - u^*(t)$ would be a periodic solution (with period 2π) of the homogeneous equation

$$a_0 v^{(n)}(t) + a_1 v^{(n-1)}(t) + \cdots + a_n v(t) = 0.$$

According to Sec. 2.17, the general solution of this equation can be expressed in terms of exponentials of the form $e^{\lambda t}$, where $p(\lambda) = 0$. But such exponentials lead to solutions with period 2π only if $\lambda = ik$ for some integer $k = 0, \pm 1, \pm 2, \ldots$ Since this is excluded by hypothesis, there is no periodic solution of (18) other than the solution identically equal to zero. This proves that $u(t)$ is the unique periodic solution of (18). ∎

b. THEOREM. *Suppose the Fourier coefficients g_k of the periodic function* (21), *with period 2π, are such that the series* (23) *converges, and suppose $p(ik) = 0$ for certain integers $k = k_1, \ldots, k_r$. Then equation* (18) *has a periodic solution, with period 2π, if*

† For example, suppose $g(t)$ is piecewise smooth (cf. Sec. 4.42a).

and only if

$$g_{k_j}=0 \qquad (j=1, \ldots, r), \tag{24}$$

and this solution is unique to within a term

$$\sum_{j=1}^{r} c_j e^{ik_j t},$$

where the c_j are arbitrary constants.

Proof. If (24) holds, then the expansion (22), with the coefficients $g_{k_j}/p(ik_j)$ (which in this case take the form 0/0) replaced by arbitrary constants c_j, is a periodic solution of equation (18), with period 2π, just as in the proof of the preceding theorem. Moreover, suppose $p(iq)=0$ for some $k=q$, while $g_q \neq 0$. Then, substituting the proposed solution into (18), multiplying the result by e^{-iqt}, and integrating with respect to t from $-\pi$ to π, we get $g_q=c_q p(iq) =0$, a contradiction showing that (18) has no periodic solutions in this case. The uniqueness part of the theorem is proved in the same way as before. ∎

***4.46.** Next we consider two of the numerous applications of Fourier series to problems of mathematical physics.

a. The vibrating string. Consider a stretched string fastened at the points 0 and π of the x-axis, with the segment $[0,\pi]$ as its equilibrium position (see Figure 23). Suppose the string is given an arbitrary initial form, described by the function $y=f(x)$, say, and is afterwards released. Then the string begins to vibrate. We are interested in finding the function $y=u(t,x)$ specifying the form of the string at the time t. By a familiar mechanical argument (involving certain simplifying assumptions),† the function $u(t,x)$ is found to satisfy the following partial differential equation, known as the *wave equation*:

$$\frac{\partial^2 u(t,x)}{\partial t^2}=a^2\frac{\partial^2 u(t,x)}{\partial x^2} \tag{25}$$

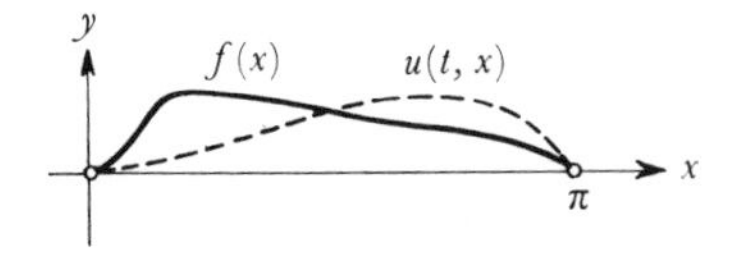

Figure 23

† See e.g., J. C. Slater and N. H. Frank, *Mechanics*, McGraw-Hill Book Co., Inc., New York (1947), Chapter 8.

($a>0$ constant). The function $u(t,x)$ must also satisfy the initial conditions

(a) $u(0,x)=f(x)$ (the initial form of the string is specified),

(b) $\dfrac{\partial u(0,x)}{\partial t}=0$ (the string is released without initial velocity),

as well as the "boundary condition"

(c) $u(t,0)=u(t,\pi)\equiv 0$ (the ends of the string are fixed).

To solve the problem, we first expand the function $u(t,x)$, defined on the interval $0\leqslant x\leqslant \pi$ for all $t\geqslant 0$, in a Fourier sine series

$$u(t,x)=\sum_{k=1}^{\infty} b_k(t)\sin kx, \tag{26}$$

where the coefficients $b_k(t)$ (functions of the time t) are still to be determined. The series (26) automatically satisfies Condition c, and the other initial conditions will be satisfied if the functions $b_k(t)$ are such that

(a') $b_k(t)=b_k$, where the numbers b_k are the Fourier coefficients of $f(x)$;
(b') $b_k'(t)=0$.

Next we require the function (26) to satisfy the wave equation (25). It follows from (26) that

$$\frac{\partial^2 u(t,x)}{\partial t^2}=\sum_{k=1}^{\infty} b_k''(t)\sin kx, \tag{27}$$

$$\frac{\partial^2 u(t,x)}{\partial x^2}=-\sum_{k=1}^{\infty} k^2 b_k(t)\sin kx, \tag{28}$$

at least formally, and therefore (25) holds if

$$b_k''(t)=-a^2k^2b_k(t) \tag{29}$$

for all $k=1,2,\ldots$ But the solution of (25) subject to Conditions a' and b' is just

$$b_k(t)=b_k\cos akt. \tag{30}$$

Hence the solution of (25) finally takes the form

$$u(t,x)=\sum_{k=1}^{\infty} b_k\cos akt\sin kx. \tag{31}$$

The formal differentiations leading to (27) and (28) will be justified if the corresponding series on the right are uniformly convergent. Taking account of (30), we see that this will be the case if the series

$$\sum_{k=1}^{\infty} b_k k^2 \cos akt \sin kx$$

is uniformly convergent on the intervals $0 \leqslant x \leqslant \pi$ and $t \geqslant 0$, which in turn is guaranteed by the convergence of the numerical series

$$\sum_{k=1}^{\infty} |b_k| k^2.$$

The latter series converges if $f(x)$ is continuous, with continuous first and second derivatives and a piecewise continuous third derivative (see Sec. 4.42b).

On the other hand, we can give the solution (31) a form in which no smoothness requirements are imposed on $f(x)$. In fact, we need only observe that

$$u(t,x) = \frac{1}{2} \sum_{k=1}^{\infty} b_k [\sin k(x+at) + \sin k(x-at)] = \frac{1}{2}[f(x+at) + f(x-at)]. \quad (32)$$

Here, whenever the arguments $x+at$ and $x-at$ of the functions $f(x+at)$ and $f(x-at)$ lie outside the original domain of definition $[0,\pi]$ of the function $f(x)$, we interpret $f(x+at)$ and $f(x-at)$ as the odd extension of $f(x)$ onto the interval $[-\pi,0]$ followed by periodic extension, with period 2π, onto the whole line $-\infty < x < \infty$. The problem now arises of the sense in which the function (32) satisfies the wave equation (25) when the function $f(x)$ does not have a second derivative. Problems of this kind are handled in mathematical physics with no particular difficulty by extending the very definition of a "solution" of an equation like (25).† Figure 24 shows successive

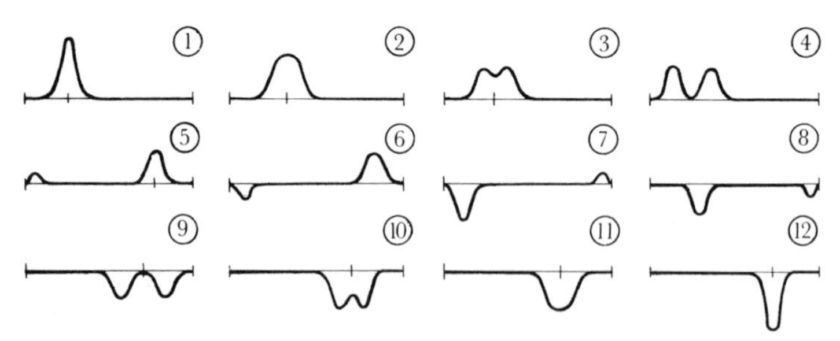

Figure 24

† Concerning this matter, which will not be gone into here, see e.g., I. G. Petrovski, *Lectures on Partial Differential Equations* (translated by A. Shenitzer), John Wiley-Interscience, New York (1954), Sec. 9.

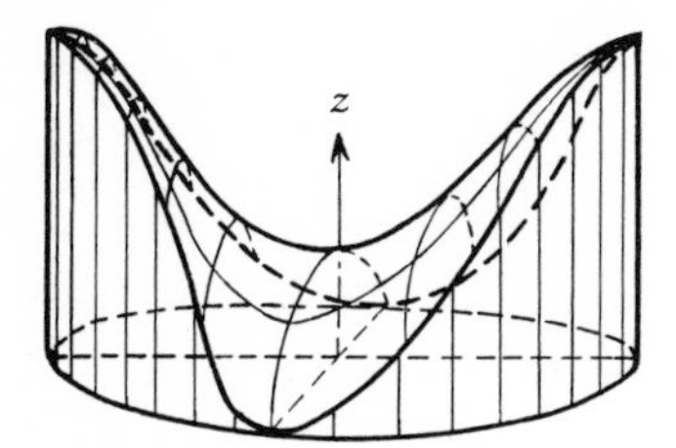

Figure 25

positions of a vibrating string, as given by formula (32), for the case where the initial form of the string is the curve labelled by the number 1.

b. The equilibrium position of a stretched circular membrane. Consider a stretched membrane lying above the closed unit disk $\overline{K}=\{(x,y): x^2+y^2\leqslant 1\}$, where the edge of the membrane is fastened to a space curve described by a continuous function $z=f(x,y)$ defined on the unit circle $\Gamma=\{(x,y): x^2+y^2=1\}$. We are interested in finding the function $z=u(x,y)$ describing the equilibrium position of the stretched membrane (see Figure 25), under the action of the elastic forces. By a familiar mechanical argument (involving certain simplifying assumptions),† the function $u(x,y)$ is found to satisfy *Laplace's equation*

$$\frac{\partial^2 u}{\partial x^2}+\frac{\partial^2 u}{\partial y^2}=0 \tag{33}$$

at every interior point of $\overline{K}$, i.e., at every point of the open unit disk $K=\{(x,y): x^2+y^2<1\}$. The function $u(x,y)$ must also satisfy the boundary condition $u(x,y)=f(x,y)$ at every point $(x,y)\in\Gamma$. In other words, our problem consists in finding a function *harmonic* on K (Vol. 1, Sec. 10.18) and continuous on $\overline{K}$ which coincides on the curve Γ with the given function $f(x,y)$. This is the celebrated *Dirichlet problem* (for the disk K).

To find such a solution, let t be the polar angle of the variable point $(x,y)\in\Gamma$. Then $f(t)$ has a Fourier series

$$f(t)\sim\frac{a_0}{2}+\sum_{k=1}^{\infty}(a_k\cos kt+b_k\sin kt), \tag{34}$$

where the sign $\sim$ means that the series (34) may not actually converge to $f(t)$ (this may well happen, as shown in Sec. 4.51). As a candidate for the solution, consider the function

$$u(r,t)=\frac{a_0}{2}+\sum_{k=1}^{\infty}r^k(a_k\cos kt+b_k\sin kt) \qquad (r<1), \tag{35}$$

† Slater and Frank, *Mechanics*, Chapter 11.

where we write the displacement u as a function of the polar coordinates r and t rather than of the rectangular coordinates x and y. The function $u(r,t)$ is the real part of the complex power series

$$\frac{a_0}{2}+\sum_{k=1}^{\infty}(a_k-ib_k)z^k, \tag{36}$$

where $z=x+iy=r(\cos t+i\sin t)$. According to Sec. 4.26a, $|a_k-ib_k|\to 0$ as $k\to\infty$, and hence, by the Cauchy-Hadamard theorem (Vol. 1, Sec. 6.62), the radius of convergence of (36) cannot be less than 1 (why?). Therefore the sum of (36) is analytic on K (Vol. 1, Sec. 10.37a), and hence $u(r,t)$ is harmonic on K (Vol. 1, Sec. 10.18), as required.

To verify that the function $u(r,t)$ can be extended by continuity onto the closed disk $\bar{K}$, and that it will then satisfy the boundary condition $u(1,t)\equiv f(t)$ on Γ, we first write (35) in the form

$$\begin{aligned}u(r,t)&=\frac{1}{2\pi}\int_{-\pi}^{\pi}f(\tau)\,d\tau+\frac{1}{\pi}\sum_{k=1}^{\infty}\int_{-\pi}^{\pi}f(\tau)r^k\cos(kt\cos k\tau+\sin kt\sin k\tau)\,d\tau\\&=\frac{1}{\pi}\int_{-\pi}^{\pi}f(\tau)\left[\frac{1}{2}+\sum_{k=1}^{\infty}r^k\cos k(t-\tau)\right]d\tau,\end{aligned}$$

where the operations of summation and integration can be interchanged because of the uniform convergence in the variable t of the series in brackets for $r<1$, which in turn follows from Weierstrass' test (Vol. 1, Sec. 6.53). But

$$\frac{1}{2}+\sum_{k=1}^{\infty}r^k\cos kt=\frac{1-r\cos t}{1-2r\cos t+r^2}-\frac{1}{2}=\frac{1-r^2}{2(1-2r\cos t+r^2)} \tag{37}$$

(Vol. 1, Sec. 6.47f), and hence

$$u(r,t)=\frac{1}{2\pi}\int_{-\pi}^{\pi}f(\tau)\frac{1-r^2}{1-2r\cos(t-\tau)+r^2}d\tau=\int_{-\pi}^{\pi}f(\tau)P_r(t-\tau)\,d\tau, \tag{38}$$

in terms of the function

$$P_r(t)=\frac{1}{2\pi}\frac{1-r^2}{1-2r\cos(t-r)+r^2}\qquad(r<1), \tag{39}$$

known as *Poisson's kernel*. The representation (38) is known as *Poisson's integral*. Examining the denominator of (39), we find that

$$1-2r\cos(t-\tau)+r^2=(1-r)^2+4r\sin^2\frac{t}{2}>0$$

if $r<1$, so that $P_r(t)$ is obviously nonnegative. Moreover, $P_r(t)$ has the other properties of a delta-like function of t as $r\nearrow 1$ (cf. Sec. 1.55g). In fact, setting

$f(t) \equiv 1$ in (38) and noting that then $u(r,t) \equiv 1$, we get

$$\int_{-\pi}^{\pi} P_r(\tau)\, d\tau = 1,$$

while, for any $\delta > 0$, we have the estimate

$$\int_{|t| \geqslant \delta} P_r(\tau)\, d\tau = \frac{1-r^2}{2\pi} \int_{|t| \geqslant \delta} \frac{d\tau}{(1-r)^2 + 4r \sin^2 \frac{\tau}{2}} \leqslant \frac{1-r^2}{4r \sin^2 \frac{\delta}{2}},$$

and hence

$$\lim_{r \nearrow 1} \int_{|t| \geqslant \delta} P_r(\tau)\, d\tau = 0.$$

It follows from the result given in Sec. 1.55g that the limit

$$\lim_{r \nearrow 1} u(r,t)$$

exists and equals $f(t)$ for every $t \in [0,2\pi]$. Therefore the function $u(r,t)$, defined by (35) for $r < 1$ and by continuity for $r = 1$, is continuous on $\overline{K}$ and satisfies the boundary condition $u(1,t) \equiv f(t)$, as required. This solves the Dirichlet problem for the disk K. It is not hard to show† that the above solution is *unique* in the class of all functions harmonic on K.

c. Let $u(r,t)$ be a function harmonic on the open disk K, taking given continuous boundary values $f(t)$ on the circle Γ. Then, as just shown (with due regard for the comment on uniqueness), $u(r,t)$ is given by the Poisson integral (38) for $r < 1$. But the Taylor series (36) has a radius of convergence $\geqslant 1$, for a reason already given, and hence represents a function analytic on K, of which the function (38), i.e., the given harmonic function, is the real part. Therefore the imaginary part of (36), which we denote by $v(r,t)$, is just the conjugate harmonic function of $u(r,t)$ (Vol. 1, Sec. 10.18). Thus every harmonic function (of the given type) has a conjugate harmonic function. To get the explicit form of $v(r,t)$, we need only replace the Poisson kernel $P_r(t)$ in (38) by *its* conjugate harmonic function. The latter is just the sum of the series whose terms are the conjugate harmonic functions of the series (37), namely the function

$$\frac{1}{2\pi} \sum_{k=1}^{\infty} r^k \sin kt = \frac{1}{2\pi} \frac{r \sin t}{1 - 2r \cos t + r^2}$$

† Petrovski, *Lectures on Partial Differential Equations*, Sec. 28.

(Vol. 1, Sec. 6.47f). It follows that

$$v(r,t)=\frac{1}{2\pi}\int_{-\pi}^{\pi}f(\tau)\frac{r\sin\tau}{1-2r\cos(t-\tau)+r^2}\,d\tau, \tag{40}$$

to within an arbitrary real constant.

4.5. Divergent Fourier Series and Generalized Summation

4.51. Let $f(t)$ be a continuous function, and let $s_n(f;t)$ be the nth symmetric partial sum of the Fourier series of $f(t)$. Then we have yet to investigate the convergence of $s_n(f;t)$ to $f(t)$ in the absence of the assumption that $f(t)$ satisfies Dini's condition (Sec. 4.34). It turns out that there exist continuous functions for which $s_n(t)$ diverges (albeit only at certain points). This is at bottom due to the fact that the Dirichlet kernels

$$D_n(h)=\frac{1}{2\pi}\frac{\sin(n+\frac{1}{2})h}{\sin\frac{h}{2}} \qquad (n=1,2,\ldots)$$

(Sec. 4.37) do not form a delta-like sequence; more exactly,

$$\sup_n\int_{-\pi}^{\pi}|D_n(h)|\,dh=\infty, \tag{1}$$

as we will see in a moment.

It will be recalled from Sec. 4.37 that

$$s_n(f;t)=\int_{-\pi}^{\pi}f(t+h)D_n(h)\,dh.$$

Setting $t=0$ for simplicity, we get

$$s_n(f;0)=\int_{-\pi}^{\pi}f(h)D_n(h)\,dh.$$

This represents a sequence of linear functionals on the Banach space $C^s(Q)$ of all continuous functions defined on $Q=[-\pi,\pi]$. We will see in a moment that *the norms of these functionals, i.e., the numbers*†

$$\int_{-\pi}^{\pi}|D_n(h)|\,dh \qquad (n=1,2,\ldots),$$

converge to infinity. It will then follow from the Banach-Steinhaus theorem

† See Sec. 1.71j.

(Sec. 1.74a) that there exists an element of the space $C^s(Q)$, i.e., a continuous function $f_0(t)$, for which the numbers $s_n(f_0;0)$ are unbounded. But then the Fourier series of $f_0(t)$ cannot converge at the point $t=0$, even in the sense of symmetric summation.

Thus everything reduces to proving formula (1). Using the inequality

$$\sin\frac{h}{2}\leqslant\frac{h}{2}\qquad(0\leqslant h\leqslant\pi),$$

we find that

$$\int_{-\pi}^{\pi}|D_n(h)|\,dh=\frac{1}{\pi}\int_0^{\pi}\frac{|\sin(n+\frac{1}{2})h|}{\sin\frac{h}{2}}\,dh\geqslant\frac{2}{\pi}\int_0^{\pi}\frac{|\sin(n+\frac{1}{2})h|}{h}\,dh.$$

Making the subsitution $(n+\frac{1}{2})h=z$, we then get

$$\int_{-\pi}^{\pi}|D_n(h)|\,dh\geqslant\frac{2}{\pi}\int_0^{(n+\frac{1}{2})h}\frac{|\sin t|}{t}\,dt.$$

But the expression on the right increases without limit as $n\to\infty$, because of the divergence of the improper integral

$$\int_0^{\infty}\frac{|\sin t|}{t}\,dt$$

(Vol. 1, Example 11.22a), which immediately implies (1).

It should be emphasized that there is a function $f_0(t)$ with a divergent Fourier series in every ball $U_\rho(g)=\{f:\|f-g\|\leqslant\rho\}\subset C^s(Q)$ (why?). The numerical series made up of the moduli of the Fourier coefficients of every such function also diverges, since otherwise Weierstrass' test would imply uniform convergence on Q of the Fourier series itself. According to a recent theorem of Carleson (1966), the points of divergence of a given piecewise continuous function $f(t)\in\mathbf{H}_C(-\pi,\pi)$ are exceptional in the sense that, given any $\varepsilon>0$, all such points can be covered by a countable family of intervals of total length less than ε.

4.52. It is now natural to ask whether this situation can be remedied by using an appropriate method of summing divergent series. First we consider Cesàro's method (Sec. 1.76b), which consists in going from a given sequence ξ_n $(n=1,2,\ldots)$ to the sequence of arithmetic means

$$\sigma_n=\frac{\xi_1+\cdots+\xi_n}{n}\qquad(n=1,2,\ldots),$$

and defining the Cesàro limit of the original sequence ξ_n as the ordinary limit of the new sequence σ_n:

$$C\text{-lim } \xi_n = \lim_{n\to\infty} \sigma_n.$$

This method leads at once to a positive result:

THEOREM (**Fejér, 1905**). *Let $f(t)$ be any function continuous on the circle $Q=[-\pi,\pi]$. Then the sequence of symmetric partial sums*

$$s_n(t) = \sum_{k=-n}^{n} c_k e^{ikt} \tag{2}$$

of the Fourier series of $f(t)$ converges uniformly to $f(t)$ on Q in Cesàro's sense, i.e.,

$$C\text{-lim } s_n(t) = \lim_{n\to\infty} \frac{s_0(t)+s_1(t)+\cdots+s_{n-1}(t)}{n} = f(t)$$

uniformly on Q.

Proof. According to Sec. 4.37,

$$s_n(t) = \int_{-\pi}^{\pi} f(t+h) D_n(h)\, dh,$$

and hence

$$\sigma_n(t) = \frac{1}{n}\sum_{k=0}^{n-1} s_k(t) = \frac{1}{n}\int_{-\pi}^{\pi} f(t+h) \sum_{k=0}^{n-1} D_k(h)\, dh.$$

But

$$\sum_{k=0}^{n-1} D_k(h) = \frac{1}{2\pi}\sum_{k=0}^{n-1} \frac{\sin(k+\frac{1}{2})h}{\sin\frac{h}{2}} = \frac{1}{2\pi}\sum_{k=0}^{n-1} \frac{\sin(k+\frac{1}{2})h\cdot\sin\frac{h}{2}}{\sin^2\frac{h}{2}}$$

$$= \frac{1}{2\pi}\sum_{k=0}^{n-1} \frac{\cos kh - \cos(k+1)h}{2\sin^2\frac{h}{2}} = \frac{1}{2\pi}\frac{1-\cos nh}{2\sin^2\frac{h}{2}} = \frac{1}{2\pi}\frac{\sin^2\frac{nh}{2}}{\sin^2\frac{h}{2}},$$

so that

$$\sigma_n(t) = \frac{1}{2\pi n}\int_{-\pi}^{\pi} f(t+h) F_n(h)\, dh, \tag{3}$$

in terms of the function

$$F_n(h)=\frac{1}{2\pi n}\frac{\sin^2\dfrac{nh}{2}}{\sin^2\dfrac{h}{2}},$$

known as the *Fejér kernel*. The function $F_n(h)$ is even, like the Dirichlet kernel $D_n(h)$, but nonnegative, unlike $D_n(h)$. Choosing $f(t)\equiv 1$ in (3) and noting that then $s_n(t)\equiv\sigma_n(t)\equiv 1$, we find that

$$\int_{-\pi}^{\pi} F_n(h)\,dh=1. \tag{4}$$

We can also write (2) in the form

$$\sigma_n(t)=\frac{1}{2\pi n}\int_{-\pi}^{\pi} f(\tau)F_n(t-\tau)\,d\tau, \tag{5}$$

bearing in mind that the integration is in effect over the circle Q, so that the values of $F_n(t-\tau)$ are to be replaced by $F_n(t-\tau+2\pi)$ or by $F_n(t-\tau-2\pi)$ whenever $t-\tau$ lies outside the original interval of integration $[-\pi,\pi]$.† Moreover, given any $\delta>0$,

$$\int_{|t-\tau|\geqslant\delta} F_n(t-\tau)\,d\tau=\int_{|h|\geqslant\delta} F_n(h)\,dh\leqslant\frac{1}{2\pi n}\int_{|h|\geqslant\delta}\frac{dh}{\sin^2\dfrac{h}{2}}=\frac{C(\delta)}{2\pi n}\to 0 \tag{6}$$

as $n\to\infty$, where $C(\delta)$ is a constant depending on δ. It follows from (4) and (6) that $F_n(t-\tau)$ $(n=1,2,\ldots)$ satisfies the properties of a delta-like sequence uniformly on Q (why?). Therefore $\sigma_n(t)$ converges uniformly to $f(t)$ on Q as $n\to\infty$, by Theorem 1.55d. ∎

4.53. Summation by the method of arithmetic means is a special case of summation with the help of a Toeplitz matrix, i.e., a matrix $T=\|t_{kn}\|$ whose elements satisfy Conditions 1–3 of Theorem 1.76c. Thus it is now natural to look for further conditions on T such that the symmetric partial sums (2) of the Fourier series of every continuous function $f(t)$ converge to $f(t)$ in the Toeplitz sense. First we note that in Sec. 1.76 the Toeplitz method was used only to find generalized limits of *bounded* sequences, whereas the partial sums of the Fourier series of a continuous function need not be bounded (Sec. 4.51). For this reason, we will assume that our Toeplitz matrix $T=\|t_{kn}\|$ is

† More exactly, by $F_n(t-\tau-2\pi)$ if $\pi<t-\tau\leqslant 2\pi$, or by $F_n(t-\tau+2\pi)$ if $-2\pi\leqslant t-\tau<-\pi$.

"triangular," i.e., that $t_{kn}=0$ if $n>k$. We can then construct the functionals

$$T_k(x)=\sum_{n=0}^{\infty} t_{kn}\xi_n=\sum_{n=0}^{k} t_{kn}\xi_n \tag{7}$$

for every numerical sequence $x=(\xi_0,\xi_1,\ldots,\xi_n,\ldots)$, whether bounded or not. By definition, the T-limit of the sequence x is just the quantity

$$T\text{-lim } \xi_n=\lim_{k\to\infty} T_k(x),$$

provided it exists.

Thus let $T=\|t_{kn}\|$ be a triangular Toeplitz matrix, let $f(t)$ be any function continuous on the circle $Q=[-\pi,\pi]$, and let $s_n(t)$ be the symmetric partial sum (2) of the Fourier series of $f(t)$. Then

$$s_n(t)=\int_{-\pi}^{\pi} f(t+h)D_n(h)\,dh,$$

as in Sec. 4.51, in terms of the Dirichlet kernel

$$D_n(h)=\frac{\sin\,(n+\frac{1}{2})h}{\sin\dfrac{h}{2}},$$

or equivalently

$$s_n(t)=\int_{-\pi}^{\pi} f(\tau)D_n(t-\tau)\,d\tau,$$

with the same proviso as made in connection with formula (5). Hence, forming the functionals (7) for the sequence $S(t)=\{s_0(t),s_1(t),\ldots,s_n(t),\ldots\}$, t being regarded as a parameter, we get

$$\begin{aligned}T_k(S(t))&=\sum_{n=0}^{k} t_{kn}s_n(t)=\sum_{n=0}^{k} t_{kn}\int_{-\pi}^{\pi} f(\tau)D_n(t-\tau)\,d\tau\\&=\int_{-\pi}^{\pi} f(\tau)\left\{\sum_{n=0}^{k} t_{kn}D_n(t-\tau)\right\}d\tau=\int_{-\pi}^{\pi} f(\tau)G_k(t-\tau)\,d\tau,\end{aligned}$$

where

$$G_k(h)=\sum_{n=0}^{k} t_{kn}D_n(h).$$

We now prove the desired generalization of Fejér's theorem:†

THEOREM (**Nikolski, 1948**). *Let $f(t)$ be any function continuous on the circle $Q=$*

† The Toeplitz matrix generating the Cesàro limit, and hence reducing Nikolski's theorem to Fejér's theorem, is given in Example 1.77b.

$[-\pi,\pi]$, *and suppose there exists a constant* $C>0$ *such that*

$$\int_{-\pi}^{\pi} |G_k(h)|\, dh < C \tag{8}$$

for all $k=0,1,2,\ldots$ *Then*

$$\lim_{k\to\infty} T_k(S(t)) = f(t), \tag{9}$$

where the convergence in uniform on Q. *However, if no such constant* C *exists, then there is a continuous function* $f(t)$ *for which the sequence* $T_k(S(t))$ $(k=0,1,2,\ldots)$ *fails to have a limit (at* $t=0$, *say).*

Proof. The crux of the proof is to show that the kernels $G_k(h)$ form a delta-like sequence. First we note that

$$\int_{-\pi}^{\pi} G_k(h)\, dh = \sum_{n=0}^{k} t_{kn} \int_{-\pi}^{\pi} D_n(h)\, dh = \sum_{n=0}^{k} t_{kn} \to 1 \tag{10}$$

as $k\to\infty$, by the properties of the Dirichlet kernel and of a Toeplitz matrix (Condition 2 of Theorem 1.76c). Moreover, given any $\delta>0$ and any continuous function $\varphi(h)$,

$$\left| \int_{|h|\geqslant\delta} Q_k(h)\varphi(h)\, dh \right| = \left| \sum_{n=0}^{k} t_{kn} \int_{|h|\geqslant\delta} D_n(h)\varphi(h)\, dh \right| \leqslant \sum_{n=0}^{k} |t_{kn}| D_{n\delta},$$

where

$$D_{n\delta} = \left| \int_{|h|\geqslant\delta} D_n(h)\varphi(h)\, dh \right|.$$

It follows from Lemma 4.32 that $D_{n\delta}\to 0$ as $n\to\infty$ for fixed δ, and hence that the sequence $D_{n\delta}$ is bounded. Let

$$D_\delta = \sup_n D_{n\delta},$$

and let

$$G = \sup_k \sum_{n=0}^{k} |t_{kn}|$$

(G is finite by Condition 1 of Theorem 1.76c). Then, given any $\varepsilon>0$, we first find an integer $n_0>0$ such that $D_{n\delta}<\varepsilon/2G$ for all $n>n_0$ and then a number $k_0\geqslant n_0$ such that

$$|t_{kn}| < \frac{\varepsilon}{2n_0 D_\delta} \qquad (n=1,2,\ldots,n_0)$$

for all $k > k_0$ (this is possible because of Condition 3 of Theorem 1.76c). It follows that

$$\left|\int_{|h|\geqslant\delta} G_k(h)\,\varphi(h)\,dh\right| \leqslant \sum_{n=0}^{n_0} |t_{kn}| D_{n\delta} + \sum_{n=n_0+1}^{k} |t_{kn}| D_{n\delta} < \frac{\varepsilon}{2} + \frac{\varepsilon}{2} = \varepsilon$$

for all $k > k_0$, and hence

$$\lim_{k\to\infty} \int_{|h|\geqslant\delta} G_k(h)\varphi(h)\,dh = 0. \tag{11}$$

Together (8), (10), and (11) imply that $G_k(t)$ $(k=0,1,2,\ldots)$ does in fact satisfy the properties of a delta-like sequence uniformly on Q (in particular, see Sec. 1.55h). Therefore Theorem 1.55d implies (9), where the convergence is uniform on Q. This proves the first part of the theorem. The second part follows from the Banach-Steinhaus theorem by the same method as used in Sec. 4.51 to prove the existence of a continuous function with a divergent Fourier series at $t=0$. ∎

If the kernel $G_k(t)$ turns out to be nonnegative, the condition (8) is automatically satisfied, because of (10). For example, the Fejér kernel is of this type.

4.54. Another kind of generalized summation of Fourier series has in effect already been used in Sec. 4.49b. In fact, if

$$\frac{a_0}{2} + \sum_{k=1}^{\infty} (a_k \cos kt + b_k \sin kt)$$

is the Fourier series of a continuous function $f(t)$, then

$$f(t) = \lim_{r\nearrow 1} \left\{\frac{a_0}{2} + \sum_{k=1}^{\infty} r^k (a_k \cos kt + b_k \sin kt)\right\},$$

where the calculation in the right-hand side can be regarded as a method of generalized summation of Fourier series. This method, known as *generalized summation in the Poisson sense*, has also been studied for discontinuous functions, but then one does not get such definitive results.

4.6. Other Orthogonal Systems

4.61. Orthogonalization. The system of trigonometric functions is a relatively rare example of a "ready-made" system of orthogonal functions. In most cases, orthogonal systems are constructed from nonorthogonal systems by resorting to the *orthogonalization theorem* of Sec. 1.43i. We now summarize the content of this theorem. Let $\mathbf{H}$ be a real or complex Hilbert space, and let $f_1,\ldots,f_n,\ldots$ be a finite or infinite system of vectors in $\mathbf{H}$ which are linearly independent in the sense that every finite subsystem $f_1,\ldots,f_n$ is linearly independent in the ordinary algebraic sense (Sec. 1.14a). Starting from $f_1,\ldots,f_n,\ldots$ we form new vectors $g_1,\ldots,g_n,\ldots$ with the help of the following "triangular table":

$$\begin{aligned}
g_1 &= f_1, \\
g_2 &= a_{21} f_1 + f_2, \\
g_3 &= a_{31} f_1 + a_{32} f_2 + f_3, \\
&\ldots \\
g_n &= a_{n1} f_1 + a_{n2} f_2 + a_{n3} f_3 + \ldots + f_n, \\
&\ldots
\end{aligned} \tag{1}$$

Then, according to the orthogonalization theorem, *there is a unique choice of coefficients a_{jk} in* (1) *such that the vectors $g_1,\ldots,g_n,\ldots$ are (pairwise) orthogonal.*

4.62. The Legendre polynomials. Let $\mathbf{H}=\mathbf{H}_R(-1,1)$ be the Hilbert space of all piecewise continuous real functions defined on the interval $[-1,1]$, and consider the following system of functions in $\mathbf{H}$:

$$f_0(t)=1,\ f_1(t)=t,\ \ldots,\ f_n(t)=t^n,\ \ldots \tag{2}$$

Since the functions $1,t,\ldots,t^n$ are linearly independent for every n (why?), we can orthogonalize the system (2). The subspace $\mathbf{L}_n \subset \mathbf{H}$ "generated" by the functions $1,t,\ldots,t^n$ is just the set of all polynomials of degree $k \leqslant n$, and the function $g_n(t)$ is a polynomial of degree n with 1 as its leading coefficient. We now derive an explicit formula for $g_n(t)$.

LEMMA. *If* $\varphi(t)=(t^2-1)^n$, *then*

$$\varphi(\pm 1)=\varphi'(\pm 1)=\cdots=\varphi^{(n-1)}(1)=0, \quad \varphi^{(n)}(\pm 1)\neq 0.$$

Proof. Use Leibniz's rule (Vol. 1, Sec. 8.12b) to differentiate

$$\varphi(t)=(t+1)^n(t-1)^n$$

$m \leqslant n$ times; in particular,†

$$\varphi^{(n)}(\pm 1)=\sum_{k=1}^{n} C_k^n[(t+1)^n]^{(k)}\Big|_{t=\pm 1}[(t-1)^n]^{(n-k)}\Big|_{t=\pm 1}=2^n n!, \tag{3}$$

since only the first or the last term of the sum ($C_0^n 2^n n!$ or $C_n^n 2^n n!$) is nonzero. ∎

† Here C_k^n denotes the binomial coefficient $n!/k!(n-k)!$.

THEOREM. *The polynomials $g_n(t)$ obtained by orthogonalizing the system (2) are given by*

$$g_n(t) = C_n[(t^2-1)^n]^{(n)}, \tag{4}$$

where

$$C_n = \frac{n!}{(2n)!}. \tag{5}$$

Proof. We will show that (4) is orthogonal to all the functions $1, t, \ldots, t^{n-1}$. It will then follow from the uniqueness of the coefficients a_{jk} in (1) that $g_n(t)$ has the representation (4), since the choice of constant (4) obviously makes the leading coefficient of $g_n(t)$ equal to 1.

To prove that the polynomial (4) is orthogonal to t^k for all $k = 1, 2, \ldots, n-1$, we form the scalar product and then integrate by parts repeatedly, using the lemma, until the exponent of t is reduced to zero. This gives

$$\begin{aligned}
(t^k, [(t^2-1)^n]^{(n)}) &= \int_{-1}^{1} t^k[(t^2-1)^n]^{(n)}\,dt \\
&= t^k[(t^2-1)^n]^{(n-1)}\Big|_{-1}^{1} - k\int_{-1}^{1} t^{k-1}[(t^2-1)^n]^{(n-1)} \\
&= -k\int_{-1}^{1} t^{k-1}[(t^2-1)^n]^{(n-1)} \\
&= -kt^{k-1}[(t^2-1)^n]^{(n-2)}\Big|_{-1}^{1} + k(k-1)\int_{-1}^{1} [(t^2-1)^n]^{(n-2)}dt \\
&= \cdots = \pm k!\int_{-1}^{1} [(t^2-1)]^{(n-k)}\,dt = \pm k![(t^2-1)^{(n-k-1)}]\Big|_{-1}^{1} \\
&= 0
\end{aligned}$$

for all $k = 1, 2, \ldots, n-1$, as required. ▮

For computational purposes, it is more convenient to choose the constant C_n in such a way as to make $g_n(1) = 1$. It follows from (3) that this can be done by replacing (5) by

$$C_n = \frac{1}{2^n n!}. \tag{5'}$$

With this choice of C_n, the polynomials (4) take the form

$$P_n(t) = \frac{1}{2^n n!}[(t^2-1)^n]^{(n)} \qquad (n = 0, 1, 2, \ldots),$$

and are known as the *Legendre polynomials*. The first few Legendre polynomials are easily calculated:

$$P_0(t)=1,\quad P_1(t)=t,\quad P_2(t)=\tfrac{3}{2}(t^2-\tfrac{1}{3}),\quad P_3(t)=\tfrac{5}{2}(t^3-\tfrac{3}{5}t),\ \ldots$$

4.63. Next we find the norm of the Legendre polynomial $P_n=P_n(t)$. As before, we form the scalar product and then integrate by parts repeatedly, using the lemma, until the exponent of the second term in the integrand is reduced to zero. This gives

$$\begin{aligned}(P_n,P_n)&=\frac{1}{2^{2n}(n!)^2}\int_{-1}^{1}[(t^2-1)^n]^{(n)}[(t^2-1)^n]^{(n)}\,dt\\&=\frac{1}{2^{2n}(n!)^2}[(t^2-1)^n]^{(n)}[(t^2-1)^n]^{(n-1)}\Big|_{-1}^{1}\\&\quad-\frac{1}{2^{2n}(n!)^2}\int_{-1}^{1}[(t^2-1)^n]^{(n+1)}[(t^2-1)^n]^{(n-1)}\,dt\\&=-\frac{1}{2^{2n}(n!)^2}\int_{-1}^{1}[(t^2-1)^n]^{(n+1)}[(t^2-1)^n]^{(n-1)}\,dt\\&=\cdots=\frac{(-1)^n}{2^{2n}(n!)^2}\int_{-1}^{1}[(t^2-1)^n]^{(2n)}(t^2-1)^n\,dt\\&=\frac{(-1)^n(2n)!}{2^{2n}(n!)^2}\int_{-1}^{1}(t-1)^n(t+1)^n\,dt.\end{aligned}$$

Next we again use repeated integration by parts to reduce the exponent of $t-1$ to zero, obtaining

$$\begin{aligned}(P_n,P_n)&=\frac{(-1)^n(2n)!}{2^{2n}(n!)^2}\left\{\left[(t-1)^n\frac{(t+1)^{n+1}}{n+1}\right]_{-1}^{1}-n\int_{-1}^{1}(t-1)^{n-1}\frac{(t+1)^{n+1}}{n+1}dt\right\}\\&=\cdots=\frac{(-1)^n(2n)!\,(-1)^n n!}{2^{2n}(n!)^2(n+1)\cdots 2n}\int_{-1}^{1}(t+1)^{2n}\,dt\\&=\frac{1}{2^{2n}}\frac{(t+1)^{2n+1}}{2n+1}\Big|_{-1}^{1}=\frac{2}{2n+1}.\end{aligned}$$

Thus, finally,

$$\|P_n\|=\sqrt{(P_n,P_n)}=\sqrt{\frac{2}{2n+1}}. \tag{6}$$

4.64. Expansion in Legendre polynomials. With every function $f(t)\in \mathbf{H}_R(-1,1)$ we can associate a *Fourier-Legendre series*

$$f(t)\sim\sum_{k=0}^{\infty}\alpha_k P_k(t). \tag{7}$$

It follows from formula (6) and the considerations of Sec. 4.17 that the

Fourier-Legendre coefficients α_k are given by

$$\alpha_k = \frac{(f,P_k)}{(P_k,P_k)} = \frac{2k+1}{2}\int_{-1}^{1} f(t)P_k(t)\,dt.$$

Just as in the case of trigonometric Fourier series (Sec. 4.24), it can be shown that the Fourier-Legendre series (7) converges to $f(t)$ in the mean square, i.e., that

$$\left\| f(t) - \sum_{k=0}^{n} \alpha_k P_k(t) \right\|^2 = \int_{-1}^{1} \left| f(t) - \sum_{k=0}^{n} \alpha_k P_k(t) \right|^2 dt \to 0$$

as $n \to \infty$. Here Parseval's theorem (Sec. 4.17) takes the form

$$\|f\|^2 = \int_{-1}^{1} |f(t)|^2\,dt = \sum_{k=0}^{\infty} \frac{2}{2k+1}|\alpha_k|^2.$$

***4.65.** The theorems of Sec. 4.3 on pointwise and uniform convergence of trigonometric Fourier series have natural analogues for Fourier-Legendre series. Let $s_n(t)$ be the nth partial sum of the series (7). Then

$$\begin{aligned} s_n(t) &= \sum_{k=0}^{n} \alpha_k P_k(t) = \sum_{k=0}^{n} \frac{(2k+1)P_k(t)}{2}\int_{-1}^{1} f(\tau)P_k(\tau)\,d\tau \\ &= \int_{-1}^{1} f(\tau) \sum_{k=0}^{n} (2k+1)\frac{P_k(\tau)P_k(t)}{2}\,d\tau. \end{aligned}$$

The function

$$L_n(t,\tau) = \sum_{k=0}^{n} (2k+1)\frac{P_k(\tau)P_k(t)}{2},$$

known as the *Fourier-Legendre kernel*, can be explicitly calculated (see Problem 11), leading to the *Christoffel-Darboux formula*

$$L_n(t,\tau) = \frac{n+1}{2}\,\frac{P_{n+1}(\tau)P_n(t) - P_n(\tau)P_{n+1}(t)}{t-\tau}.$$

By operating with the Fourier-Legendre kernel in the same way as with the Dirichlet kernel, it can be shown† that *if a function* $f(t) \in \mathbf{H}_R(-1,1)$ *has finite left-hand and right-hand derivatives* $f'(t_0-0)$ *and* $f'(t_0+0)$ *at* $t=t_0 \in (-1,1)$, *then the Fourier-Legendre series* (7) *converges to the value* $f(t_0)$ *if* $f(t)$ *is continuous at* $t=t_0$ (*where the convergence is uniform on every interval* $[-1+\delta,\ 1-\delta]$ *on which the indicated derivatives are bounded*) *and to the value*

$$\tfrac{1}{2}[f(t_0-0) + f(t_0+0)]$$

† See e.g., D. Jackson, *Fourier Series and Orthogonal Polynomials*, The Mathematical Association of America, Buffalo, N. Y. (1941), Chapter 2.

if $f(t)$ is discontinuous at $t=t_0$.

4.66. As an example of the application of Legendre polynomials to mathematical physics, let

$$\frac{\partial^2 u}{\partial x^2}+\frac{\partial^2 u}{\partial y^2}+\frac{\partial^2 u}{\partial z^2}=0 \tag{8}$$

be *Laplace's equation* in three dimensions, and consider the problem of finding a solution of (8) in the unit ball $r^2=x^2+y^2+z^2\leqslant 1$ which takes given continuous boundary values $u=f(\theta)$ on the boundary of the ball (the sphere $r=1$), where θ is the angle between the z-axis and the vector (x,y,z). This is the natural three-dimensional generalization of the Dirichlet problem for the disk, considered in Sec. 4.46b. It turns out† that if we first expand $f(\theta)$ in Legendre polynomials of the argument $\cos\theta$, writing

$$f(\theta)=\sum_{k=0}^{\infty}\alpha_k P_k(\cos\theta),$$

then the desired function $u=u(r,\theta)$ is just

$$u(r,\theta)=\sum_{k=0}^{\infty}\alpha_k r^k P_k(\cos\theta).$$

4.67. Orthogonal systems with a weight function. We now indicate a few of the many other systems of orthogonal polynomials encountered in mathematical physics. These systems are all obtained by the following general method: Starting from a given nonnegative function $p(x)$, called the *weight function*, defined on a finite or infinite interval $-\infty\leqslant a\leqslant x\leqslant b\leqslant +\infty$,‡ we first construct the Hilbert space $\mathbf{H}_{p(x)}(a,b)$ with scalar product

$$(f,g)_{p(x)}=\int_a^b f(x)\overline{g(x)}\,p(x)\,dx,$$

and then use the method of Sec. 4.61 to orthogonalize the functions $1,x,\ldots,x^n,\ldots$

a. If

$$a=-1,\qquad b=1,\qquad p(x)=1,$$

we obviously get the Legendre polynomials.

b. If

$$a=-1,\qquad b=1,\qquad p(x)=\frac{1}{\sqrt{1-x^2}},$$

† See e.g., N. N. Lebedev, *Special Functions and Their Applications* (translated by R. A. Silverman), Dover Publications, Inc., New York (1972), p. 208.

‡ We write $a<x$ instead of $a\leqslant x$ if $a=-\infty$ and $x<b$ instead of $x\leqslant b$ if $b=+\infty$.

we get the *Chebyshev polynomials*

$$T_n(x)=\cos\,(n\ \text{arc}\cos x);$$

under the change of variables $x=\cos t$, $T_n(x)$ reduces to $\cos nt$, with the space $\mathbf{H}_{p(x)}(-1,1)$ being isomorphic to the space $\mathbf{H}_1(0,\pi)$.

c. If

$$a=0,\qquad b=1,\qquad p(x)=x^{q-1}(1-x)^{p-q},$$

we get the *Jacobi* (or *hypergeometric*) *polynomials.*

d. If

$$a=-\infty,\qquad b=+\infty,\qquad p(x)=e^{-x^2},$$

we get the *Hermite polynomials*

$$H_n(x)=C_ne^{x^2}(e^{-x^2})^{(n)}.$$

e. If

$$a=0,\qquad b=+\infty,\qquad p(x)=e^{-x},$$

we get the *Laguerre polynomials*

$$L_n(x)=C_ne^x(x^ne^{-x})^{(n)}.$$

For a further discussion of these and many other systems of orthogonal functions, the reader is referred to the abundant literature on the subject.†

Problems

1. Let $f(t)$ be the odd function equal to $\pi/4$ for $0<t<\pi$. By expanding $f(t)$ in Fourier series, prove the formulas

$$1-\frac{1}{3}+\frac{1}{5}-\frac{1}{7}+\cdots=\frac{\pi}{4},$$

$$1+\frac{1}{5}-\frac{1}{7}-\frac{1}{11}+\frac{1}{13}+\frac{1}{17}-\cdots=\frac{\pi}{4},$$

† See e.g., R. Courant and D. Hilbert, *Methods of Mathematical Physics*, Vol. 1, John Wiley-Interscience, New York (1953), Chapter 2; G. Szegö, *Orthogonal Polynomials*, revised edition, American Mathematical Society, Providence, R. I. (1959); G. Alexits, *Convergence Problems of Orthogonal Series*, Pergamon Press, New York (1961); G. Sansone, *Orthogonal Functions* (translated by A. H. Diamond), John Wiley-Interscience, New York (1959); A. Erdélyi (editor), *Higher Transcendental Functions*, Vol. II (Bateman Manuscript Project), McGraw-Hill Book Co., New York (1953), Chapter 10. More on Legendre, Hermite, and Laguerre polynomials can also be found in Lebedev, *Special Functions and Their Applications*, Chapter 2.

$$1-\frac{1}{5}+\frac{1}{7}-\frac{1}{11}+\frac{1}{13}-\cdots=\frac{\pi}{2\sqrt{3}}.$$

(Euler)

2. Let $f(t)$ be the even function equal to t for $0<t<\pi$. By expanding $f(t)$ in Fourier series, prove the formulas

$$1+\frac{1}{4}+\frac{1}{9}+\frac{1}{16}+\cdots=\frac{\pi^2}{6},$$

$$1-\frac{1}{4}+\frac{1}{9}-\frac{1}{16}+\cdots=\frac{\pi^2}{12}.$$

(Euler)

3. Sum the following Fourier series:

(a) $1+\dfrac{\cos t}{1}+\dfrac{\cos 2t}{1\cdot 2}+\cdots+\dfrac{\cos nt}{n!}+\cdots;$

(b) $\dfrac{\sin t}{1}+\dfrac{\sin 2t}{1\cdot 2}+\cdots+\dfrac{\sin nt}{n!}+\cdots.$

4. Prove that a Fourier series is uniformly convergent if its partial sums form a precompact set (Sec. 1.24a) in the space $C^s(-\pi,\pi)$.

5. Suppose $f(t)$ is continuous and monotonic in a neighborhood of a point t_0. Prove that the symmetric partial sums of the Fourier series of $f(t)$ converge at $t=t_0$.

6 (*Continuation*). Suppose $f(t)$ is continuous and monotonic on an open interval I. Prove that the symmetric partial sums of the Fourier series of $f(t)$ converge uniformly to $f(t)$ on every closed subinterval of I.

7. Let $f(t)$ be a function satisfying the following conditions:

(1) $f(-t)\equiv -f(t),\ f(0)=0,\ f(\pi-t)\equiv f(t),\ f(t+2\pi)\equiv f(t)$;
(2) $f(t)$ is continuous;
(3) $f'(t)$ is continuous and nonincreasing for $0<t\leqslant\pi/2$;
(4) $\lim\limits_{t\searrow 0}\dfrac{tf'(t)}{f(t)}=1$, i.e., as $t\searrow 0$ the segment cut from the y-axis by the tangent to the curve $y=f(t)$ is equivalent (Vol. 1, Sec. 4.38a) to the ordinate at the point of tangency (see Figure 26).

Prove that the coefficients b_k of the Fourier sine series

$$\sum_{k=1}^{\infty} b_k \sin kt$$

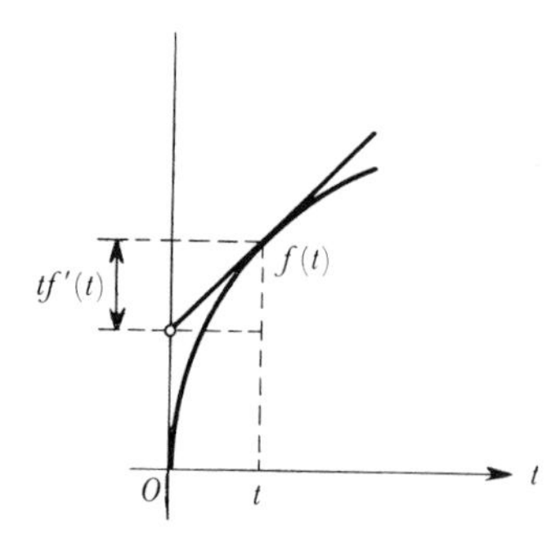

Figure 26

are of the form

$$b_k = \begin{cases} 0 & \text{if } k \text{ is even}, \\ \dfrac{\theta_k}{k} f\left(\dfrac{\pi}{k}\right) + \varepsilon_k & \text{if } k \text{ is odd}, \end{cases}$$

where $\theta_k \to 4/\pi$ as $k \to \infty$ and

$$\sum_{k=1}^{\infty} |\varepsilon_k| < \infty.$$

8. Using Problem 7, give an example of a continuous function $f(t)$ whose Fourier series is uniformly convergent on $[-\pi,\pi]$, although the series made up of its Fourier coefficients is not absolutely convergent.

9. Using Problem 7, give an example of a function $f(t)$ whose Fourier series

$$\sum_{k=-\infty}^{\infty} c_k e^{ikt}$$

is uniformly convergent on $[-\pi,\pi]$, at the same time that each of the series

$$\sum_{k=0}^{\infty} c_k e^{ikt}, \qquad \sum_{k=-\infty}^{0} c_k e^{ikt}$$

has points of divergence.

10. Let $p(x)$ be a weight function (Sec. 4.67), and let

$$Q_n(x) = \alpha_n x^n + \beta_n x^{n-1} + \gamma_{n-2}^{(n)} x^{n-2} + \cdots + \gamma_0^{(n)}$$

be the corresponding sequence of orthonormal polynomials. Prove the recurrence relation

$$xQ_n(x) = \frac{\alpha_n}{\alpha_{n+1}} Q_{n+1}(x) + \frac{\beta_n - \beta_{n+1}}{\alpha_n} Q_n(x) + \frac{\alpha_{n-1}}{\alpha_n} Q_{n-1}(x).$$

11 (*Continuation*). Prove the *Christoffel-Darboux formula*

$$\sum_{k=0}^{n} Q_k(x)\, Q_k(t) = \frac{\alpha_n}{\alpha_{n+1}} \frac{Q_n(x)Q_{n+1}(t) - Q_n(t)Q_{n+1}(x)}{t-x}.$$

12 (*Continuation*). Prove that the polynomial $Q_n(t)$ $(n \geqslant 1)$ has zeros in $[a,b]$, and in fact precisely n zeros.

5 The Fourier Transform

5.1. The Fourier Integral and Its Inversion

5.11. To represent a periodic function $\varphi(x)$, with period 2π, as a superposition of pure harmonic oscillations, we resort to a Fourier series

$$\varphi(x) = \sum_{k=-\infty}^{\infty} a_k e^{ikx}.$$

Somewhat more generally, if $\varphi(x)$ is periodic, with period $2\pi l$, the Fourier series of $\varphi(x)$ takes the form

$$\varphi(x) = \sum_{k=-\infty}^{\infty} a_k e^{ikx/l}, \tag{1}$$

where the coefficients a_k are given by the formula

$$a_k = \frac{1}{2\pi l} \int_{-\pi}^{\pi} \varphi(\xi) e^{-ik\xi/l}\, d\xi. \tag{2}$$

Note that (2) is obtained formally by multiplying (1) by $e^{-ikx/l}$ and then integrating term by term from $-\pi l$ to πl. Together (1) and (2) imply

$$\varphi(x) = \frac{1}{2\pi} \sum_{k=-\infty}^{\infty} \frac{1}{l} \int_{-\pi l}^{\pi l} \varphi(\xi) e^{ik(x-\xi)/l}\, d\xi. \tag{3}$$

At this point, it is natural to try taking the limit as $l \to \infty$ in formula (3), with the aim of representing an arbitrary function $\varphi(x)$ defined on the whole real line $-\infty < x < \infty$ as a superposition of harmonic oscillations. Taking this limit formally, we get

$$\varphi(x) = \frac{1}{2\pi} \int_{-\infty}^{\infty} d\sigma \left\{ \int_{-\infty}^{\infty} \varphi(\xi) e^{i\sigma(x-\xi)} \right\} d\xi, \tag{4}$$

where σ denotes the continuous argument "generated" by the discrete argument $\sigma_k = k/l$. Thus the desired formula for expanding $\varphi(x)$ in harmonic oscillations is of the form

$$\varphi(x) = \frac{1}{2\pi} \int_{-\infty}^{\infty} \psi(\sigma) e^{i\sigma x}\, d\sigma, \tag{5}$$

where

$$\psi(\sigma) = \int_{-\infty}^{\infty} \varphi(\xi) e^{-i\sigma\xi}\, d\xi. \tag{6}$$

Integrals of the type appearing in (5) and (6) are called *Fourier integrals*, and have already been encountered in our discussion of improper integrals (Vol.

1, Sec. 11.42). We now distinguish (5) and (6) by calling ψ the *Fourier transform* of φ, and φ the *inverse Fourier transform* of ψ. Formula (5) is also called the *formula for inversion of the Fourier transform*. Note that (5) and (6) differ only by the factor $1/2\pi$ and by the sign of the argument of the exponential.

5.12. Instead of trying to justify the legitimacy of the passage to the limit leading to formula (4), we will now prove (4) *directly* under certain assumptions on the function $\varphi(x)$. The first assumption is that $\varphi(x)$ *is piecewise continuous*† *and absolutely integrable on the whole real line* $-\infty < x < \infty$. This not only guarantees the existence of the integral (6) for all σ in the interval $-\infty < \sigma < \infty$ (Vol. 1, Sec. 11.21a), but also implies the following

LEMMA. *The function* $\psi(\sigma)$ *is bounded and continuous for all* $\sigma \in (-\infty, \infty)$, *and approaches the limit* 0 *as* $|\sigma| \to \infty$.

Proof. The boundedness of $\psi(\sigma)$ follows from the estimate

$$|\psi(\sigma)| \leqslant \int_{-\infty}^{\infty} |\varphi(\xi)| \, d\xi.$$

Moreover, it follows from the absolute integrability of the function $\varphi(x)$ and the majorant test (Vol. 1, Sec. 11.57a) that the Fourier integral (6) is uniformly convergent in σ for all $\sigma \in (-\infty, \infty)$. But the uniform convergence of (6) and the continuity of $e^{-i\sigma\xi}$ together imply the continuity of the function $\psi(\sigma)$ (cf. Vol. 1, Sec. 11.53). Finally, given any $\varepsilon > 0$, let the number $A > 0$ be such that

$$\int_{-\infty}^{-A} |\varphi(\xi)| \, d\xi + \int_{A}^{\infty} |\varphi(\xi)| \, d\xi < \frac{\varepsilon}{2}.$$

Applying Lemma 4.32 to the interval $[-A, A]$, we then find a number $\sigma_0 > 0$ such that

$$\left| \int_{-A}^{A} \varphi(\xi) e^{-i\sigma\xi} \, d\xi \right| < \frac{\varepsilon}{2}$$

whenever $|\sigma| > \sigma_0$. But then $|\sigma| > \sigma_0$ implies

$$\left| \int_{-\infty}^{\infty} \varphi(\xi) e^{-i\sigma\xi} \, d\xi \right| \leqslant \int_{-\infty}^{-A} |\varphi(\xi)| \, d\xi + \left| \int_{-A}^{A} \varphi(\xi) e^{-i\sigma\xi} \, d\xi \right| + \int_{A}^{\infty} |\varphi(\xi)| \, d\xi < \varepsilon,$$

which proves the last part of the lemma. ∎

5.13. The proof of formula (6) will involve the following special improper

† More exactly, piecewise continuous on every finite interval.

integral of the third kind:

$$I_{pq}=\int_{-\infty}^{\infty}\frac{e^{ipt}-e^{-iqt}}{t}\,dt \qquad (p,q>0).$$

The integrand is continuous on the whole real line (the indeterminacy at $t=0$ is easily removed by noting that the integrand approaches $(p+q)i$ as $t\to 0$). The convergence of I_{pq} is an immediate consequence of the Abel-Dirichlet test for improper integrals (Vol. 1, Sec. 11.23c). To evaluate I_{pq}, we write

$$\begin{aligned}I_{pq}&=\int_{-\infty}^{\infty}\left(\frac{\cos pt-\cos qt}{t}+i\,\frac{\sin pt}{t}+i\,\frac{\sin qt}{t}\right)dt\\&=\int_{-\infty}^{\infty}\frac{\cos pt-\cos qt}{t}\,dt+i\int_{-\infty}^{\infty}\frac{\sin pt}{t}\,dt+i\int_{-\infty}^{\infty}\frac{\sin qt}{t}\,dt.\end{aligned}$$

But the first integral vanishes because of the oddness of the integrand, while to evaluate the second and third integrals, we need only use the integral

$$\int_{-\infty}^{\infty}\frac{\sin \sigma x}{x}\,dx=\pi$$

(Vol. 1, Sec. 11.42h). It follows that

$$I_{pq}=2\pi i. \tag{7}$$

Similarly, we have

$$\int_{-T}^{T}\frac{e^{ipt}-e^{-iqt}}{t}\,dt=i\int_{-T}^{T}\frac{\sin pt}{t}\,dt+i\int_{-T}^{T}\frac{\sin qt}{t}\,dt,$$

where the improper integral

$$\int_{-\infty}^{\infty}\frac{\sin x}{x}\,dx$$

is uniformly convergent on every interval $\sigma\geqslant\sigma_0>0$ (Vol. 1, Sec. 11.59b). Therefore, given any $\varepsilon>0$, we can find a T such that

$$\left|\int_{|t|\geqslant T}\frac{e^{ipt}-e^{-iqt}}{t}\,dt\right|<\varepsilon \tag{8}$$

for all $p\geqslant 1$, $q\geqslant 1$ (say) whenever $T\geqslant T_0$.

5.14. We are now in a position to prove formula (4) with suitable conditions on the function $\varphi(x)$:

THEOREM. *Let $\varphi(x)$ be piecewise continuous and absolutely integrable on the real line*

$-\infty<x<\infty$, *and suppose* $\varphi(x)$ *satisfies Dini's condition at* $x=x_0$, *i.e., suppose there exists a* $\delta>0$ *such that*

$$\int_{|t|<\delta}\left|\frac{\varphi(x_0+t)-\varphi(x_0)}{t}\right|dt<\infty.$$

Then

$$\varphi(x_0)=\lim_{\substack{p\to\infty\\ q\to\infty}}\frac{1}{2\pi}\int_{-p}^{q}\left\{\int_{-\infty}^{\infty}\varphi(\xi)e^{i\sigma(x_0-\xi)}\,d\xi\right\}d\sigma, \tag{9}$$

where the limit on the right exists as p and q approach (plus) infinity independently.

Proof. Let

$$\varphi_{p,q}(x_0)=\frac{1}{2\pi}\int_{-p}^{q}\left\{\int_{-\infty}^{\infty}\varphi(\xi)e^{i\sigma(x_0-\xi)}\,d\xi\right\}d\sigma$$

for arbitrary $p\geqslant 0$, $q\geqslant 0$. Since the "inner integral" converges uniformly in σ, we can reverse the order of the integrations with respect to ξ and σ (cf. Vol. 1, Sec. 11.54), obtaining

$$\begin{aligned}\varphi_{p,q}(x_0)&=\frac{1}{2\pi}\int_{-\infty}^{\infty}\varphi(\xi)\left\{\int_{-p}^{q}e^{i\sigma(x_0-\xi)}\,d\sigma\right\}d\xi\\ &=\frac{1}{2\pi i}\int_{-\infty}^{\infty}\varphi(\xi)\frac{e^{iq(x_0-\xi)}-e^{-ip(x_0-\xi)}}{x_0-\xi}d\xi\\ &=\frac{1}{2\pi i}\int_{-\infty}^{\infty}\varphi(x_0+t)\frac{e^{ipt}-e^{-iqt}}{t}dt\end{aligned} \tag{10}$$

after making the substitution $x_0-\xi=t$. It follows from (7) that

$$\begin{aligned}\varphi_{p,q}(x_0)-\varphi(x_0)&=\frac{1}{2\pi i}\int_{-\infty}^{\infty}[\varphi(x_0+t)-\varphi(x_0)]\frac{e^{ipt}-e^{-iqt}}{t}dt\\ &=\frac{1}{2\pi i}\int_{|t|\leqslant T}\frac{\varphi(x_0+t)-\varphi(x_0)}{t}(e^{ipt}-e^{-iqt})\,dt\\ &\quad+\frac{1}{2\pi i}\int_{|t|\geqslant T}\varphi(x_0+t)\frac{e^{ipt}-e^{-iqt}}{t}dt\\ &\quad-\frac{\varphi(x_0)}{2\pi i}\int_{|t|\geqslant T}\frac{e^{ipt}-e^{-iqt}}{t}dt,\end{aligned} \tag{11}$$

where we choose $T\geqslant 1$. Since $\varphi(x_0+t)$ is absolutely integrable as a function of t and since

$$\left|\frac{e^{ipt}-e^{-iqt}}{t}\right|\leqslant 2$$

if $|t| \geqslant T \geqslant 1$, the second term in the right-hand side of (11) approaches 0 as $T \to \infty$ independently of the values of p and q, say for $p \geqslant 1$, $q \geqslant 1$. The same is true of the third term, because of the inequality (8). As for the first term, it approaches 0 as $p \to \infty$, $q \to \infty$, by Lemma 5.12, since the function

$$\frac{\varphi(x_0+t)-\varphi(x_0)}{t}$$

is absolutely integrable on the interval $|t| \leqslant T$, by Dini's condition. Thus, finally,

$$\lim_{\substack{p\to\infty \\ q\to\infty}} \varphi_{p,q}(x_0) = \varphi(x_0), \tag{12}$$

as required. ∎

In much the same way, it can be shown that the convergence in (9) is uniform on every bounded subset E of the real line $-\infty < x < \infty$, provided Dini's condition holds uniformly on E. The proof resembles that of Theorem 4.35.

5.15. If $\varphi(x)$ does not satisfy Dini's condition at the point $x = x_0$, Theorem 5.14 no longer works and the Fourier integral may well fail to converge at $x = x_0$. As in the case of Fourier series, we can then only expect formula (12) to hold for some kind of generalized method of passing to the limit. First we consider the symmetric "partial integral"

$$\varphi_{p,p}(x) = \frac{1}{2\pi} \int_{-p}^{p} \left\{ \int_{-\infty}^{\infty} \varphi(\xi) e^{i\sigma(x-\xi)}\, d\xi \right\} d\sigma,$$

which we henceforth denote simply by $\varphi_p(x)$. According to (10), $\varphi_p(x)$ can be written in the form

$$\varphi_p(x) = \frac{1}{2\pi i} \int_{-\infty}^{\infty} \varphi(x+t) \frac{e^{ipt}-e^{-ipt}}{t}\, dt = \frac{1}{\pi} \int_{-\infty}^{\infty} \varphi(x+t) \frac{\sin pt}{t}\, dt. \tag{13}$$

THEOREM. *Let $\varphi(x)$ be piecewise continuous and absolutely integrable on the real line $-\infty < x < \infty$, with a discontinuity point of the first kind at $x = x_0$.*† *Suppose $\varphi(x)$ satisfies "one-sided Dini conditions" at $x = x_0$, i.e., suppose both integrals*

$$\int_{x_0-\delta}^{x_0} \left| \frac{\varphi(x)-\varphi(x_0-0)}{x} \right| dx, \qquad \int_{x_0}^{x_0+\delta} \left| \frac{\varphi(x)-\varphi(x_0+0)}{x} \right| dx$$

converge for some $\delta > 0$. Then

$$\lim_{p\to\infty} \varphi_p(x_0) = \tfrac{1}{2}[\varphi(x_0-0)+\varphi(x_0+0)].$$

† Thus the one-sided limits $\varphi(x_0-0)$ and $\varphi(x_0+0)$ exist.

Proof. Completely analogous to that of Theorem 4.38. ▮

5.16. Next we drop the requirement that Dini's condition be satisfied, and investigate the analogue for Fourier integrals of Fejér's theorem for Fourier series (Sec. 4.52). Here, of course, instead of the arithmetic mean of the symmetric partial sums of the Fourier series, we consider the integral mean value

$$\sigma_N(x) = \frac{1}{N} \int_0^N \varphi_\nu(x)\, d\nu \tag{14}$$

of the symmetric partial integrals (13). Substituting (13) into (14), we get

$$\begin{aligned}\sigma_N(x) &= \frac{1}{\pi N} \int_0^N \left\{ \int_{-\infty}^{\infty} \varphi(x+t) \frac{\sin \nu t}{t} dt \right\} d\nu \\ &= \frac{1}{\pi N} \int_{-\infty}^{\infty} \frac{\varphi(x+t)}{t} \left\{ \int_0^N \sin \nu t \, d\nu \right\} dt \\ &= \frac{1}{\pi N} \int_{-\infty}^{\infty} \frac{\varphi(x+t)}{t} \frac{1-\cos Nt}{t} dt \\ &= \frac{2}{\pi N} \int_{-\infty}^{\infty} \varphi(x+t) \frac{\sin^2 \frac{Nt}{2}}{t^2} dt. \end{aligned} \tag{15}$$

The expression

$$F_N(t) = \frac{2}{\pi N} \frac{\sin^2 \frac{Nt}{2}}{t^2},$$

known as the *Fejér kernel for the Fourier integral*, has the following properties:

(1) $F_N(t) \geqslant 0$;

(2) $\displaystyle\int_{-\infty}^{\infty} F_N(t)\, dt = 1$;

(3) $\displaystyle\int_{|t| \geqslant \delta} F_N(t)\, dt \to 0$ as $N \to \infty$ for every fixed $\delta > 0$.

Property 1 is obvious, and Property 2 follows from Vol. 1, Sec. 11.59b. As for Property 3, it is implied by the estimate

$$\int_{|t| \geqslant \delta} F_N(t)\, dt \leqslant \frac{2}{\pi N} \int_{|t| \geqslant \delta} \frac{dt}{t^2} = \frac{4}{\pi N \delta}.$$

THEOREM. *Let $\varphi(x)$ be bounded, piecewise continuous, and absolutely integrable on the real line R_1 and uniformly continuous on a set $E \subset R_1$.*† *Then the integral mean value*

† This means that given any $\varepsilon > 0$, there is a $\delta > 0$ such that $|t| < \delta$ implies $|\varphi(x+t) - \varphi(x)| < \varepsilon$ for all $x \in E$, $t \in R_1$. It should be emphasized that the point $x+t$ need not lie in the set E.

$\sigma_N(x)$ of the symmetric partial integrals of the Fourier transform of $\varphi(x)$ converges uniformly to $\varphi(x)$ on E.

Proof. Given any $\varepsilon > 0$, let $\delta > 0$ be such that $|t| < \delta$ implies

$$|\varphi(x+t) - \varphi(x)| < \frac{\varepsilon}{2}$$

for all $x \in E$, $t \in R_1$. Then, since

$$\sigma_N(x) - \varphi(x) = \int_{-\infty}^{\infty} [\varphi(x+t) - \varphi(x)] F_N(t)\, dt,$$

because of (15) and Property 2 of the Fejér kernel, we have

$$\begin{aligned} |\sigma_N(x) - \varphi(x)| &\leqslant \int_{|t| \leqslant \delta} |\varphi(x+t) - \varphi(x)| F_N(t)\, dt \\ &\quad + \int_{|t| \geqslant \delta} |\varphi(x+t) - \varphi(x)| F_N(t)\, dt \\ &\leqslant \frac{\varepsilon}{2} \int_{-\infty}^{\infty} F_N(t)\, dt + 2 \sup_{-\infty < x < \infty} |\varphi(x)| \int_{|t| \geqslant \delta} F_N(t)\, dt. \end{aligned}$$

But the first term on the right does not exceed $\varepsilon/2$, while the second term becomes less than $\varepsilon/2$ for sufficiently large N, say for $N > N_0$. Therefore $|\sigma_N(x) - \varphi(x)| < \varepsilon$ for all $N > N_0$, $x \in E$. ∎

5.17. COROLLARY (**Uniqueness of the Fourier transform**). *Let $\varphi(x)$ be bounded, piecewise continuous, and absolutely integrable on the real line $-\infty < x < \infty$. Suppose $\psi(\sigma)$, the Fourier transform of $\varphi(x)$, vanishes for all σ. Then $\varphi(x)$ vanishes for all x, with the possible exception of a set having no finite limit points.*

Proof. In this case $\psi(\sigma) \equiv 0$, $\varphi_\nu(x) \equiv 0$, $\sigma_N(x) \equiv 0$, and hence

$$\varphi(x) = \lim_{N \to \infty} \sigma_N(x) = 0$$

at every continuity point of $\varphi(x)$, since every such point belongs to an interval of (uniform) continuity of $\varphi(x)$. Moreover, every finite interval contains no more than a finite number of discontinuity points of $\varphi(x)$. ∎

5.2. Further Properties of the Fourier Transform

5.21. Relation between the rate of decrease of a function and the smoothness of its Fourier transform. According to Lemma 5.12, the Fourier transform $\psi(\sigma)$ of a piecewise continuous, absolutely integrable function $\varphi(x)$ is bounded and continuous for all σ, and approaches 0 as

$|\sigma|\to\infty$. Suppose now that $x\varphi(x)$, as well as $\varphi(x)$, is absolutely integrable on the real line $-\infty<x<\infty$. Then $\psi(\sigma)$ is differentiable. In fact, formal differentiation of the Fourier integral

$$\int_{-\infty}^{\infty} \varphi(x)e^{-i\sigma x}\,dx=\psi(\sigma)$$

with respect to the parameter σ leads to the integral

$$-i\int_{-\infty}^{\infty} x\varphi(x)e^{-i\sigma x}\,dx,$$

which is absolutely convergent and uniformly convergent in the parameter σ. But then, by the theorem on differentiation of a parameter-dependent improper integral (cf. Vol. 1, Sec. 11.55a), the function $\psi(\sigma)$ is differentiable, with derivative

$$\psi'(\sigma)=-i\int_{-\infty}^{\infty} x\varphi(x)e^{-i\sigma x}\,dx.$$

Hence, if $\mathbf{F}$ denotes the (linear) operator carrying a function into its Fourier transform, so that

$$\mathbf{F}\varphi(x)=\int_{-\infty}^{\infty} \varphi(x)e^{-i\sigma x}\,dx,$$

we have just proved the important formula

$$\mathbf{F}(x\varphi)=i(\mathbf{F}\varphi)', \tag{1}$$

showing that *multiplying a function* $\varphi(x)$ *by* x *has the effect of differentiating its Fourier transform* $\psi(\sigma)=\mathbf{F}\varphi(x)$ *and multiplying the result by* i. Being the Fourier transform of an absolutely integrable function, $\psi'(\sigma)$ is again bounded and continuous, and approaches 0 as $|\sigma|\to\infty$.

Suppose further that not only $\varphi(x)$, but also the n products

$$x\varphi(x),\ x^2\varphi(x),\ \ldots,\ x^n\varphi(x),$$

are all absolutely integrable on the real line. Then, repeating the above argument n times, we find that the function $\psi(\sigma)=\mathbf{F}\varphi(x)$ has derivatives up to order n inclusive, all of which are bounded and continuous, and approach 0 as $|\sigma|\to\infty$. In this case, (1) is replaced by the more general formula

$$\mathbf{F}(x^k\varphi)=i^k(\mathbf{F}\varphi)^{(k)} \qquad (k=0,1,\ldots,n). \tag{1'}$$

If *all* the products

$$x^k\varphi(x) \qquad (k=0,1,2,\ldots)$$

are absolutely integrable, then $\psi(\sigma)=\mathbf{F}\varphi(x)$ has derivatives of *all* orders, and each of these derivatives is bounded and continuous, and approaches 0 as $|\sigma|\to\infty$.

Thus, roughly speaking, the faster the function $\varphi(x)$ falls off at infinity, the smoother the function $\psi(\sigma)=\mathbf{F}\varphi(x)$.

5.22. Next we consider the improvements in the smoothness properties of the function $\psi(\sigma)$ that result when further restrictions are imposed on the behavior of the function $\varphi(x)$ at infinity.

a. THEOREM. *Suppose the product $\varphi(x)e^{b|x|}$ is absolutely integrable for some constant $b>0$. Then the Fourier transform $\psi(\sigma)=\mathbf{F}\varphi(x)$ is not only infinitely differentiable, but also analytic (in a suitable strip).*

Proof. In this case, the Fourier integral

$$\psi(\sigma)=\int_{-\infty}^{\infty}\varphi(x)e^{-i\sigma x}\,dx$$

is defined not only for real values of σ, but also for certain complex values of σ. Going over to a complex variable $s=\sigma+i\tau$, where σ and τ are real, we have

$$\psi(s)=\int_{-\infty}^{\infty}\varphi(x)e^{-isx}\,dx=\int_{-\infty}^{\infty}\varphi(x)e^{-i\sigma x}e^{\tau x}\,dx,$$

where the integral converges for $|\tau|\leqslant b$, i.e., in a whole horizontal strip in the s-plane. The resulting function of a complex variable is analytic at every interior point s_0 of this strip. In fact, the integral is uniformly convergent in some neighborhood of s_0 (entirely contained in the strip), and the analyticity then follows from a familiar theorem of complex analysis (cf. Vol. 1, Sec. 11.55b). ∎

b. THEOREM. *The function $\psi(s)=\psi(\sigma+i\tau)$ is bounded in the strip $|\tau|\leqslant b$. Moreover, $\psi(s)$ approaches 0 uniformly for all τ in the strip as $|\sigma|\to\infty$.*

Proof. The first assertion follows from the obvious estimate

$$|\psi(s)|\leqslant\int_{-\infty}^{\infty}|\varphi(x)|e^{|\tau||x|}\,dx\leqslant\int_{-\infty}^{\infty}|\varphi(x)|e^{b|x|}\,dx.$$

The proof of the second assertion requires a slight sharpening of the argument given in the proof of Lemma 4.12. Since the function $\varphi(x)e^{b|x|}$ is absolutely integrable, given any $\varepsilon>0$, we can find a number $A>0$ such that

$$\int_{|x|\geqslant A}|\varphi(x)|e^{b|x|}\,dx=\int_{-\infty}^{-A}|\varphi(x)|e^{b|x|}\,dx+\int_{A}^{\infty}|\varphi(x)|e^{bx}\,dx<\frac{\varepsilon}{2}\cdot \qquad (2)$$

As for the integral

$$\int_{-A}^{A} \varphi(x)e^{-isx}\,dx = \int_{-A}^{A} \varphi(x)e^{\tau x}e^{-i\sigma x}\,dx,$$

it satisfies the estimate

$$\left|\int_{-A}^{A} \varphi(x)e^{-isx}\,dx\right| \leqslant 2A\omega\left\{\varphi(x)e^{\tau x}, \frac{2\pi}{|\sigma|}\right\} + N_A M_A \frac{2\pi}{|\sigma|} \tag{3}$$

(cf. formula (8), p. 234), where $\omega(f,\delta)$ is the modulus of continuity of the function $f(x)$, N_A is the number of intervals of continuity of $f(x)$ in the interval $[-A,A]$, and

$$M_A = \sup_{|x|\leqslant A} |\varphi(x)|e^{\tau x}.$$

The first term in the right-hand side of (3) does not exceed the quantity

$$2A\omega\{\varphi(x), 2\pi/|\sigma|\}e^{Ab} + 2A\omega\{e^{bx}, 2\pi/|\sigma|\} \sup_{|x|\leqslant A} |\varphi(x)|$$

(Vol. 1, Sec. 5.17d), which approaches 0 as $|\sigma|\to\infty$ independently of the value of τ in the strip $|\tau|\leqslant b$, and the same is obviously true of the second term. Thus we can find a $\sigma_0>0$ such that

$$\left|\int_{-A}^{A} \varphi(x)e^{-isx}\,dx\right| < \frac{\varepsilon}{2}$$

whenever $|\sigma|>\sigma_0$, $|\tau|\leqslant b$. But

$$\left|\int_{|x|\geqslant A} \varphi(x)e^{-isx}\,dx\right| \leqslant \int_{|x|\geqslant A} |\varphi(x)|e^{b|x|}\,dx < \frac{\varepsilon}{2},$$

because of (2), and hence, just as in the proof of Lemma 5.12,

$$\left|\int_{-\infty}^{\infty} \varphi(x)e^{-isx}\,dx\right| < \varepsilon,$$

whenever $|\sigma|>\sigma_0$, $|\tau|\leqslant b$. ∎

c. Suppose further that the product $\varphi(x)e^{b|x|}$ is integrable for *every* $b>0$. Then the function $\psi(s)$ is defined and analytic in every strip $|\tau|\leqslant b$, i.e., $\psi(s)$ is an *entire* function (Vol. 1, Sec. 10.38). As just shown, in every strip $|\tau|\leqslant b$ this entire function is bounded (with a bound depending on b) and approaches 0 as $|\sigma|\to\infty$.

5.23.a. We can also consider functions $\varphi(x)$ which fall off even faster at infinity, namely those for which the product $\varphi(x)e^{M(x)}$ is absolutely integrable, where $M(x)$ grows more rapidly than any linear function. It will be convenient to represent $M(x)$ in the form

$$M(x) = \int_0^x \mu(\xi)\, d\xi \qquad (0 \leqslant x < \infty),$$

where $\mu(\xi)$ is an increasing continuous function with $\mu(0)=0$, $\mu(\infty)=\infty$, and we define $M(x)$ for negative x by the formula $M(x) = M(-x)$. To describe the properties of the Fourier transform of $\varphi(x)$, we introduce another function $\Omega(\tau)$, known as the *dual of* $M(x)$, *in Young's sense*, defined by

$$\Omega(\tau) = \int_0^\tau \lambda(t)\, dt \qquad (0 \leqslant \tau < \infty), \qquad \Omega(-\tau) = \Omega(\tau),$$

where $\lambda(t)$ is the inverse of the function $\mu(\xi)$. Functions which are duals of each other, in Young's sense, are related by *Young's inequality*

$$x\tau \leqslant M(x) + \Omega(\tau) \qquad (x \geqslant 0,\ \tau \geqslant 0)$$

(Vol. 1, Sec. 9.62g).

b. THEOREM. *Let* $\varphi(x)$ *be a piecewise continuous function for which the integral*

$$\int_{-\infty}^{\infty} |\varphi(x)| e^{M(x)}\, dx$$

is finite. Then the Fourier transform $\psi(s) = \mathbf{F}\varphi(x)$ *is an entire function satisfying the inequality*

$$|\psi(\sigma + i\tau)| \leqslant C e^{\Omega(\tau)}.$$

Proof. The fact that $\psi(s)$ is entire follows from Sec. 5.22b, since

$$\int_{-\infty}^{\infty} |\varphi(x)| e^{b|x|}\, dx \leqslant \int_{-\infty}^{\infty} |\varphi(x)| e^{M(x)}\, dx < \infty$$

for every $b > 0$ (why?). Furthermore, we have

$$|\psi(\sigma + i\tau)| = \left| \int_{-\infty}^{\infty} \varphi(x) e^{-i(\sigma + i\tau)x}\, dx \right| \leqslant \int_{-\infty}^{\infty} |\varphi(x)| e^{M(x)} e^{|\tau||x| - M|x|}\, dx.$$

But

$$|x||\tau| - M(x) \leqslant \Omega(\tau),$$

by Young's inequality, with x and τ replaced by $|x|$ and $|\tau|$, and hence

$$|\psi(\sigma + i\tau)| \leqslant e^{\Omega(\tau)} \int_{-\infty}^{\infty} |\varphi(x)| e^{M(x)}\, dx = C e^{\Omega(\tau)}. \quad \blacksquare$$

c. Example. Suppose $\varphi(x)$ is such that

$$\int_{-\infty}^{\infty} |\varphi(x)| e^{(1/p)|x|^p}\, dx \qquad (p>1).$$

Then the corresponding function $\psi(s) = \mathbf{F}\varphi(x)$ satisfies the condition

$$|\psi(\sigma+i\tau)| \leqslant C^{(1/q)|\tau|^q} \qquad \left(\frac{1}{p}+\frac{1}{q}=1\right),$$

since the function $(1/q)|\tau|^q$ is the dual, in Young's sense, of the function $(1/p)|x|^p$ (Vol. 1, Sec. 9.62g). Note that the numbers p and q both exceed 1, but change in opposite directions. Thus q decreases as p increases, and $q \to 1$ as $p \to \infty$.

5.24. Suppose finally that the product of $\varphi(x)$ and *every* increasing function of $|x|$ is absolutely integrable. Functions which vanish outside some interval $|x| \geqslant a$ obviously have this property, and moreover, it is easy to see that *only* such functions have the property. Thus suppose $\varphi(x)$ vanishes outside some interval $|x| \geqslant a$. Then the Fourier transform

$$\psi(s) = \int_{-a}^{a} \varphi(x) e^{-isx}\, dx$$

is an entire function of s, satisfying the estimate

$$|\psi(\sigma+i\tau)| \leqslant \int_{-a}^{a} |\varphi(x)| e^{|\tau||x|}\, dx \leqslant Ce^{a|\tau|} \tag{4}$$

in the s-plane, where

$$C = \int_{-a}^{a} |\varphi(x)|\, dx.$$

An entire function $\psi(s)$ satisfying the inequality (4) is called an *entire function of finite exponential type* $\leqslant a$.

Thus the faster the function $\varphi(x)$ falls off at infinity, the "smoother" its Fourier transform $\psi(\sigma) = \mathbf{F}\varphi(x)$. Starting with continuous functions $\psi(\sigma)$, we have considered in turn functions which are finitely differentiable, infinitely differentiable, analytic in a strip, and analytic in the whole plane (i.e., entire), arriving finally at entire functions of finite exponential type. In this sense, the latter represent the "limiting case of smoothness" for functions approaching 0 in both directions along the real axis, a property shared by the Fourier transforms of all absolutely integrable functions (see Lemma 5.12). It can be shown (see Vol. 1, Chapter 10, Problem 24) that there are no nonzero entire functions which approach 0 in both directions along the real axis and grow more slowly at infinity than $e^{a|\tau|}$ for every $a>0$.

5.25. We now pursue our investigation in another direction, i.e., instead of requiring the function $\varphi(x)$ to fall off more and more rapidly, we require that it be more and more smooth. In view of the results of Secs. 5.21–5.24, we then have every reason to believe that the Fourier transform $\psi(\sigma) = \mathbf{F}\varphi(x)$ will fall off more and more rapidly.

Suppose $\varphi(x)$ is continuous and absolutely integrable, and has a piecewise continuous derivative which is also absolutely integrable (on the whole real line $-\infty<x<\infty$). Then, first of all,

$$\varphi(x)=\varphi(0)+\int_0^x \varphi'(\xi)\,d\xi$$

has a limit as $x\to\infty$, and this limit can only be 0, since otherwise $\varphi(x)$ would not be absolutely integrable. The same is true of the case $x\to-\infty$. Moreover, integrating by parts, we get

$$\mathbf{F}\varphi' = \int_{-\infty}^{\infty} \varphi'(x)e^{-i\sigma x}\,dx = \varphi(x)e^{-i\sigma x}\Big|_{-\infty}^{\infty} + i\sigma\int_{-\infty}^{\infty}\varphi(x)e^{-i\sigma x}\,dx.$$

The integrated term vanishes, as just shown, and therefore

$$\mathbf{F}\varphi' = i\sigma\mathbf{F}\varphi. \tag{5}$$

In other words, *differentiating a function $\varphi(x)$ has the effect of multiplying its Fourier transform $\psi(\sigma)=\mathbf{F}\varphi(x)$ by $i\sigma$.* Since $\mathbf{F}\varphi'$ is a bounded function of σ (and even approaches 0 as $|\sigma|\to\infty$), being a Fourier transform of an integrable function, $\mathbf{F}\varphi$ satisfies the estimate

$$|\mathbf{F}\varphi| = \frac{|\mathbf{F}\varphi'|}{|\sigma|} \leqslant \frac{C}{|\sigma|}. \tag{6}$$

Thus, in the present case, $\psi(\sigma)$ not only approaches 0 as $|\sigma|\to\infty$, but actually approaches 0 faster than $1/|\sigma|$.

Suppose further that not only $\varphi(x)$, but also the first n derivatives

$$\varphi'(x),\ \varphi''(x),\ldots,\varphi^{(n)}(x)$$

are absolutely integrable on the real line. Then, integrating by parts n times, we find that

$$\mathbf{F}\varphi^{(k)} = (i\sigma)^k\mathbf{F}\varphi \qquad (k=0,1,\ldots,n). \tag{5$'$}$$

In this case, (6) is replaced by the more general formula

$$|\mathbf{F}\varphi| = \frac{|\mathbf{F}\varphi^{(k)}|}{|\sigma|^k} \leqslant \frac{C}{|\sigma|^k} \qquad (k=0,1,\ldots,n). \tag{6$'$}$$

In particular, if the absolutely integrable function $\varphi(x)$ is "suitably smooth," then its Fourier transform $\psi(\sigma) = \mathbf{F}\varphi(x)$ is also absolutely integrable. It is clear from (6′) that this will certainly be the case if the functions φ, φ', and φ'' exist and are absolutely integrable.

If *all* the derivatives

$$\varphi^{(k)}(x) \qquad (k = 0,1,2,\ldots)$$

exist and are absolutely integrable, the function $\psi(\sigma) = \mathbf{F}\varphi(x)$ falls off more rapidly than *any* function $1/|\sigma|^k$ as $|\sigma| \to \infty$.

5.26. In this section, we will investigate the behavior of the Fourier transform of a function which is not only infinitely differentiable, but also analytic in some strip $|y| < b$ in the plane of the complex variable $z = x + iy$.

a. LEMMA. *Let $\varphi(z)$ be analytic in the strip $|y| < b$, and suppose $\varphi(z)$ satisfies the inequality*

$$|\varphi(x+iy)| \leqslant \Phi(x) \tag{7}$$

in the strip, where $\Phi(x) \geqslant 0$ is a function such that

$$\lim_{|x| \to \infty} \Phi(x) = 0 \tag{8}$$

and

$$\int_{-\infty}^{\infty} \Phi(x)\, dx < \infty. \tag{9}$$

Then the value of the integral

$$\int_{-\infty}^{\infty} \varphi(x+iy)\, dx \tag{10}$$

is independent of the point $y \in (-b,b)$.

Proof. The existence of (10) follows at once from the continuity of the function $\varphi(x+iy)$ in x and the estimates (7) and (9). Let y_1 and y_2 $(y_1 < y_2)$ be any two points in the interval $(-b,b)$, and consider the closed contour $L = ABCD$ shown in Figure 27. By Cauchy's theorem (Vol. 1, Sec. 10.32),

$$\int_L \varphi(z)\, dz = \int_A^B \varphi(z)\, dz + \int_B^C \varphi(z)\, dz + \int_C^D \varphi(z)\, dz + \int_D^A \varphi(z)\, dz = 0. \tag{11}$$

Let $-R$ and R be the abscissas of the points A and B. Then

$$\left| \int_B^C \varphi(z)\, dz \right| \leqslant \int_B^C |\varphi(z)|\, dz \leqslant \int_{y_1}^{y_2} \Phi(R)\, dy = \Phi(R)(y_2 - y_1),$$

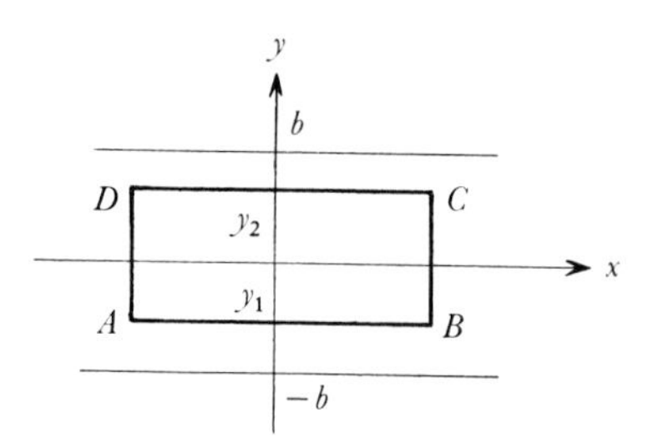

Figure 27

where the expression on the right approaches 0 as $R \to \infty$, because of (8). In just the same way, the integral

$$\int_D^A \varphi(z)\, dz$$

approaches 0 as $R \to \infty$. The limits as $R \to \infty$ of the remaining two integrals are just

$$\int_{-\infty}^{\infty} \varphi(x+iy_1)\, dx$$

and

$$-\int_{-\infty}^{\infty} \varphi(x+iy_2)\, dx.$$

Thus, finally, taking the limit as $R \to \infty$ in (11), we get

$$\int_{-\infty}^{\infty} \varphi(x+iy_1)\, dx = \int_{-\infty}^{\infty} \varphi(x+iy_2)\, dx. \quad \blacksquare$$

b. THEOREM. *Let $\varphi(z)$ and $\Phi(x)$ be the same as in the lemma. Then the Fourier transform $\psi(\sigma) = \mathbf{F}\varphi(x)$ satisfies the inequality*

$$|\psi(\sigma)| \leqslant Ce^{-b|\sigma|}, \tag{12}$$

where

$$C = \int_{-\infty}^{\infty} \Phi(x)\, dx.$$

Proof. Applying the lemma to the analytic function

$$\varphi^*(z) = \varphi(z)e^{-i\sigma z},$$

with $\Phi(x)$ replaced by

$$\Phi^*(x) = \Phi(x)e^{b|\sigma|},$$

we find that

$$\psi(\sigma)=\int_{-\infty}^{\infty}|\varphi(x)e^{-i\sigma x}\,dx=\int_{-\infty}^{\infty}\varphi(x+iy)e^{-i\sigma(x+iy)}\,dx \tag{13}$$

for every $y\in(-b,b)$. For fixed y, we have the estimate

$$|\psi(\sigma)|\leqslant\int_{-\infty}^{\infty}|\varphi(x+iy)|e^{\sigma y}\,dx=e^{\sigma y}\int_{-\infty}^{\infty}|\varphi(x+iy)|\,dx\leqslant e^{\sigma y}\int_{-\infty}^{\infty}\Phi(x)\,dx.$$

Taking the limit as $y\to-b$ if $\sigma>0$ and as $y\to b$ if $\sigma<0$, we get (12). ∎

c. Let $\varphi(z)$ be entire, and suppose that for every strip $|y|<b$, there is a function $\Phi(x)$ satisfying the conditions (7)–(9), where $\Phi(x)$ may depend on b. Then, according to the theorem, *the Fourier transform $\psi(\sigma)=\mathbf{F}\varphi(x)$ satisfies an inequality of the form*

$$|\psi(\sigma)|\leqslant C_b e^{-b|\sigma|}$$

for every $b>0$.

5.27. THEOREM. *Let $\varphi(z)$ be an entire function satisfying the inequality*

$$|\varphi(x+iy)|\leqslant\Phi_b(x)e^{\Omega(y)}$$

in every strip $|y|<b$, where $\Phi_b(x)\geqslant0$ is a function such that

$$\lim_{|x|\to\infty}\Phi_b(x)=0,\qquad\int_{-\infty}^{\infty}\Phi_b(x)\,dx<\infty$$

for every $b>0$, and

$$\Omega(y)=\int_0^y\lambda(\eta)\,d\eta\qquad(0\leqslant y<\infty),\qquad\Omega(-y)=\Omega(y),$$

where $\lambda(\eta)$ is an increasing continuous function with $\lambda(0)=0$, $\lambda(\infty)=\infty$. Let $M(\sigma)$ be the dual, in Young's sense, of $\Omega(y)$, so that

$$M(\sigma)=\int_0^\sigma\mu(\xi)\,d\xi,$$

where $\mu(\xi)$ is the inverse of the function $\lambda(\eta)$. Then the Fourier transform $\psi(\sigma)=\mathbf{F}\varphi(x)$ satisfies an inequality of the form

$$|\psi(\sigma)|\leqslant C_b e^{-M(\sigma)} \tag{14}$$

in every strip $|y|<b$, where

$$C_b=\int_{-\infty}^{\infty}\Phi_b(x)\,dx. \tag{15}$$

Proof. We have

$$|\varphi(x+iy)| \leqslant \Phi_b(x) e^{\Omega(y)} \leqslant \Phi_b(x) e^{\Omega(b)}$$

in every strip $|y|<b$, and hence

$$\psi(\sigma) = \int_{-\infty}^{\infty} \varphi(x) e^{-i\sigma x}\, dx = \int_{-\infty}^{\infty} \varphi(x+iy) e^{-i\sigma(x+iy)}\, dx$$

for every $y \in (-b,b)$, just as in (13). It follows that

$$|\psi(\sigma)| \leqslant \int_{-\infty}^{\infty} \Phi_b(x) e^{\Omega(y)} e^{\sigma y}\, dx = C_b e^{\Omega(y)+\sigma y} \qquad (|y|<b), \tag{16}$$

where C_b is given by (15). Choosing the sign of y such that $\sigma y = -|\sigma||y|$ and the absolute value of y such that Young's inequality (Sec. 5.23a) reduces to an equality

$$|\sigma||y| = M(\sigma) + \Omega(y)$$

(cf. Vol. 1, Sec. 9.62g), we find that (16) reduces at once to (14). ∎

5.28. THEOREM. *Let $\varphi(z)$ be an entire function satisfying the inequality*

$$|\varphi(x+iy)| \leqslant \Phi(x) e^{a|y|},$$

where $\Phi(x) \geqslant 0$ is a function (independent of y) such that

$$\lim_{|x|\to\infty} \Phi(x) = 0, \qquad \int_{-\infty}^{\infty} \Phi(x)\, dx < \infty.$$

Then the Fourier transform $\psi(\sigma) = \mathbf{F}\varphi(x)$ vanishes if $|\sigma| > a$.

Proof. Again

$$\psi(\sigma) = \int_{-\infty}^{\infty} \varphi(x) e^{-i\sigma x}\, dx = \int_{-\infty}^{\infty} \varphi(x+iy) e^{-i\sigma(x+iy)}\, dx$$

for every fixed y, as in (13). Choosing the sign of y such that $\sigma y = -|\sigma||y|$, we find that

$$|\psi(\sigma)| \leqslant e^{\sigma y} \int_{-\infty}^{\infty} \Phi(x) e^{a|y|}\, dx \leqslant C e^{(a-|\sigma|)|y|}. \tag{17}$$

If $|\sigma| > a$, then, taking the limit as $|y| \to \infty$ in (17), we find that $\psi(\sigma) = 0$. ∎

5.29. It should be noted that the theorems of Secs. 5.25–5.28 are not the exact converses of those of Secs. 5.21–5.24, and in fact require certain supplementary conditions (e.g., the existence of a function $\Phi(x)$ satisfying the conditions of Sec. 5.26a). We now consider the problem of constructing classes of functions $\varphi(x)$ for which the class of corresponding Fourier transforms $\psi(\sigma)=\mathbf{F}\varphi(x)$ has an exact description. Some classes of this kind can be constructed by combining the theorems of Secs. 5.21–5.28.

a. Let S be the class of all infinitely differentiable functions $\varphi(x)$ $(-\infty< x<\infty)$ satisfying inequalities of the form

$$|x^k\varphi^{(q)}(x)|\leqslant C_{pq} \qquad (k,q=0,1,2,\ldots), \tag{18}$$

where the C_{pq} are constants (depending on the choice of the function $\varphi(x)$). Each of the functions $x^k\varphi^{(q)}(x)$ is not only bounded on the real line, but also absolutely integrable on the real line, since besides (18) we also have

$$|x^k\varphi^{(q)}(x)|\leqslant\frac{C_{k+2,q}}{x^2} \qquad (k,q=0,1,2,\ldots)$$

if $x\neq 0$, so that

$$|x^k\varphi^{(q)}(x)|\leqslant \min\left\{C_{kq},\frac{C_{k+2,q}}{x^2}\right\}\leqslant\frac{C'_{kq}}{1+x^2},$$

where C'_{kq} is a new constant.

Along with $\varphi(x)$ itself, every function $x^k\varphi(x)$ is infinitely differentiable, and the successive derivatives of $x^k\varphi(x)$ are all absolutely integrable on the real line, since, by Leibniz's rule (Vol. 1, Sec. 8.12b), the qth derivative of $x^k\varphi(x)$ is a linear combination of the absolutely integrable functions

$$x^{k-j}\varphi^{(q-j)}(x) \qquad (j=0,1,\ldots,q).$$

The function $\psi(\sigma)=\mathbf{F}\varphi(x)$ is infinitely differentiable, by Sec. 5.21. Moreover, it follows from formulas (1′) and (5′) that

$$\mathbf{F}[x^q\varphi(x)]^{(k)}=(i\sigma)^k\mathbf{F}[x^q\varphi(x)]=(i\sigma)^k i^q\psi^{(q)}(\sigma),$$

where the right-hand side, being the Fourier transform of the absolutely integrable function $[x^q\varphi(x)]^{(k)}$, is bounded for all k and q, i.e.,

$$|\sigma^k\psi^{(q)}(\sigma)|\leqslant B_{kq} \qquad (k,q=0,1,2,\ldots)$$

for suitable constants B_{kq}. Thus, *if $\varphi(x)$ belongs to S, then so does $\psi(\sigma)$.*

Conversely, *any given function $\psi(\sigma)\in S$ is the Fourier transform of a function $\varphi(x)\in S$.* In fact, let

$$\varphi(x)=\frac{1}{2\pi}\int_{-\infty}^{\infty}\psi(\sigma)e^{i\sigma x}\,dx. \tag{19}$$

Then the function $2\pi\varphi(-x)$ is the direct Fourier transform of $\psi(\sigma)$, and hence belongs to S. But then obviously $\varphi(x)$ also belongs to S, and moreover, taking the inverse Fourier transform, we have

$$\psi(\sigma)=\frac{1}{2\pi}\int_{-\infty}^{\infty} 2\pi\varphi(-x)e^{i\sigma x}\,dx=\int_{-\infty}^{\infty}\varphi(x)e^{-i\sigma x}\,dx,$$

so that $\psi(\sigma)=\mathbf{F}\varphi(x)$.

Thus, finally, $\mathbf{F}S=S$, i.e., *the Fourier transform operator* $\mathbf{F}$ *maps the class* S *onto the whole class* S.

***b.** Let m_{kq} $(k,q=0,1,2,\ldots)$ be any double sequence of constants, and let $S_{\langle m_{kq}\rangle}$ be the class of all infinitely differentiable functions $\varphi(x)$ $(-\infty<x<\infty)$ satisfying inequalities of the form

$$|x^k\varphi^{(q)}(x)|\leqslant CA^kB^qm_{kq} \qquad (k,q=0,1,2,\ldots),$$

where the constants A, B, and C can depend on the function $\varphi(x)$. Then it turns out that

$$\mathbf{F}S_{\langle m_{kq}\rangle}=S_{\langle m_{kq}\rangle},$$

under appropriate conditions on the sequence m_{kq}.†

c. Let $M(x)$ and $\Omega(t)$ be dual functions, in Young's sense (Sec. 5.32a), and let W_M be the class of all infinitely differentiable functions $\varphi(x)$ $(-\infty<x<\infty)$ satisfying the inequalities

$$|\varphi^{(q)}(x)|\leqslant C_qe^{-M(x)} \qquad (q=0,1,2,\ldots).$$

If $\mathbf{F}\varphi(x)=\psi(s)$, then

$$\mathbf{F}\varphi^{(q)}(s)=(is)^q\psi(s)$$

(Sec. 5.25), and hence, by Theorem 5.23b,

$$|s^q\psi(\sigma+i\tau)|\leqslant C_q'e^{\Omega(\tau)} \qquad (q=0,1,2,\ldots). \tag{20}$$

Let W^Ω be the class of all entire functions $\psi(s)$ satisfying inequalities of the form (20). Then we have just shown that $\mathbf{F}W_M\subset W^\Omega$. To show that $\mathbf{F}W_M=W^\Omega$, let $\psi(s)$ be any function in W^Ω. Then it follows from (20) and the same inequalities with q replaced by $q+2$ that

$$|s^q\psi(\sigma+i\tau)|\leqslant e^{\Omega(\tau)}\min\left\{C_q',\frac{C_{q+2}'}{|s|^2}\right\}=\Phi_{\tau q}(\sigma)e^{\Omega(\tau)},$$

where

$$\Phi_{\tau q}(\sigma)=\min\left\{C_q',\frac{C_{q+2}'}{|\sigma+i\tau|^2}\right\}\leqslant\frac{C_{\tau q}}{1+|\sigma|^2}$$

† See I. M. Gelfand and G. E. Shilov, *Generalized Functions*, Vol. 2 (translated by A. Feinstein et al.), Academic Press, New York (1968), Chapter 4, Sec. 6.

is an integrable function. Applying Theorem 5.27, we arrive at the conclusion that the function $\varphi(x)$ defined by (19) satisfies the inequalities

$$|\varphi^{(q)}(x)| \leqslant C_q'' e^{-M(x)} \qquad (q=0,1,2,\ldots),$$

i.e., $\varphi(x) \in W_M$. Therefore $\mathbf{F} W_M = W^\Omega$. It is easy to see that $\mathbf{F} W^\Omega = W_M$ as well (why?).

5.3. Examples and Applications

5.31. The Fourier transform of the rational function

$$\varphi(x) = \frac{P(x)}{Q(x)} = \frac{a_0 + a_1 x + \cdots + a_n x^n}{b_0 + b_1 x + \cdots + b_m x^m} \qquad (a_n \neq 0,\ b_m \neq 0),$$

where $n < m-1$ and $Q(x)$ has no real zeros, can readily be calculated by contour integration (Vol. 1, Sec. 11.42). Using Theorem 5.26b and the analyticity of $Q(z)$ in some strip $|y| < b$ containing no zeros of $Q(z)$, we can predict that $\mathbf{F}\varphi(x)$ will fall off exponentially as $|\sigma| \to \infty$ (give the details), without bothering to calculate $\mathbf{F}\varphi(x)$ explicitly.

5.32. Example. Find the Fourier transform $\psi(\sigma)$ of the function $\varphi(x) = e^{-ax^2}$ $(a > 0)$.

Solution. The function $\varphi(x)$ can be continued analytically into the whole z-plane $(z = x + iy)$, where it satisfies the relation

$$|\varphi(z)| = |e^{-az^2}| = |e^{-a(x+iy)^2}| = e^{ay^2} e^{-ax^2}.$$

Hence, by Lemma 5.26a, in calculating the Fourier transform we can replace the x-axis by any parallel line, obtaining

$$\begin{aligned} \psi(\sigma) &= \int_{-\infty}^{\infty} e^{-a(x+iy)^2} e^{-i\sigma(x+iy)}\, dx \\ &= \int_{-\infty}^{\infty} e^{-ax^2 + ay^2 + \sigma y - 2aixy - i\sigma x}\, dx \\ &= \int_{-\infty}^{\infty} e^{ay^2 + \sigma y} \int_{-\infty}^{\infty} e^{-ax^2 - ix(2ay + \sigma)}\, dx. \end{aligned}$$

Choosing $y = -\sigma/2a$, we find that

$$\psi(\sigma) = e^{-\sigma^2/4a} \int_{-\infty}^{\infty} e^{-ax^2}\, dx.$$

But

$$\int_{-\infty}^{\infty} e^{-ax^2}\, dx = \sqrt{\frac{\pi}{a}}$$

(Vol. 1, Sec. 11.66), and hence

$$\psi(\sigma)=e^{-\sigma^2/4a}\sqrt{\frac{\pi}{a}}.$$

In particular, if

$$\varphi(x)=e^{-x^2/2}$$

$(a=\frac{1}{2})$, then

$$\psi(\sigma)=\sqrt{2\pi}\,e^{-\sigma^2/2}.$$

5.33. Fourier transforms and convolutions. By the *convolution* of two functions $f(x)$ and $g(x)$ defined for $-\infty<x<\infty$, we mean the function

$$h(x)=f(x)*g(x)=\int_{-\infty}^{\infty} f(\xi)g(x-\xi)\,d\xi.$$

If $f(x)$ and $g(x)$ are bounded, continuous, and absolutely integrable on $(-\infty,\infty)$, then $h(x)$ exists for every x and is also bounded, continuous, and absolutely integrable on $(-\infty,\infty)$, and satisfies the relation

$$\int_{-\infty}^{\infty} h(x)\,dx=\int_{-\infty}^{\infty} f(x)\,dx\int_{-\infty}^{\infty} g(x)\,dx$$

(Vol. 1, Sec. 11.58a). The convolution of the functions $f(x)e^{-i\sigma x}$ and $g(x)e^{-i\sigma x}$ is just

$$\int_{-\infty}^{\infty} f(\xi)e^{-i\sigma\xi}g(x-\xi)e^{-i\sigma(x-\xi)}\,d\xi=e^{-i\sigma x}\int_{-\infty}^{\infty} f(\xi)g(x-\xi)\,d\xi, \tag{1}$$

and hence, applying (1) to these functions instead of to $f(x)$ and $g(x)$, we get

$$\int_{-\infty}^{\infty} e^{-i\sigma x}\left\{\int_{-\infty}^{\infty} f(\xi)g(x-\xi)\,d\xi\right\}dx=\int_{-\infty}^{\infty} f(x)e^{-i\sigma x}\,dx\int_{-\infty}^{\infty} g(x)e^{-i\sigma x}\,dx,$$

or, in terms of Fourier transforms,

$$\mathbf{F}(f*g)=\mathbf{F}f\cdot\mathbf{F}g. \tag{2}$$

In other words, *the Fourier transform of the convolution of two functions $f(x)$ and $g(x)$ equals the product of the Fourier transforms of the separate functions* (provided $f(x)$ and $g(x)$ satisfy the indicated conditions).

5.34. Solution of the heat flow equation. As an application of Fourier transforms, consider the problem of finding the solution of the *heat flow* *equation*†

$$\frac{\partial u(x,t)}{\partial t}=\frac{\partial^2 u(x,t)}{\partial x^2} \qquad (-\infty<x<\infty,\ t\geqslant 0), \tag{3}$$

reducing to a given function $u_0(x)$ for $t=0$. Physically, this means finding the temperature distribution of a one-dimensional homogeneous continuum (an infinite rod) at any given time $t>0$, given its initial temperature distribution at time $t=0$. The following conditions will be imposed on the function $u(x,t)$ and its derivatives:

(a) For every fixed $t>0$, the function $u(x,t)$ and its first and second partial derivatives $u_x(x,t)$ and $u_{xx}(x,t)$ with respect to x are continuous and absolutely integrable with respect to x on the real line $-\infty<x<\infty$;
(b) On every interval $0\leqslant t\leqslant T$, the function $u_t(x,t)$ (the partial derivative of $u(x,t)$ with respect to t) has an "integrable majorant" (Vol. 1, Sec. 11.57a):

$$|u_t(x,t)|\leqslant\Phi(x), \qquad \int_{-\infty}^{\infty}\Phi(x)\,dx<\infty.$$

To solve equation (3), we go over to Fourier transforms, i.e., we multiply both sides by $e^{-i\sigma x}$ and then integrate with respect to x from 0 to ∞. It follows from Condition b and the theorem on differentiation of a parameter-dependent improper integral (cf. Vol. 1, Secs. 11.55a, 11.57a) that

$$\int_{-\infty}^{\infty}u_t(x,t)e^{-i\sigma x}\,dx=\frac{\partial}{\partial t}\int_{-\infty}^{\infty}u(x,t)e^{-i\sigma x}\,dx=v_t(\sigma,t),$$

where

$$v(x,t)=\mathbf{F}u(x,t)=\int_{-\infty}^{\infty}u(x,t)e^{-i\sigma x}\,dx$$

is the Fourier transform of the desired function $u(x,t)$. Moreover, it follows from Condition a and formula (5′), p. 286, that

$$\mathbf{F}u_{xx}(x,t)=-\sigma^2\mathbf{F}u(x,t)=-\sigma^2 v(x,t).$$

As a result, (3) leads to the ordinary differential equation

$$v_t(\sigma,t)=-\sigma^2 v(\sigma,t).$$

Our problem is to find a solution of this equation which reduces to

$$v_0(\sigma)=\mathbf{F}u_0(x)=\int_{-\infty}^{\infty}u_0(x)e^{-i\sigma x}\,dx$$

† For the derivation of equation (3), see e.g., G. P. Tolstov, *Fourier Series* (translated by R. A. Silverman), Prentice-Hall, Inc., Englewood Cliffs, N. J. (1962), p. 296.

at $t=0$. Such a solution must obviously be of the form

$$v(\sigma,t)=e^{-\sigma^2 t}v_0(\sigma).$$

Choosing $a=1/4t$ in Example 5.32, we find that

$$e^{-\sigma^2 t}=\mathbf{F}\left[\frac{1}{2\sqrt{\pi t}}e^{-x^2/4t}\right],$$

and then, by formula (2) for the Fourier transform of a convolution,

$$v(\sigma,t)=\mathbf{F}\left[\frac{1}{2\sqrt{\pi t}}e^{-x^2/4t}\right]\mathbf{F}u_0=\mathbf{F}\left[\frac{1}{2\sqrt{\pi t}}e^{-x^2/4t}*u_0(x)\right].$$

But $v(\sigma,t)=\mathbf{F}u(x,t)$, and hence, finally,

$$u(x,t)=\frac{1}{2\sqrt{\pi t}}e^{-x^2/4t}*u_0(x)=\frac{1}{2\sqrt{\pi t}}\int_{-\infty}^{\infty}e^{-\xi^2/4t}u_0(x-\xi)\,d\xi.$$

The uniqueness of this solution of the heat flow equation in a large class of functions is proved in the theory of partial differential equations.†

5.4. The Laplace Transform

5.41. Let $\varphi(x)$ be a piecewise continuous function on the real line, such that the product $\varphi(x)e^{-\gamma x}$ (γ real) is absolutely integrable. Then the Fourier transform $\psi(s)=\mathbf{F}\varphi(x)$, which may not exist for real s, exists for certain complex values of $s=\sigma+i\tau$. In particular,

$$\psi(s)=\int_{-\infty}^{\infty}\varphi(x)e^{-isx}\,dx=\int_{-\infty}^{\infty}\varphi(x)e^{\tau x}e^{-i\sigma x}\,dx$$

exists on the line $\tau=-\gamma$, since, on this line, $\psi(s)$ is just the Fourier transform of the absolutely integrable function $\varphi(x)e^{-\gamma x}$.

The most important case of this kind occurs when

$$\begin{aligned} |\varphi(x)| &\leqslant Ce^{\alpha x} \quad \text{if} \quad x\geqslant 0, \\ \varphi(x) &=0 \quad \text{if} \quad x<0. \end{aligned} \tag{1}$$

Here the Fourier transform

$$\psi(s)=\int_{0}^{\infty}\varphi(x)e^{-isx}\,dx=\int_{0}^{\infty}\varphi(x)e^{\tau x}e^{-i\sigma x}\,dx \tag{2}$$

exists for all $\tau<-\alpha$, i.e., in the whole half-plane of the complex variable

† See I. G. Petrovski, *Lectures on Partial Differential Equations* (translated by A. Shenitzer), John Wiley-Interscience, New York (1954), Chapter 4.

$s=\sigma+i\tau$ bounded from above by the line $\tau=-\alpha$. Suppose we make the change of variable $is=p$ in (2), noting that as s varies over the half-plane $\operatorname{Im} s<-\alpha$, p varies over the half-plane $\operatorname{Re} p>\alpha$. Then it is easy to see (give the details) that the function

$$\Phi(p)\equiv\psi(s)=\int_0^\infty \varphi(x)e^{-px}\,dx$$

is defined and analytic on the half-plane $\operatorname{Re} p>\alpha$, that it converges to 0 as $\operatorname{Im} p\to\infty$ on every vertical line in this half-plane, and that this convergence is uniform on every finite closed interval of values of $\operatorname{Re} p$. Moreover, in the half-plane $\operatorname{Re} p>\alpha$ the function $\Phi(p)$ satisfies the estimate

$$|\Phi(p)|\leqslant\int_0^\infty |\varphi(x)|e^{-\xi x}\,dx\leqslant C\int_0^\infty e^{(\alpha-\xi)x}\,dx=\frac{C}{\xi-\alpha},$$

which shows that $\Phi(p)$ is bounded and approaches 0 as $\xi\to\infty$ in every closed half-plane $\operatorname{Re} p\geqslant\beta>\alpha$.

The function $\Phi(p)$ is called the *Laplace transform* of the function $\varphi(x)$. The Laplace transform, as we see, differs from the Fourier transform (considered in the complex plane) only by a rotation through 90° in the plane of the complex variable s.

5.42.a. The following simple theorem gives sufficient (but hardly necessary) conditions for a function $\Phi(p)$ to be the Laplace transform of a function $\varphi(x)$ satisfying the conditions (1):

THEOREM. *Suppose the function $\Phi(p)$ $(p=\xi+i\eta)$ satisfies the following conditions:*

(a) *$\Phi(p)$ is analytic on the half-plane* $\operatorname{Re} p>\alpha\geqslant 0$;
(b) *There exist a constant C and a nonnegative function $B(\eta)$ integrable on the line $-\infty<\eta<\infty$ such that*

$$\left|\Phi(p)-\frac{C}{p}\right|\leqslant B(\eta)$$

for all $\xi>\alpha$.

Then $\Phi(p)$ is the Laplace transform of a piecewise continuous function $\varphi(x)$ which vanishes for $x<0$, and is continuous and satisfies the inequality

$$|\varphi(x)|\leqslant Ce^{\alpha x} \tag{3}$$

for $x\geqslant 0$.

Proof. The function C/p is obviously the Laplace transform of the function $\varphi_0(x)$ equal to 0 for $x<0$ and to C for $x\geqslant 0$. The function $\varphi_0(x)$ satisfies the conditions and the conclusion of the theorem. Therefore, by subtracting out $\varphi_0(x)$, we can assume without loss of generality that the function $\Phi(p)$ itself satisfies the inequality

$$|\Phi(p)| \leqslant B(\eta) \tag{4}$$

for all $\xi > \alpha$. We then define the function $\varphi(x)$ by the formula

$$\varphi(x) = \frac{1}{2\pi i} \int_{\gamma - i\infty}^{\gamma + i\infty} \Phi(p) e^{px}\, dp \qquad (\gamma > \alpha). \tag{5}$$

It follows from the analyticity of $\Phi(p)$, the estimate (4), and an argument like that made in proving Lemma 5.26a that the integral (5) is independent of γ. On the other hand, we have the estimate

$$|\varphi(x)| \leqslant \frac{1}{2\pi} \int_{-\infty}^{\infty} |\Phi(\xi + i\eta)| e^{\gamma x}\, d\eta \leqslant \frac{B}{2\pi} e^{\gamma x}.$$

If $x \geqslant 0$, we let γ approach α, obtaining the inequality (3), while if $x < 0$, we let γ approach $+\infty$, obtaining $\varphi(x) \equiv 0$. Moreover, writing (5) in the form

$$\varphi(x) = \frac{1}{2\pi i} \int_{-\infty}^{\infty} \Phi(\xi + i\eta) e^{(\xi + i\eta)x} i\, d\eta = \frac{e^{\xi x}}{2\pi} \int_{-\infty}^{\infty} \Phi(\xi + i\eta) e^{i\eta x}\, d\eta,$$

we find that $2\pi\varphi(-x)e^{\xi x}$ is the Fourier transform (in the variable η) of the absolutely integrable function $\Phi(\xi + i\eta)$ (ξ fixed). Taking the inverse Fourier transform, we have

$$\Phi(\xi + i\eta) = \frac{1}{2\pi} \int_{-\infty}^{\infty} 2\pi\varphi(-x) e^{(\xi + i\eta)x}\, dx = \int_0^{\infty} \varphi(x) e^{-px}\, dx,$$

so that $\Phi(\xi + i\eta)$ is actually the Laplace transform of $\varphi(x)$. ∎

Formula (5) plays an important role in the theory of the Laplace transform, and is known as the *formula for inversion of the Laplace transform.*

b. It is interesting to consider what properties the function $\varphi(x)$ must have to make its Laplace transform satisfy the conditions of the above theorem. Suppose $\varphi(x)$ has continuous derivatives up to order $m-1$ and a piecewise continuous mth derivative satisfying the conditions (1). Then, integrating by parts m times in the formula

$$\Phi(p) = \int_0^{\infty} \varphi(x) e^{-px}\, dx,$$

we get

$$|\Phi(p)| = \left| \frac{1}{p^m} \int_0^{\infty} \varphi^{(m)}(x) e^{-px}\, dx \right| \leqslant \frac{C}{|p|^m} = \frac{C}{|\xi^2 + \eta^2|^{m/2}} \leqslant \frac{C}{|\gamma^2 + \eta^2|^{m/2}} \tag{6}$$

for $\xi = \text{Re } p \geqslant \gamma > \alpha$. Clearly the conditions of the theorem are satisfied if $m=2$, i.e., if $\varphi(x)$ has a continuous derivative and a piecewise continuous second derivative.

5.43. The Laplace transform is often used to solve differential equations, both ordinary and partial, arising in "nonstationary problems." In such problems, the unknown function $f(t)$ vanishes for $t<0$, while for $t \geqslant 0$, the function $f(t)$ must be the solution of some differential equation, satisfying some initial condition for $t=0$.

First we consider the ordinary nonhomogeneous linear differential equation

$$a_0 y^{(n)}(t) + a_1 y^{(n-1)}(t) + \cdots + a_n y(t) = b(t) \tag{7}$$

of order n with constant coefficients and initial conditions

$$y(0) = y_0, \quad y'(0) = y_1, \quad \ldots, \quad y^{(n-1)}(0) = y_{n-1}, \tag{8}$$

where $b(t)$ is such that

$$\begin{aligned} |b(t)| &\leqslant Ce^{\alpha t} \quad \text{if} \quad t \geqslant 0, \\ b(t) &= 0 \quad\quad\;\; \text{if} \quad t<0. \end{aligned} \tag{9}$$

Suppose we multiply equation (7) by e^{-pt} and integrate with respect to t from 0 to ∞, letting

$$Y(p) = \int_0^\infty y(t) e^{-pt}\, dt \tag{10}$$

denote the Laplace transform of $y(t)$. Integrating (10) by parts repeatedly with the help of (8), we get

$$\int_0^\infty y'(t) e^{-pt}\, dt = y(t) e^{-pt}\Big|_0^\infty + p \int_0^\infty y(t) e^{-pt}\, dt = -y_0 + pY(p),$$

$$\begin{aligned} \int_0^\infty y''(t) e^{-pt}\, dt = y'(t) e^{-pt}\Big|_0^\infty + p \int_0^\infty y'(t) e^{-pt}\, dt &= -y_1 + p(-y_0 + pY(p)) \\ &= -y_1 - py_0 + p^2 Y(p), \end{aligned}$$

$$\ldots$$

$$\begin{aligned} \int_0^\infty y^{(n)}(t) e^{-pt}\, dt &= y^{(n-1)}(t) e^{-pt}\Big|_0^\infty + p \int_0^\infty y^{(n-1)}(t) e^{-pt}\, dt \\ &= -y_{n-1} + p(-y_{n-2} - py_{n-3} - \cdots - p^{n-2} y_0 + p^{n-1} Y(p)) \\ &= -y_{n-1} - py_{n-2} - \cdots - p^{n-1} y_0 + p^n Y(p). \end{aligned}$$

Then, multiplying each of these equations by the corresponding coefficient a_k and adding the results, we get an equation of the form

$$R_0(p) + R(p) Y(p) = B(p),$$

where $R_0(p)$ is a polynomial in p of degree no greater than $n-1$, $R(p)$ is a polynomial in p of degree n, and $B(p)$ is the Laplace transform of the function $b(t)$. Thus the unknown function $Y(p)$ satisfies a purely algebraic equation. Solving this equation, we find that

$$Y(p)=\frac{B(p)-R_0(p)}{R(p)}.$$

The function $R_0(p)/R(p)$ satisfies Condition b of Theorem 5.42a. In fact, setting

$$C=\lim_{p\to\infty}\frac{pR_0(p)}{R(p)},$$

we find that

$$\frac{R_0(p)}{R(p)}=\frac{C}{p}-\frac{pR_0(p)-CR(p)}{pR(p)}$$

is a rational function of p, such that the degree of the denominator, equal to $n+1$, exceeds the degree of the numerator by at least 2, since the terms of degree n in the numerator cancel out, by the very definition of C. As for the function $B(p)/R(p)$, whether or not it satisfies the conditions of the theorem depends, naturally, on the nature of $B(p)$. If the degree of the polynomial $R(p)$ exceeds 1, then the conditions of the theorem are satisfied for any function $b(t)$ satisfying (9), because of the boundedness of $B(p)$; if the degree of $R(p)$ equals 1, then, as shown by the inequality (6), it is sufficient that $b(t)$ be continuous and have a piecewise continuous derivative satisfying the conditions (9), along with $b(t)$ itself.

In any event, if $B(p)/R(p)$ also satisfies the conditions of Theorem 5.42a, we can use formula (5) to write the desired solution $y(t)$ in the form

$$y(t)=\frac{1}{2\pi i}\int_{\gamma-i\infty}^{\gamma+i\infty}\frac{B(p)-R_0(p)}{R(p)}e^{pt}\,dp. \tag{11}$$

If the function $B(p)$ can be continued analytically into the whole p-plane (with only isolated singular points), then, as a rule, the evaluation of (11) is accomplished with the help of contour integration and the residue theorem, just as is done in calculating Fourier integrals of rational functions (Vol. 1, Sec. 11.42). Note that the function e^{pt} is bounded in the left half-plane $\operatorname{Re} p<\gamma$ for $t>0$, but not in the right half-plane, so that the semicircle making up part of the contour of integration must be constructed on the left side of the line $\operatorname{Re} p=\gamma$, rather than on the right side. For γ we can choose any real number satisfying the condition that all the singular points of the integrand lie to the left of the line $\operatorname{Re} p=\gamma$.

5.44. Example. Consider the second-order differential equation

$$a_0 y'' + a_1 y' + a_2 y = b \sin kt \tag{12}$$

with initial conditions

$$y_0 = 0, \qquad y_1 = 1,$$

and suppose the equation†

$$a_0 \lambda^2 + a_1 \lambda + a_2 = 0 \tag{13}$$

has a pair of conjugate complex roots

$$\lambda = \alpha + i\beta, \qquad \lambda = \alpha - i\beta \qquad (\alpha < 0,\ \beta \neq 0)$$

(the sign of α is important). In electrical engineering an equation like (12) describes the forced oscillations of a circuit with resistance, inductance, and capacitance, under the action of an applied emf of (angular) frequency k. Going over to Laplace transforms, we get the equation

$$(a_0 p^2 + a_1 p + a_2) Y(p) = \int_0^\infty b e^{-pt} \sin kt \, dt = \frac{bk}{k^2 + p^2}. \tag{14}$$

Solving (14), we find that

$$Y(p) = \frac{bk}{(a_0 p^2 + a_1 p + a_2)(k^2 + p^2)},$$

and hence, by the inversion formula,

$$y(t) = \frac{bk}{2\pi i} \int_{\gamma - i\infty}^{\gamma + i\infty} f(p) \, dp, \tag{15}$$

where

$$f(p) = \frac{e^{pt}}{(a_0 p^2 + a_1 p + a_2)(k^2 + p^2)}.$$

The denominator of $f(p)$ has four simple zeros, at the points $\pm ik$ and $\alpha \pm i\beta$. To evaluate the integral (15), let Γ_R be the contour made up of the segment $\operatorname{Re} p = \gamma$, $-R \leqslant \operatorname{Im} p \leqslant R$, and a circular arc C_R of radius R in the left half-plane centered at the origin,‡ where R is large enough so that Γ_R contains all four points $\pm ik$, $\alpha \pm i\beta$ (Figure 28). Then, by the residue theorem (Vol. 1, Sec. 10.43),

† Note that (13) is the characteristic equation (Sec. 2.17b) of the homogeneous differential equation corresponding to (12).

‡ Note that C_R is somewhat larger than an exact semicircle.

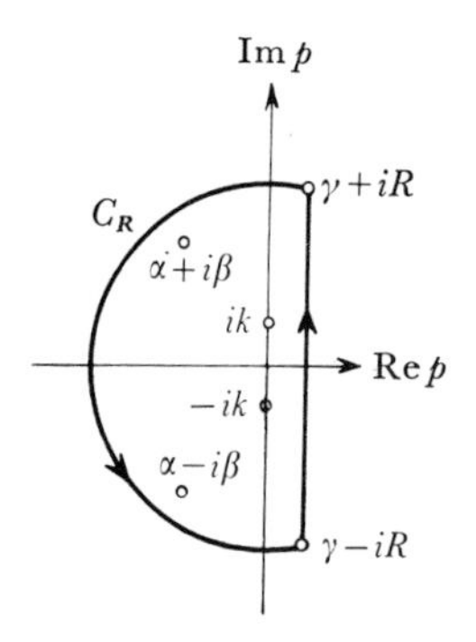

Figure 28

$$\int_{\Gamma_R} f(p)\,dp = \int_{C_R} f(p)\,dp + \int_{-\gamma-iR}^{\gamma+iR} f(p)\,dp$$
$$= 2\pi i\left[\operatorname*{Res}_{p=ik} f(p) + \operatorname*{Res}_{p=-ik} f(p) + \operatorname*{Res}_{p=\alpha+i\beta} f(p) + \operatorname*{Res}_{p=\alpha-i\beta} f(p)\right].$$

Taking the limit as $R\to\infty$ and noting that

$$\lim_{R\to\infty}\int_{C_R} f(p)\,dp = 0,$$

by a slight modification of Jordan's lemma (see Problem 8), we find that

$$y(t) = \frac{bk}{2\pi i}\int_{\gamma-i\infty}^{\gamma+i\infty} f(p)\,dp$$
$$= bk\left[\operatorname*{Res}_{p=ik} f(p) + \operatorname*{Res}_{p=-ik} f(p) + \operatorname*{Res}_{p=\alpha+i\beta} f(p) + \operatorname*{Res}_{p=\alpha-i\beta} f(p)\right].$$

The residue at each of these points is calculated by the general formula (Vol. 1, Sec. 10.42a) for the residue at a simple pole:

$$\operatorname*{Res}_{p=p_0}\frac{A(p)}{B(p)} = \frac{A(p_0)}{B'(p_0)}.$$

As a result, we get

$$y(t) = bk\left[\frac{e^{(\alpha+i\beta)t}}{(\lambda^2+k^2)2i\beta a_0} - \frac{e^{(\alpha-i\beta)t}}{(\bar{\lambda}^2+k^2)2i\beta a_0}\right.$$
$$\left. + \frac{e^{ikt}}{(-a_0k^2+a_1ik+a_2)2ik} - \frac{e^{-ikt}}{(-a_0k^2-a_1ik+a_2)2ik}\right].$$

The resulting waveform is the superposition of a periodic oscillation with the frequency k of the applied emf and a damped oscillation with the "characteristic frequency" β of the system; the rate of damping is determined by the number α, i.e., by the abscissa of the roots of equation (13).

The phenomenon of resonance occurs if $\alpha=0$, $\beta=k$. In this case, the original differential equation takes the form

$$y''+k^2y=b\sin kt, \tag{12'}$$

with solution

$$y(t)=\frac{bk}{2\pi i}\int_{\gamma-i\infty}^{\gamma+i\infty}\frac{e^{pt}}{(p^2+k^2)^2}\,dp.$$

The points $p=\pm ik$ are now poles of order 2 of the integrand. The residue at each of these points is calculated by the general formula (Vol. 1, Sec. 10.42b) for the residue at a pole of order 2:

$$\operatorname*{Res}_{p=p_0}\frac{A(p)}{B(p)}=\left[\frac{A(p)(p-p_0)^2}{B(p_0)}\right]'_{p=p_0}.$$

As a result, we get the solution

$$\begin{aligned}y(t)&=bk\left[e^{ikt}\left(-\frac{t}{4k^2}+\frac{1}{4ik^3}\right)+e^{-ikt}\left(-\frac{t}{4k^2}-\frac{1}{4ik^3}\right)\right]\\&=-\frac{bt}{2k}\cos kt+\frac{b}{2k^2}\sin kt,\end{aligned}$$

corresponding to an oscillation whose amplitude increases without limit.

5.45.a. The same method can be used to solve partial differential equations. Just as taking Laplace transforms with respect to t carries an ordinary differential equation into an algebraic equation in the unknown function (the Laplace transform of the solution), so taking Laplace transforms with respect to t carries an equation containing derivatives not only with respect to t, but also with respect to other variables $x, y, \ldots$ into an equation in which the derivatives with respect to t vanish, while the derivatives with respect to $x, y, \ldots$ remain.

***b. Example.** Consider the heat flow equation

$$\frac{\partial u}{\partial t}=\frac{\partial^2 u}{\partial x^2}$$

(Sec. 5.34) in a finite interval $0\leqslant x\leqslant l$ with boundary and initial conditions

$$u_x(0,t)=0, \qquad u(l,t)=u_1, \qquad u(x,0)=u_0.$$

Physically, these conditions mean that heat does not leave the end $x=0$, while a constant temperature u_1 is maintained at the end $x=l$ by the addition of heat from the outside; moreover, the temperature has the constant value u_0 at the initial time $t=0$.

To solve the problem, we take Laplace transforms with respect to t, i.e., we go from the function $u(x,t)$ to the function

$$v(x,p)=\int_0^{\infty} e^{-pt}u(x,t)\,dt.$$

This gives the ordinary differential equation

$$\frac{d^2v(x,p)}{dx^2}-pv(x,p)=-u_0 \tag{16}$$

of order 2, with boundary conditions

$$v_x(0,p)=0, \qquad v(l,p)=\frac{u_1}{p}.$$

Solving (16), we get

$$v(x,p)=\frac{u_0}{p}+\frac{u_1-u_0}{p}\frac{\cosh x\sqrt{p}}{\cosh l\sqrt{p}},$$

and hence, by the inversion formula,

$$u(x,t)=u_0+\frac{u_1-u_0}{2\pi i}\int_{\gamma-i\infty}^{\gamma+i\infty} f(p)\,dp, \tag{17}$$

where

$$f(p)=\frac{e^{pt}}{p}\frac{\cosh x\sqrt{p}}{\cosh l\sqrt{p}}$$

is a single-valued function of p with poles $p_0=0$ and

$$p_n=-\frac{\pi^2}{l^2}\left(n-\frac{1}{2}\right)^2 \qquad (n=1,2,\ldots).$$

We now show that the integral in (17) equals $2\pi i$ times the (infinite) sum of the residues of $f(p)$ at all the poles $p_0,p_1,p_2,\ldots$ To this end, we construct a circular arc C_n of radius $n^2\pi^2/l^2$ in the left half-plane centered at the origin, cutting the real axis between two consecutive poles (Figure 29). As we will see in a moment, the ratio

$$\frac{\cosh x\sqrt{p}}{\cosh l\sqrt{p}} \tag{18}$$

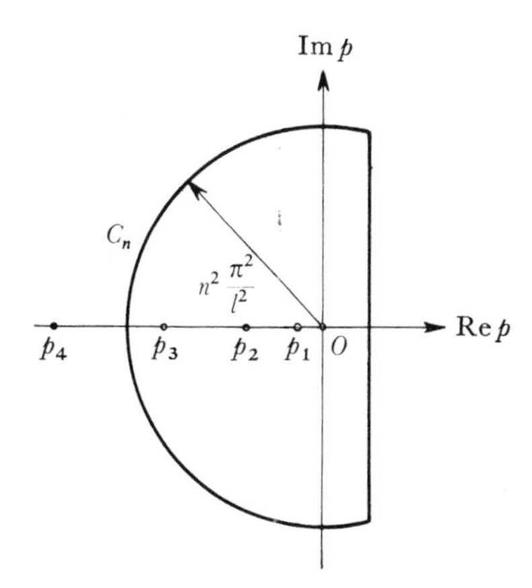

Figure 29

is bounded (in absolute value) on C_n. It then follows from Jordan's lemma (Problem 8) that the integral of $f(p)$ along C_n approaches 0 as $n \to \infty$, and hence, by the usual argument (cf. Example 5.44), that (17) is just

$$u(x,t) = u_0 + (u_1 - u_0) \sum_{n=0}^{\infty} \operatorname*{Res}_{p=p_n} f(p). \tag{19}$$

Instead of considering the ratio (18) on the arc C_n, where $|p| = n^2\pi^2/l^2$, we can replace $\sqrt{p}$ by ζ and p by ζ^2, and then consider the ratio

$$\frac{\cosh x\zeta}{\cosh l\zeta} \tag{18$'$}$$

on the new circular arc L_n of radius $n\pi/l$, with argument varying from $\frac{1}{2}\pi - \varepsilon_R$ to $\frac{3}{4}\pi + \varepsilon_R$, where ε_R is a small quantity depending on R which approaches 0 as $R \to \infty$ (Figure 30).† Writing $\zeta = \xi + i\eta$ ($\eta > 0$), we have

$$\begin{aligned}\left|\frac{\cosh x\zeta}{\cosh l\zeta}\right|^2 &= \left|\frac{\cosh x(\xi + i\eta)}{\cosh l(\xi + i\eta)}\right|^2 \\ &= \left|\frac{\cosh x\xi \cos x\eta + i \sinh x\xi \sin x\eta}{\cosh l\xi \cos l\eta + i \sinh l\xi \sin l\eta}\right|^2 \\ &= \frac{\cosh^2 x\xi \cos^2 x\eta + \sinh^2 x\xi \sin^2 x\eta}{\cosh^2 l\xi \cos^2 l\eta + \sinh^2 l\xi \sin^2 l\eta} \\ &= \frac{\cosh^2 l\xi}{\cosh^2 l\xi \cos^2 l\eta + \sinh^2 l\xi \sin^2 l\eta}.\end{aligned} \tag{20}$$

† Note that L_R is somewhat larger than an exact "quarter-circle."

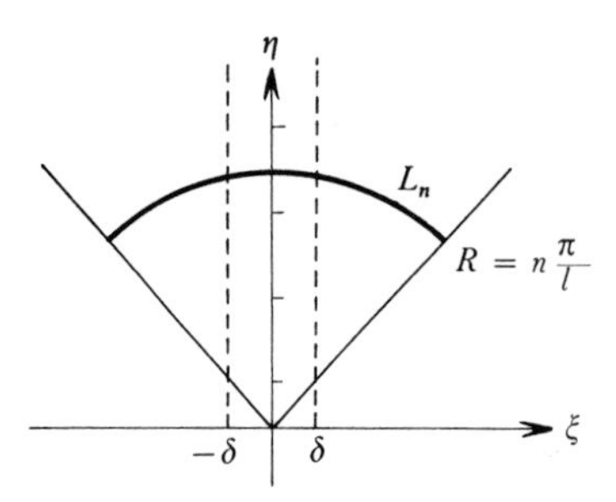

Figure 30

If $|\xi| < \delta$, then on the arc L_n, for sufficiently large n, we have

$$\left|\eta - \frac{n\pi}{l}\right| < \alpha$$

and hence $\cos^2 l\eta > 1 - \beta$, where α and β are arbitrarily small. Therefore

$$\left|\frac{\cosh^2 x\zeta}{\cosh^2 l\zeta}\right| \leqslant \frac{\cosh^2 l\xi}{(1-\beta)\cosh^2 l\xi} = \frac{1}{1-\beta} \tag{21}$$

on the corresponding part of L_n. On the other hand, if $|\xi| \geqslant \delta$, then, replacing $\cosh^2 l\xi$ by $\sinh^2 l\xi$ in the denominator in the right-hand side of (20), we get

$$\left|\frac{\cosh x\zeta}{\cosh l\zeta}\right|^2 \leqslant \frac{\cosh^2 l\xi}{\sinh^2 l\xi} = \coth^2 l\xi \leqslant \coth^2 l\delta. \tag{22}$$

Together (21) and (22) imply that the ratio (18) is bounded on the whole arc L_n, and, as already noted, this implies the validity of formula (19). Clearly,

$$\operatorname*{Res}_{p=p_0} f(p) = 1,$$

while it is easy to see that

$$\operatorname*{Res}_{p=p_n} f(p) = \frac{4(-1)^n}{(2n-1)\pi} \exp\left\{-\frac{\pi^2}{l^2}\left(n-\frac{1}{2}\right)^2 t\right\} \cos\left(n-\frac{1}{2}\right)\frac{\pi x}{l} \qquad (n=1,2,\ldots).$$

Thus, finally, we can express the solution $u(x,t)$ as the sum of the following infinite series:

$$u(x,t) = u_1 + \frac{4}{\pi}(u_1 - u_0) \sum_{n=1}^{\infty} \frac{(-1)^n}{2n-1} \exp\left\{-\frac{\pi^2}{l^2}\left(n-\frac{1}{2}\right)^2 t\right\} \cos\left(n-\frac{1}{2}\right)\frac{\pi x}{l}.$$

*5.5. Quasi-Analytic Classes of Functions

5.51. The method of the Laplace transform can also be used successfully to solve theoretical problems. As an example of such an application, we now present the basic theorem of the theory of quasi-analytic functions.†

Let $f(x)$ be a function of a real variable x. Then $f(x)$ is said to be *analytic at a point* x_0 if $f(x)$ has a Taylor series expansion

$$f(x) = \sum_{n=0}^{\infty} \frac{f^{(n)}(x_0)}{n!}(x-x_0)^n$$

convergent in a neighborhood of x_0. Even if $f(x)$ is infinitely differentiable in a neighborhood of x_0, $f(x)$ need not be analytic at x_0 (see Vol. 1, Chapter 8, Problem 1). But if the successive derivatives of $f(x)$ do not grow "too rapidly," in the sense that

$$\max_{|x-x_0|<\delta} |f^{(n)}(x)| \leqslant An!B^n \qquad (n=0,1,2,\ldots) \tag{1}$$

for suitable constants $A>0$ and $B>0$, then $f(x)$ is analytic at x_0 (Vol. 1, Sec. 8.52). Conversely, it follows from the formula

$$f^{(n)}(x) = \frac{n!}{2\pi i}\oint_{|\zeta-x|=r} \frac{f(\zeta)}{(\zeta-x)^{n+1}}\,d\zeta \qquad (n=0,1,2,\ldots)$$

for the derivatives of an analytic function (Vol. 1, Sec. 10.34) that if $f(x)$ is analytic at x_0, then the conditions (1) hold (show this).

Now let $m_0, m_1, \ldots, m_n, \ldots$ be an arbitrary sequence of positive numbers, and let $C_{\langle m_n \rangle}$ be the class of functions $f(x)$ defined on $-\infty < x < \infty$ satisfying the inequalities

$$|f^{(n)}(x)| \leqslant Am_nB^n \quad (n=0,1,2,\ldots)$$

for all x, where the constants $A>0$ and $B>0$ can depend on the choice of the function $f(x)$. If the numbers m_n grow faster than $n!$, the class $C_{\langle m_n \rangle}$ may contain nonanalytic functions as well as analytic functions. However, as shown by Denjoy in 1921, there are classes $C_{\langle m_n \rangle}$ containing both analytic and nonanalytic functions, but which nevertheless have the *uniqueness property*, i.e., which are such that *if* $f(x) \in C_{\langle m_n \rangle}$, $g(x) \in C_{\langle m_n \rangle}$, *and if*

$$f^{(n)}(x_0) = g^{(n)}(x_0) \qquad (n=0,1,2,\ldots)$$

at some point x_0, *then* $f(x)$ *and* $g(x)$ *coincide for all values of* x. The fact that ana-

† Following S. Mandelbrojt, *Séries de Fourier et Classes Quasi-Analytiques de Fonctions*, Gauthier-Villars, Paris (1935).

lytic functions have the uniqueness property is well known (cf. Vol. 1, Sec. 10.39e).

5.52. A class $C_{\langle m_n \rangle}$ is said to be *quasi-analytic* if it has the uniqueness property. A complete description of quasi-analytic classes was given by Carleman in 1926, and a somewhat simpler formulation of Carleman's result was given by Ostrowski in 1930. To understand the formulation of the Carleman-Ostrowski theorem (the central result of this section), we now carry out some preliminary constructions.

It will be assumed that as $n \to \infty$ the sequence m_n grows more rapidly than any function of the form r^n, where $r > 0$ (we will see later that if this is not the case, then the problem can be solved very simply). Because of this assumption, the sequence r^n/m_n approaches 0 as $n \to \infty$ for every $r > 0$, and hence is bounded from above. Hence the function

$$T(r) = \sup_{n \geqslant 0} \frac{r^n}{m_n}$$

exists and is finite for every $r > 0$.

The function $T(r)$, which plays a key role in our subsequent considerations, has a useful geometric interpretation: Consider the sequence of points with coordinates

$$x_n = n, \qquad y_n = -\ln m_n,$$

known as *Valiron points*, lying in the right-hand xy-plane. Since $r^n/m_n \to 0$ as $n \to \infty$, we have $n \ln r - \ln m_n \to -\infty$ as $n \to \infty$, and therefore, given any $r > 0$, $b > 0$, only a finite number of Valiron points satisfy the inequality

$$n \ln r - \ln m_n \geqslant b. \tag{2}$$

The equation $x \ln r + y = b$, or equivalently

$$y = -x \ln r + b,$$

corresponds to a half-line in the right-hand xy-plane of slope $-\ln r$, intersecting the y-axis in the point with ordinate b. According to the inequality (2), only a finite number of Valiron points can lie above any such half-line. Therefore, given any $r > 0$, we can find a $b = b(r)$ such that there are no Valiron points at all above the half-line

$$y = -x \ln r + b(r),$$

while at the same time there is at least one Valiron point on the half-line itself (Figure 31). Such a half-line is known as a *Valiron half-line*. By construc-

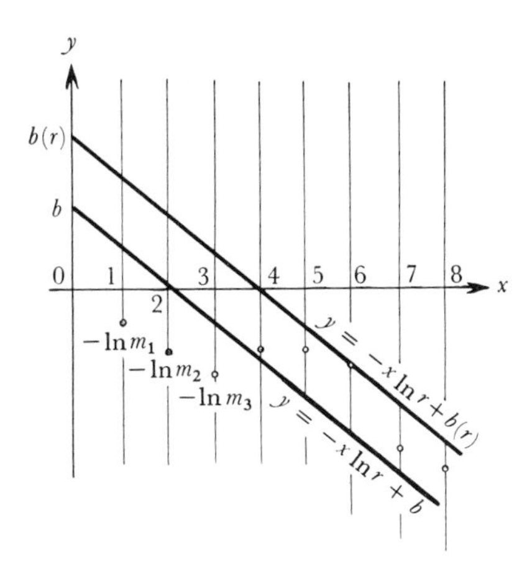

Figure 31

tion, we have

$$-n \ln r + b(r) \geqslant -\ln m_n \tag{3}$$

for $b = b(r)$, and hence

$$b(r) \geqslant \sup_n \{n \ln r - \ln m_n\} = \sup_n \ln \frac{r^n}{m_n}.$$

Actually

$$b(r) = \sup_n \frac{r^n}{m_n},$$

since the inequality (3) becomes an equality for at least one value of n. But then

$$b(r) = \ln\ T(r).$$

Some of the properties of the function $\ln\ T(r)$ can be inferred from its geometric meaning. In the first place, it follows from the definition of $b(r)$ that $b(r)$ is a nondecreasing function of r. In fact, if $b(r_2) < b(r_1)$ for any $r_2 > r_1$, then the whole half-line $y = -x \ln r_2 + b(r_2)$ would lie below the half-line $y = -x \ln r_1 + b(r_1)$, which is incompatible with the fact that there are Valiron points on the latter half-line. Moreover, we can always construct a Valiron half-line with any preassigned value of b ($> \inf b(r)$), by considering the family of all half-lines with y-intercept b and using an argument like that given above. This means that the nondecreasing function $b(r) = \ln\ T(r)$ takes *all* values ($> \inf b(r)$), and hence is continuous (Vol. 1, Sec. 5.34). In addition, it can be shown that $b(r)$ is a piecewise linear function of $\ln r$, but this fact will not be needed here.

5.53. Before proceeding to the statement and proof of the Carleman-Ostrowski theorem itself, we digress to establish an auxiliary proposition from the theory of analytic functions:

LEMMA. *Suppose $f(z)$ is analytic and satisfies the conditions*

$$|f(z)| \leqslant 1, \qquad f(z_0) \neq 0$$

on the open disk $|z-z_0|<h$. Suppose further that $f(z)$ is continuous on the closed disk $|z-z_0| \leqslant h$ and has one and only one zero on the circle $|z-z_0|=h$, at the point z^*. Then the integral

$$-\int_0^{2\pi} \ln |f(z_0+he^{i\theta})| \, d\theta$$

exists and is finite.

Proof. There is no loss of generality in assuming that $z_0=0$, $h=1$, $z^*=1$. The function $f(z)$ is analytic on every disk $|z| \leqslant r < 1$, in which it can have only a finite number of zeros $z_1,\ldots,z_m$; it can be assumed that r is such that there are no zeros on the circle $|z|=r$. Consider the closed contour C shown in Figure 32, made up of the circle $|z|=r$ traversed in the positive (counter-clockwise) direction, circles C_j $(j=1,2,\ldots,m)$ of small radius ε centered at the points z_j and traversed in the negative direction, and arcs $L_j (j=1,2,\ldots,m)$ traversed twice in opposite directions, where the arc L_j, with initial point

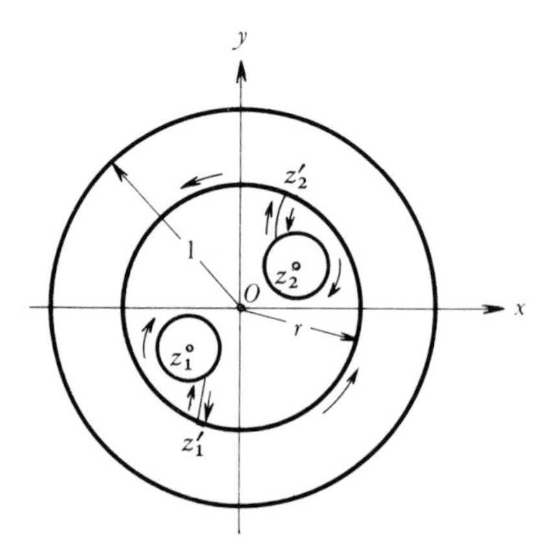

Figure 32

z_j' and final point z_j'', joins the circles C and C_j. The function $\ln f(z)$ is analytic inside C, and hence, by Cauchy's formula (Vol. 1, Sec. 10. 33c),

$$\ln f(0) = \frac{1}{2\pi i} \oint_C \ln f(z) \frac{dz}{z}. \tag{4}$$

Now consider the part of the contour C consisting of the circle C_j of radius ε centered at the point z_j and traversed in the negative direction. The contribution to the integral (4) due to C_j is of the form

$$\frac{1}{2\pi i} \int_{2\pi}^{0} \ln f(z) \frac{i\varepsilon e^{i\theta}}{z_j + \varepsilon e^{i\theta}} d\theta. \tag{5}$$

If k_j is the order of the zero z_j, then

$$f(z) = (z - z_j)^{k_j} f_j(z),$$

where $f_j(z_j) \neq 0$, and

$$\begin{aligned} |\ln f(z)| &= |\ln (z - z_j)^{k_j} f_j(z)| = |k_j \ln (z - z_j) + \ln f_j(z)| \\ &\leqslant k_j |\ln (z - z_j)| + |\ln f_j(z)| \\ &= k_j (|\ln |z - z_j|| + 2\pi) + |\ln f_j(z)| \leqslant k_j |\ln \varepsilon| + M_j, \end{aligned}$$

where M_j is a constant. It follows from this estimate that the integral (5) approaches 0 as $\varepsilon \to 0$. In other words, the contribution to (4) due to the part of the contour C consisting of the circles C_j $(j = 1, 2, \ldots, m)$ vanishes as $\varepsilon \to 0$.

In making a circuit around the point z_j in the negative direction, the function

$$\ln f(z) = \ln |f(z)| + i \arg f(z)$$

acquires an increment $-2\pi k_j i$, and hence the contribution to the integral (4) due to the arc L_j, traversed twice in opposite directions, equals

$$k_j \int_{L_j} \frac{dz}{z} = k_j (\ln z_j'' - \ln z_j').$$

By the same token, on every successive part of the circle $|z| = r$, the function $\ln f(z)$ acquires an increment $-2\pi k_j i$, thereby making a contribution of the form

$$k_j \int_{\arg z_j'}^{\arg z_{j+1}'} i \, d\theta \tag{6}$$

to the integral (4). But (6) is obviously purely imaginary. Taking all this into account, and going over to real parts in (4), we find that

$$\ln |f(0)| = \sum_{j=1}^{m} k_j \ln \frac{|z_j|}{r} + \frac{1}{2\pi} \int_0^{2\pi} \ln |f(re^{i\theta})| \, d\theta$$

in the limit as $\varepsilon \to 0$. But $|z_j| < r$, and hence

$$\frac{|z_j|}{r} < 1, \qquad \ln \frac{|z_j|}{r} < 0.$$

Therefore

$$\frac{1}{2\pi} \int_0^{2\pi} \ln |f(re^{i\theta})| \, d\theta \geqslant \ln |f(0)|,$$

or equivalently,

$$-\frac{1}{2\pi} \int_0^{2\pi} \ln |f(re^{i\theta})| \, d\theta \leqslant -\ln |f(0)|.$$

By hypothesis, there is one and only one zero of $f(z)$ on the circle $|z| = 1$, at the point $z^* = 1$. Given any $\delta > 0$, we obviously have

$$-\frac{1}{2\pi} \int_\delta^{2\pi - \delta} \ln |f(re^{i\theta})| \, d\theta \leqslant -\frac{1}{2\pi} \int_0^{2\pi} \ln |f(re^{i\theta})| \, d\theta \leqslant -\ln |f(0)|$$

($|f(z)| \leqslant 1$). Holding δ fixed, we now take the limit as $r \to 1$, obtaining

$$-\frac{1}{2\pi} \int_0^{2\pi - \delta} \ln |f(e^{i\theta})| \, d\theta \leqslant -\ln |f(0)|.$$

This inequality holds for every $\delta > 0$. Taking the limit as $\delta \to 0$, we finally find that the integral

$$-\frac{1}{2\pi} \int_0^{2\pi} \ln |f(e^{i\theta})| \, d\theta$$

exists and is finite. ∎

5.54. We are now equipped to prove the following key

THEOREM (**Carleman-Ostrowski**). *Let*

$$T(r) = \sup_{n \geqslant 0} \frac{r^n}{m_n}.$$

Then the class $C_{\langle m_n \rangle}$ *is quasi-analytic if and only if*

$$\int_1^\infty \frac{\ln T(r)}{r^2} \, dr = \infty. \tag{7}$$

Proof. The proof will be given in several steps (Secs. 5.55–5.57).

5.55. *Step 1.* First we use the Laplace transform to reduce the problem of describing quasi-analytic classes to another problem, involving analytic functions in the half-plane. Suppose the class $C_{\langle m_n\rangle}$ is not quasi-analytic. This means that $C_{\langle m_n\rangle}$ contains functions $f(x)$ and $g(x)$ such that $f^{(n)}(x_0)=g^{(n)}(x_0)$ for all $n=0,1,2,\dots$ while $f(x)\not\equiv g(x)$. There is no loss of generality in assuming that $x_0=0$ and $f(x)\not\equiv g(x)$ for $x\geqslant 0$; these conditions can always be achieved by making a shift and replacing x by $-x$, i.e., by operations that can be performed without leaving the class $C_{\langle m_n\rangle}$. Consider the function

$$\varphi(x)=\begin{cases} f(x)-g(x) & \text{if } x\geqslant 0,\\ 0 & \text{if } x<0,\end{cases}$$

which obviously belongs to the class $C_{\langle m_n\rangle}$. This function vanishes for $x<0$ and is bounded for $x\geqslant 0$; hence it has a Laplace transform

$$\Phi(p)=\int_0^\infty \varphi(x)e^{-px}\,dx, \tag{8}$$

which is analytic in the half-plane $\operatorname{Re} p>0$.

To find other conditions satisfied by $\Phi(p)$, we integrate (8) by parts n times, obtaining

$$p^n\Phi(p)=\int_0^\infty \varphi^{(n)}(x)e^{-px}\,dx,$$

$$|p^n\Phi(p)|\leqslant Am_nB^n\int_0^\infty e^{-px}\,dx=Am_nB^n\frac{1}{|p|}\leqslant A_\alpha m_nB^n,$$

if $|p|>\alpha>0$. Conversely, suppose there is a function $\Phi(p)\not\equiv 0$ analytic on a half-plane $\operatorname{Re} p>\alpha>0$ satisfying inequalities of the form

$$|p^n\Phi(p)|\leqslant Am_nB^n \qquad (n=0,1,2,\dots).$$

Then obviously $\Phi(p)/p^2$ satisfies the conditions of Theorem 5.42a; in fact, for the integrable majorant $B(\eta)$ we need only choose the function $Cm_0/|\alpha+i\eta|^2$. Applying the theorem, we find that the function

$$\varphi(x)=\frac{1}{2\pi i}\int_{\gamma-i\infty}^{\gamma+i\infty}\frac{\Phi(p)}{p^2}e^{px}\,dp \qquad (\gamma>\alpha)$$

vanishes for $x<0$. Since $\Phi(p)\not\equiv 0$, we also have $\varphi(x)\not\equiv 0$ for $x\geqslant 0$. Moreover, $\varphi(x)$ has derivatives of all orders, and

$$\begin{aligned}|[\varphi(x)e^{-\alpha x}]^{(n)}|&=\frac{1}{2\pi}\left|\int_{\gamma-i\infty}^{\gamma+i\infty}\frac{\Phi(p)}{p^2}(p-\alpha)^n e^{(p-\alpha)x}\,dp\right|\\ &\leqslant\frac{A}{2\pi}m_nB^n\int_{\gamma-i\infty}^{\gamma+i\infty}\left|\frac{p-\alpha}{p}\right|^n\frac{|dp|}{|p^2|}\leqslant\frac{A}{2\pi}m_nB^n\int_{\gamma-i\infty}^{\gamma+i\infty}\frac{|dp|}{|p|^2}=A'm_nB^n.\end{aligned}$$

It follows that $\varphi(x)e^{-\alpha x}$ belongs to the class $C_{\langle m_n\rangle}$. Since $\varphi(x)=0$ for $x<0$

and $\varphi(x) \not\equiv 0$, the class $C_{\langle m_n \rangle}$ is not quasi-analytic. Thus *the problem of whether or not a given class $C_{\langle m_n \rangle}$ is quasi-analytic is equivalent to the problem (known as* **Watson's problem**) *of whether or not there exists a function which is analytic on a half-plane* $\operatorname{Re} p > \alpha$ *and satisfies the inequalities*

$$|p^n \Phi(p)| \leqslant A m_n B^n \qquad (n = 0,1,2,\ldots).$$

5.56. *Step 2.* The transformation $p = 2\alpha/s$ carries the half-plane $\operatorname{Re} p > \alpha$ into the disk $|z-1| < 1$, and correspondingly, Watson's problem takes the following form: *Find conditions on the sequence m_n such that there exists a function $F(s) \not\equiv 0$ analytic on the disk $|s-1| < 1$ satisfying the inequalities*

$$|F(s)| \leqslant A m_n B^n |s|^n \qquad (n = 0,1,2,\ldots). \tag{9}$$

For example, if $m_{n_k} \leqslant A_1 r_0^{n_k}$ for some increasing sequence of positive integers $n_1, n_2, \ldots, n_k, \ldots$, then no such function $F(s)$ exists. In fact, if

$$|F(s)| \leqslant A A_1 r_0^{n_k} B^{n_k} |s|^{n_k} = A A_1 (r_0 B |s|)^{n_k} \qquad (k = 1,2,\ldots),$$

then, choosing $|s| < 1/r_0 B$ and taking the limit as $k \to \infty$, we find that $F(s) \equiv 0$, contrary to hypothesis. Thus, under the indicated conditions on the sequence m_n, the class $C_{\langle m_n \rangle}$ is quasi-analytic. On the other hand, in this case,

$$T(r) = \sup_n \frac{r^n}{m_n} = \infty$$

if $r > r_0$, so that the condition (7) is satisfied in an obvious way. The fact that the solution of the problem is particularly simple if $m_{n_k} \leqslant A_1 r_0^{n_k}$ $(k = 1,2,\ldots)$ has already been noted in Sec. 5.52.

Returning to the general case, suppose there exists a function $F(s) \not\equiv 0$ analytic on the disk $|s-1| < 1$ satisfying the inequalities (9). Then we can find a number ρ between 0 and 1 such that $F(\rho) \neq 0$ and $|F(\rho + \rho e^{i\theta})| \leqslant 1$ for all real θ, where the only zero of the function $F(s)$ on the circle $s = \rho + \rho e^{i\theta}$ is at the point $s = 0$ (why?). The rest of the construction will take place in the disk $|s - \rho| \leqslant \rho$. According to (9),

$$|F(\rho + \rho e^{i\theta})| \leqslant A m_n B^n \, |1 + e^{i\theta}|^n = A m_n B^n \left| 2\rho \cos \frac{\theta}{2} \right|^n.$$

Taking the minimum of the right-hand side, we get

$$|F(\rho + \rho e^{i\theta})| \leqslant \frac{A}{\displaystyle \max_n \frac{1}{m_n B^n \left| 2\rho \cos \frac{\theta}{2} \right|^n}}.$$

Therefore, by the definition of the function $T(r)$,

$$|F(\rho+\rho e^{i\theta})| \leqslant \frac{A}{T\left(\dfrac{1}{2B\rho\left|\cos\dfrac{\theta}{2}\right|}\right)},$$

which implies

$$\ln|F(\rho+\rho e^{i\theta})| \leqslant \ln A - \ln T\left(\frac{1}{2B\rho\left|\cos\dfrac{\theta}{2}\right|}\right).$$

Applying Lemma 5.53 to the function $F(z)$, we now deduce that the integral

$$-\int_0^{2\pi} \ln|F(\rho+\rho e^{i\theta})|\, d\theta$$

exists and is finite. Hence the same is true of the integral

$$\int_0^{2\pi} \ln T\left(\frac{1}{2B\rho\left|\cos\dfrac{\theta}{2}\right|}\right) d\theta \leqslant 2\pi \ln A - \int_0^{2\pi} \ln|F(\rho+\rho e^{i\theta})|\, d\theta,$$

which can be written in the form

$$2\int_0^{\pi} \ln T\left(\frac{1}{2B\rho\cos\dfrac{\theta}{2}}\right) d\theta.$$

Making the substitution

$$r = \frac{1}{2B\rho\cos\dfrac{\theta}{2}},$$

we find that the integral

$$2\int_0^{\pi} \ln T\left(\frac{1}{2B\rho\cos\dfrac{\theta}{2}}\right) d\theta = 4\int_{1/2B\rho}^{\infty} \frac{\ln T(r)}{r^2}\frac{r}{\sqrt{4B^2\rho^2r^2-1}}\, dr$$

is finite. But then the integral

$$\int_1^{\infty} \frac{\ln T(r)}{r^2}\, dr$$

is also finite (why?). Thus we have shown that *the condition* (7) *fails if the class* $C_{\langle m_n \rangle}$ *is not quasi-analytic*, thereby proving the sufficiency of the condition.

5.57. *Step 3.* We now prove the necessity of the condition (7). Suppose

$$\int_1^\infty \frac{\ln T(r)}{r^2} dr < \infty.$$

Then

$$\int_0^{2\pi} \ln T\left(\frac{1}{2B\rho\left|\cos\frac{\theta}{2}\right|}\right) d\theta < \infty,$$

and hence we can use Poisson's integral (Sec. 4.46b) to construct a function

$$G(re^{i\varphi}) = \frac{1}{2\pi}\int_0^{2\pi} \ln T\left(\frac{1}{2\left|\cos\frac{\theta}{2}\right|}\right) \frac{1-r^2}{1-2r\cos(\theta-\varphi)+r^2} d\theta,$$

which is harmonic on the disk $r<1$ (justify setting $B\rho=1$). Let $G(s-1)=P(s)$, and let $Q(s)$ be the conjugate harmonic function of $P(s)$ on the disk $|s-1|<1$. Then the function

$$F(s) = e^{-P(s)-iQ(s)}$$

is analytic on the disk $|s-1|<1$, and we assert that

$$|F(s)| \leqslant m_n |s|^n \qquad (n=0,1,2,\ldots). \tag{10}$$

To prove (10), we note that these inequalities are equivalent to

$$e^{-P(s)} \leqslant m_n |s|^n \qquad (n=0,1,2,\ldots), \tag{10'}$$

or to

$$-G(s) = -P(s+1) \leqslant \ln m_n + n \ln (s+1) \qquad (n=0,1,2,\ldots). \tag{11}$$

Both terms on the right can be represented as Poisson integrals:†

$$\ln m_n = \frac{1}{2\pi}\int_0^{2\pi} \frac{(1-r^2)\ln m_n}{1-2r\cos(\theta-\varphi)+r^2} d\theta,$$

$$n \ln |s+1| = \frac{1}{2\pi}\int_0^{2\pi} \frac{n(1-r^2)\ln |e^{i\theta}+1|}{1-2r\cos(\theta-\varphi)+r^2} d\theta. \tag{12}$$

† The perceptive reader will note that the integral (12) is improper, since $\ln|e^{i\theta}+1|$ becomes infinite for $\theta=\pi$. Nevertheless, the validity of (12) is not hard to prove. See A. I. Markushevich, *Theory of Functions of a Complex Variable*, Vol. II (translated by R. A. Silverman), Prentice-Hall, Inc., Englewood Cliffs, N. J. (1965), Sec. 35.

Therefore (11) will be proved if we succeed in showing that

$$\int_0^{2\pi} \ln\left[T\left(\frac{1}{2\left|\cos\frac{\theta}{2}\right|}\right) m_n |1+e^{i\theta}|^n\right] \frac{1-r^2}{1-2r\cos(\theta-\varphi)+r^2}\, d\theta \geqslant 0. \tag{13}$$

Since

$$T(r) = \sup_{n\geqslant 0} \frac{r^n}{m^n},$$

we have

$$T(r) \geqslant \frac{r^n}{m_n}, \qquad T(r) m_n r^{-n} \geqslant 1$$

for every n, and moreover

$$|1+e^{i\theta}| = 2\left|\cos\frac{\theta}{2}\right|.$$

It follows that the integrand in (13) is a nonnegative function, and hence that the inequality (13) actually holds. But (13) implies (10), and if (10) holds, then the class $C_{\langle m_n \rangle}$ is not quasi-analytic, as shown in Sec. 5.55. Thus we have shown that *the class* $C_{\langle m_n \rangle}$ *is not quasi-analytic if the condition* (7) *fails*, thereby proving the necessity of the condition. The proof of the Carleman-Ostrowski theorem is now complete! ∎

Problems

1. Suppose $\varphi(x)$ is continuous and monotonic on an open interval I. Prove that the Fourier integral

$$\varphi(x) = \lim_{p\to\infty} \frac{1}{\pi} \int_{-\infty}^{\infty} \varphi(x+t) \frac{\sin pt}{t}\, dt$$

is convergent at every point of I, and uniformly convergent on every closed subinterval of I.

2. Give an example of a continuous function $\varphi(x)$ such that the Fourier integral

$$\psi(\sigma) = \int_{-\infty}^{\infty} \varphi(x) e^{i\sigma x}\, dx$$

is uniformly convergent on the whole line $-\infty < \sigma < \infty$, while $\psi(\sigma)$ fails to be absolutely integrable on $-\infty < \sigma < \infty$.

3. Give an example of a function $\varphi(x)$ such that the Fourier integral

$$\lim_{p\to\infty}\frac{1}{\pi}\int_{-\infty}^{\infty}\varphi(x+t)\frac{\sin pt}{t}\,dt=\lim_{p\to\infty}\frac{1}{2\pi}\int_{-p}^{p}\left\{\int_{-\infty}^{\infty}\varphi(\xi)e^{i\sigma(x-\xi)}\,d\xi\right\}d\sigma$$

is uniformly convergent on the whole line $-\infty<x<\infty$, while each of the integrals

$$\lim_{p\to\infty}\int_{0}^{p}\left\{\int_{-\infty}^{\infty}\varphi(\xi)\,e^{i\sigma(x-\xi)}\,d\xi\right\}d\sigma,$$

$$\lim_{p\to\infty}\int_{-p}^{0}\left\{\int_{-\infty}^{\infty}\varphi(\xi)\,e^{i\sigma(x-\xi)}\,d\xi\right\}d\sigma$$

has points of divergence.

4. Prove *Parseval's theorem*

$$\int_{-\infty}^{\infty}|g(\sigma)|^2\,d\sigma=2\pi\int_{-\infty}^{\infty}|f(x)|^2\,dx,$$

where $g(\sigma)$ is the Fourier transform of a function $f(x)$ which satisfies the conditions of Theorem 5.14 and has an integrable square on the whole line $-\infty<x<\infty$. (Plancherel)

5. Prove the "uncertainty relation"

$$\int_{-\infty}^{\infty}x^2|f(x)|^2\,dx\cdot\int_{-\infty}^{\infty}\sigma^2|g(\sigma)|^2\,d\sigma\geqslant\frac{\pi}{2},$$

assuming that the functions $xf(x)$ and $f'(x)$ satisfy the conditions of Problem 4 and that

$$\int_{-\infty}^{\infty}|f(x)|^2\,dx=1.$$

6. Given that $F(p)$ is the Laplace transform of $f(t)$, find the Laplace transforms of the following functions:

(a) $e^{at}f(t)$; (b) $f'(t)$; (c) $\int_0^t f(\tau)\,d\tau$; (d) $tf(t)$; (e) $f(t)/t$.

7. Find the Laplace transforms of the following functions:

(a) e^{at}; (b) $t^{\alpha-1}$ $(\alpha>0)$; (c) $t^{\alpha-1}e^{at}$; (d) $\sin at$; (e) $\cos at$; (f) $(\sin at)/t$.

8. Prove that Jordan's lemma (Vol. 1, Sec. 11.42d) remains valid if the condition $\operatorname{Im} z\geqslant 0$ is replaced by $\operatorname{Im} z\geqslant -a$, where a is any fixed real number. Discuss the modification of Jordan's lemma for the case where the integrand is $f(p)e^{pt}$ instead of $f(z)e^{i\sigma z}$, and the path of integration lies in the half-plane $\operatorname{Re} p \leqslant \gamma$.

9 (*The Mellin transform and its inversion*). Prove that if

$$F(s) = \int_0^\infty f(x) x^{s-1}\, dx,$$

then

$$f(x) = \frac{1}{2\pi i} \int_{c-i\infty}^{c+i\infty} F(s) x^{-s}\, ds.$$

10. Prove that the class $C_{\langle (n!)^\alpha \rangle}$ is quasi-analytic if $\alpha \leqslant 1$, but not if $\alpha > 1$.

Comment. This class contains only analytic functions if $\alpha \leqslant 1$ (why?). On the other hand, it can be shown that there exist quasi-analytic classes containing nonanalytic functions as well as analytic functions (e.g., if $m_n = n!\ln^n n$).

Hints and Answers

Chapter 1

1. *Ans.* (a) Complete; (b) Complete; (c) Incomplete.
2. *Hint.* The set of all increasing sequences of positive integers has the power of the continuum (Vol. 1, Chapter 2, Problem 8). For every such sequence $n_1 < n_2 < \cdots$ choose a function $x(t) \in R^s(0,\infty)$ equal to 1 at the points $n_1, n_2, \ldots$ and to 0 at the other integral points.
3. *Hint.* If an extremal function existed, it would have to equal 1 for $0 < x < \frac{1}{2}$ and -1 for $\frac{1}{2} < x < 1$.
4. *Hint.* We must verify that (x,y) satisfies the axioms for a scalar product (Sec. 1.41). To verify Axiom d, apply the parallelogram formula to the parallelograms constructed on the four pairs of vectors

$$x+z, y; \quad x-z, y; \quad y+z, x; \quad y-z, x.$$

Verify Axiom c first for integral α and then for fractional α, afterwards taking the limit for arbitrary real α.
5. *Hint.* By Cauchy's theorem,

$$\oint_{|z|=1} p(z)\, dz = 0,$$

and this formula continues to hold after passage to the limit.
6. *Hint.* Let

$$F = \prod_{f \in J} \{x \in Q : f(x) = 0\}.$$

Show that every function $g(x) \in R^s(Q)$ equal to zero in a neighborhood of F belongs to the ideal J. Moreover, every function $f(x) \in R^s(Q)$ equal to zero on F is a limit of functions of the form $g(x)$.
7. *Hint.* Let $\delta > 0$ be the number corresponding to $\varepsilon > 0$ in the equicontinuity condition for the set E. Then the values of the functions $x(t) \in E$ at the points of a finite δ-net for the compactum Q form a precompact 2ε-net for the set E.
8. *Hint.* It can be assumed without loss of generality that

$$\left| \sum_{n=1}^{\infty} t_{kn} - 1 \right| < \delta$$

for every $k = 1, 2, \ldots$ Find two sequences $N_1, N_2, \ldots$ and $k_1, k_2 \ldots$ such that

$$\sum_{n=N_1}^{\infty} |t_{1n}| < \delta, \quad \sum_{n=1}^{N_1} |t_{k_1 n}| < \delta, \quad \sum_{n=N_2}^{\infty} |t_{k_1 n}| < \delta, \quad \sum_{n=1}^{N_2} |t_{k_2 n}| < \delta, \ldots$$

Then let

$$\xi_n = \begin{cases} 1 & \text{if} \quad N_{2p-1} \leqslant n < N_{2p}, \\ -1 & \text{if} \quad N_{2p} \leqslant n < N_{2p+1} \end{cases}$$

$(p=1,2,\ldots)$.

9. *Hint.* It is sufficient to consider sequences $x=(\xi_1,\ldots,\xi_n,\ldots)$ such that

$$\sup \xi_n = \overline{\lim}\ \xi_n = -\underline{\lim}\ \xi_n = -\inf \xi_n.$$

10. *Hint.* Use the uniqueness theorem (Vol. 1, Sec. 10.39a) and the maximum modulus principle (Vol. 1, Chapter 10, Problem 3) for analytic functions.

11. *Hint.* The inverse of the operator of multiplication by $z-\lambda$ can only be the operator of multiplication by $1/(z-\lambda)$.

12. *Hint.* Use Theorem 1.87 to show that every point of the spectrum of $\mathbf{A}$ is a generalized eigenvalue. Deduce from $(\mathbf{A}-\lambda\mathbf{E})x_n \to 0$, $|x_n|=1$ that $[p(\mathbf{A})-p(\lambda)\mathbf{E}]x_n \to 0$, and then use the continuity of $p(\mathbf{A})$ and the condition $p(\lambda)\neq 0$.

13. *Hint.* Use Problem 12.

14. *Hint.* Integrate Young's inequality

$$|x(t)y(t)| \leqslant \frac{1}{p}|x(t)|^p + \frac{1}{q}|y(t)|^q$$

(Vol. 1, Sec. 9.62g), assuming that

$$\int_a^b |x(t)|^p\,dt = \int_a^b |y(t)|^q\,dt = 1$$

(why is this assumption permissible?).

15. *Hint.* Integrate the inequality

$$\begin{aligned} |f(x)+g(x)|^p &\leqslant (|f(x)|+|g(x)|)^p \\ &= |f(x)|(|f(x)|+|g(x)|)^{p-1} + |g(x)|(|f(x)|+|g(x)|)^{p-1}, \end{aligned}$$

afterwards applying Hölder's inequality (Problem 14) to the right-hand side.

16. *Hint.* The solution is analogous to that of Problem 14, with integration replaced by summation.

17. *Hint.* The solution is analogous to that of Problem 15.

18. *Hint.* If $x^{(n)}=(\xi_1^{(n)},\ldots,\xi_k^{(n)},\ldots)$ is a fundamental sequence in l_p, then every numerical sequence $\xi_k^{(n)}$ $(k=1,2,\ldots)$ is also fundamental and hence convergent. Let

$$\xi_k = \lim_{n\to\infty} \xi_k^{(n)}.$$

Then, given any $\varepsilon > 0$, there is an integer N such that

$$\sum_{k=1}^{\infty} |\xi_k^{(m)} - \xi_k^{(n)}| < \varepsilon$$

for all $m,n > N$. Replacing ∞ by p, we first take the limit as $m \to \infty$ and then as $p \to \infty$.

19. *Hint.* The set $\{x \in R_n: |x|_p \leqslant 1\}$ is not convex.

20. *Hint.* Use Arzelà's theorem (Sec. 1.24b) to find a uniformly convergent subsequence, then use Arzelà's theorem again to find a "finer" subsequence whose first derivatives are also uniformly convergent, and so on. Then choose the "diagonal" subsequence.

21. *Hint.* If the set $E \subset P(Q)$ consists of a single function $x(t)$, let $Q_k = \{t \in Q: k\varepsilon \leqslant x(t) < (k+1)\varepsilon\}$. In the general case, use Hausdorff's criterion (Vol. 1, Sec. 3.93c).

Chapter 2

1. *Hint.* Try to verify the Lipschitz condition.

2. *Hint.* (a) A parabola; (b) A semicubical parabola.

3. *Hint.* Use the expression for the Wronskian (Sec. 2.74).

4. *Hint.* In the Jordan basis the system decomposes into a number of unrelated subsystems, one for each root of the characteristic equation. Each of these subsystems is equivalent to an equation of the form

$$\left(\frac{d}{dt} - \lambda\right)^m u(t) = 0,$$

and the whole system is equivalent to the equation

$$\prod_k \left(\frac{d}{dt} - \lambda_k\right)^{m_k} u(t) = 0.$$

5. *Hint.* The operators Ω_0^t and Ω_T^{T+t} satisfy the same equation with the same initial condition.

6. *Hint.* Define the operator $\mathbf{A}(t)$ by the conditions

$$u_k'(t) = \mathbf{A}(t) u_k(t) \qquad (k = 1, \ldots, n)$$

on the subspace $\mathbf{X}_t$ spanned by the vectors $u_1(t), \ldots, u_n(t)$, setting $\mathbf{A}(t)$ equal to zero on the orthogonal complement of $\mathbf{X}_t$ (Lin. Alg., Sec. 8.36b).

7. *Ans.* Yes. For example, if $n = 2$, choose the functions

$$y_1(t) = \begin{cases} t^3 & \text{for } t > 0, \\ 0 & \text{for } t < 0, \end{cases} \qquad y_2(t) = y_1(-t).$$

8. *Ans.* For example,

$$\begin{vmatrix} y^{(n)}(t) & y^{(n-1)}(t) & \cdots & y(t) \\ y_1^{(n)}(t) & y_1^{(n-1)}(t) & \cdots & y_1(t) \\ \cdot & \cdot & \cdots & \cdot \\ y_n^{(n)}(t) & y_n^{(n-1)}(t) & \cdots & y_n(t) \end{vmatrix} = 0.$$

9. *Hint.* The result follows from the existence theorem for $n=0$ and more generally by induction.
10. *Hint.* Let $y(t)=e^{kt}z(t)$.
11. *Hint.* Let

$$y(t)=k\int_0^t w(s)\, ds.$$

12. *Hint.* An ε-almost-solution also satisfies the inequality

$$\left\| y(t)-y(0)-\int_0^t f(s,y(s))\, ds \right\| \leqslant \varepsilon t.$$

Now use the solution of Problem 11.
13. *Hint.* For $t_k \leqslant t \leqslant t_{k+1}$,

$$y_\Pi(t)=\left\{[\mathbf{E}+(t-t_k)\mathbf{A}(t_k)]\prod_{j=k-1}^{0}[\mathbf{E}+\mathbf{A}(t_j)\Delta t_j]\right\}y_0,$$

$$\begin{aligned} y'_\Pi(t) &= \left\{\mathbf{A}(t_k)\prod_{j=k-1}^{0}[\mathbf{E}+\mathbf{A}(t_j)\Delta t_j]\right\}y_0 \\ &= \mathbf{A}(t_k)\,[\mathbf{E}+(t-t_k)\mathbf{A}(t_k)]^{-1}y_\Pi(t) \\ &= \mathbf{A}(t_k)[\mathbf{E}+\mathbf{B}_k(t)]y_\Pi(t)\mathbf{A}(t)y_\Pi(t)+\mathbf{C}_k(t)y_\Pi(t), \end{aligned}$$

where the operators $\mathbf{B}_k(t)$ and $\mathbf{C}_k(t)$ approach zero as

$$d(\Pi)=\max\{t_1-t_0,\ldots,t_n-t_{n-1}\}\to 0.$$

14. *Hint.* Use the solutions of Problems 12 and 13.
15. *Hint.* Use the method of solution of Problem 13.
16. *Hint.* Use the solutions of Problems 12 and 15.

Chapter 3

1. *Hint.* The projection onto the horizontal plane of the helix consisting of the centers of curvature of the helix $r=(a\cos t, a\sin t, bt)$ is of radius b^2/a.
2. *Hint.* Orthogonal self-congruent transformations commute and hence have the same canonical basis.
3. *Hint.* The indicated ratios all equal ds/ds^*.

4. *Hint.* $S_{m-1} \subset S_m$.
5. *Hint.* Use the decimal expansions of both coordinates of a given point of the square.
6. *Hint.* Show that the vector

$$\frac{Z(t+h)-Z(t)}{h}$$

has norm $1/\sqrt{h}$.

7. *Hint.* Show that the scalar product $(Z(t+h)-Z(t),\ Z(s+k)-Z(s))$ vanishes if the intervals $[t,t+h]$ and $[s,s+k]$ are nonoverlapping.
8. *Hint.* Choose $Z(t)$ to be the result of shifting any fixed smooth function $\varphi_0(\tau)$ by the amount t along the τ-axis.

Chapter 4

1. *Hint.* Choose appropriate numerical values of the argument in the Fourier series.
2. *Hint.* Do the same as in Problem 1.
3. *Ans.* (a) $s(t)=e^{\cos t}\cos(t\sin t)$; (b) $s(t)=e^{\cos t}\sin(t\sin t)$.
4. *Hint.* This precompact set has a unique limit point.
5. *Hint.* Use the second mean value theorem for integrals (Vol. 1, Chapter 9, Problem 3).
6. *Hint.* Suitably sharpen the proof of Problem 5.
7. *Hint.* For odd k,

$$\frac{4}{\pi}b_k=\int_0^{\pi/2} f(t)\sin kt\,dt=\int_0^{\pi/k} f(t)\sin kt\,dt+\int_{\pi/k}^{\pi/2} f(t)\sin kt\,dt,$$

and it is enough to show that

$$I_1=\int_0^{\pi/k} f(t)\sin kt\,dt=\frac{\beta_k}{k}f\left(\frac{\pi}{k}\right),\qquad \beta_k\to 2,$$

$$I_2=\int_{\pi/k}^{\pi/2} f(t)\sin kt\,dt=-\frac{1}{k}f\left(\frac{\pi}{k}\right)+\gamma_k,\qquad \sum_{k=1}^{\infty}|\gamma_k|<\infty.$$

To prove the first assertion, change kt to τ; to prove the second assertion, integrate by parts, and apply the mean value theorem and Vol. 1, Chapter 7, Problem 16.
8. *Hint.* Besides Conditions 1–4 of Problem 7, impose the extra condition

$$\sum_{k=1}^{\infty}\frac{1}{k}f\left(\frac{\pi}{k}\right)=\infty.$$

Then use Problem 6.
9. *Hint.* Write the Fourier series of the function $f(t)$ of Problem 8 in complex form.
10. *Hint.* $(xQ_n, Q_k) = (Q_n, xQ_k)$, and xQ_k $(k < n-1)$ is a polynomial of degree less than n. Therefore the representation

$$xQ_n(x) = \gamma_{n+1} Q_{n+1}(x) + \gamma_n Q_n(x) + \gamma_{n-1} Q_{n-1}(x)$$

holds for suitable $\gamma_{n+1}, \gamma_n, \gamma_{n-1}$. Compare the coefficients of x^{n+1}, x^n and calculate (xQ_n, Q_{n-1}).
11. *Hint.* Multiply the above formula by $Q_n(t)$ and from the result subtract the analogous formula with t and x interchanged. Now sum over n.
12. *Hint.* The first assertion follows from the orthogonality to 1, the second from the orthogonality to the polynomial

$$\prod_{k=1}^{m} (x - x_k),$$

under the assumption that $x_1, \ldots, x_m$ are *all* the zeros of the polynomial $Q_n(t)$ and $m < n$.

Chapter 5

1. *Hint.* The solution is analogous to that of Chapter 4, Problems 5 and 6.
2. *Hint.* The solution is analogous to that of Chapter 4, Problem 8.
3. *Hint.* The solution is analogous to that of Chapter 4, Problem 9.
4. *Hint.* Verify the assertion for functions of class S (Sec. 5.29a), and then for the function $f_N(x)$ which equals $f(x)$ for $|x| \leqslant N$ and vanishes for $|x| > N$, approximating $f_N(x)$ by functions of class S. Then take the limit as $N \to \infty$.
5. *Hint.* Consider the integral

$$\int_{-\infty}^{\infty} |tx\varphi(x) + \varphi'(x)|^2 \, dx \geqslant 0$$

as a quadratic polynomial in the parameter t.
6. *Ans.* (a) $F(p-a)$; (b) $pF(p)$; (c) $F(p)/p$; (d) $-F'(p)$; (e) $\int F(s)\,ds$, where the integral is over any path going from p to ∞ in the half-plane of analyticity of $F(p)$.
7. *Ans.* (a) $1/(p-a)$; (b) $\Gamma(\alpha)/p^\alpha$; (c) $\Gamma(\alpha)/(p-a)^\alpha$; (d) $a/(p^2+a^2)$; (e) $p/(p^2+a^2)$; (f) arc tan (a/p).
8. *Hint.* If $\varphi(R) = \arcsin(a/R)$, then $\varphi(R) \to 0$, $R\varphi(R) \to a$ as $R \to \infty$.
9. *Hint.* Let $x = e^t$.
10. *Hint.* Use Theorem 5.54 and Stirling's formula (Vol. 1, Sec. 11.67b).

Bibliography

Arnold, V. I., *Ordinary Differential Equations* (translated by R. A. Silverman), The MIT Press, Cambridge, Mass. (1973).

Bochner, S. and K. Chandrasekharan, *Fourier Transforms*, Princeton University Press, Princeton, N. J. (1949).

Carslaw, H. S. and J. C. Jaeger, *Operational Methods in Applied Mathematics*, Dover Publications, Inc., New York (1963).

Churchill, R. V., *Fourier Series and Boundary Value Problems*, McGraw-Hill Book Co., Inc., New York (1941).

Churchill, R. V., *Operational Mathematics*, third edition, McGraw-Hill Book Co., Inc., New York (1972).

Dym, H. and H. P. McKean, *Fourier Series and Integrals*, Academic Press, New York (1972).

Edwards, R. E., *Functional Analysis, Theory and Applications*, Holt, Rinehart and Winston, New York (1965).

Edwards, R. E., *Fourier Series, A Modern Introduction* (in two volumes), Holt, Rinehart and Winston, New York (1967).

Franklin, P., *An Introduction to Fourier Methods and the Laplace Transformation*, Dover Publications, Inc., New York (1958).

Goldberg, R. R., *Fourier Transforms*, Cambridge University Press, New York (1961).

Halmos, P. R., *Introduction to Hilbert Space*, second edition, Chelsea Publishing Co., New York (1957).

Jackson, D., *Fourier Series and Orthogonal Polynomials*, The Mathematical Association of America, Buffalo, N. Y. (1941).

Kolmogorov, A. N. and S. V. Fomin, *Introductory Real Analysis* (translated by R. A. Silverman), Prentice-Hall, Inc., Englewood Cliffs, N. J. (1970).

Liusternik, L. A. and V. I. Sobolev, *Elements of Functional Analysis* (translated by A. E. Labarre, Jr. et al.), Frederick Ungar Publishing Co., New York (1961).

Petrovski, I. G., *Ordinary Differential Equations* (translated by R. A. Silverman), Dover Publications, Inc., New York (1972).

Pontryagin, L. S., *Ordinary Differential Equations* (translated by L. Kacinskas and W. B. Counts), Addison-Wesley Publishing Co., Inc., Reading, Mass. (1962).

Riesz, F. and B. Sz.-Nagy, *Functional Analysis* (translated by L. F. Boron), Frederick Ungar Publishing Co., New York (1955).

Sneddon, I. A., *Fourier Transforms*, McGraw-Hill Book Co., Inc., New York (1951).

Taylor, A. E., *Introduction to Functional Analysis*, John Wiley, New York (1958).

Titchmarsh, E. C., *Introduction to the Theory of Fourier Integrals*, second edition, Oxford University Press, New York (1948).

Tolstov, G. P., *Fourier Series* (translated by R. A. Silverman), Prentice-Hall, Inc., Englewood Cliffs, N. J. (1962).

Widder, D. V., *The Laplace Transform*, Princeton University Press, Princeton, N. J. (1941).

Wiener, N., *The Fourier Integral and Certain of Its Applications*, Dover Publications, Inc., New York (1959).

Yosida, K., *Functional Analysis*, second edition, Springer-Verlag, New York (1968).

Zygmund, A., *Trigonometric Series*, second edition (in two volumes), Cambridge University Press, New York (1959).

Index